3D PRINTING AND ADDITIVE MANUFACTURING OF ELECTRONICS

Principles and Applications

World Scientific Series in 3D Printing

Series Editor: Chee Kai Chua *(Nanyang Technological University, Singapore)*

Published:

Vol. 3 *3D Printing and Additive Manufacturing of Electronics:*
Principles and Applications
by Chee Kai Chua, Wai Yee Yeong, Hong Yee Low,
Tuan Tran and Hong Wei Tan

Vol. 2 *Lasers in 3D Printing and Manufacturing*
by Chee Kai Chua, Murukeshan Vadakke Matham and Young-Jin Kim

Vol. 1 *Bioprinting: Principles and Applications*
by Chee Kai Chua and Wai Yee Yeong

World Scientific Series in 3D Printing

3D PRINTING AND ADDITIVE MANUFACTURING OF ELECTRONICS

Principles and Applications

Chee Kai Chua
Singapore University of Technology and Design, Singapore

Wai Yee Yeong
Nanyang Technological University, Singapore

Hong Yee Low
Singapore University of Technology and Design, Singapore

Tuan Tran
Nanyang Technological University, Singapore

Hong Wei Tan
Singapore University of Technology and Design, Singapore

World Scientific

NEW JERSEY · LONDON · SINGAPORE · BEIJING · SHANGHAI · HONG KONG · TAIPEI · CHENNAI · TOKYO

Published by

World Scientific Publishing Co. Pte. Ltd.

5 Toh Tuck Link, Singapore 596224

USA office: 27 Warren Street, Suite 401-402, Hackensack, NJ 07601

UK office: 57 Shelton Street, Covent Garden, London WC2H 9HE

Library of Congress Cataloging-in-Publication Data
Names: Chua, Chee Kai, author.
Title: 3D printing and additive manufacturing of electronics : principles and applications /
    Chee Kai Chua, Singapore University of Technology and Design, Singapore,
    Wai Yee Yeong, Nanyang Technological University, Singapore,
    Hong Yee Low, Singapore University of Technology and Design, Singapore,
    Tuan Tran, Nanyang Technological University, Singapore,
    Hong Wei Tan, Singapore University of Technology and Design, Singapore.
Description: Singapore ; Hackensack, NJ ; London : World Scientific, [2021] | Series:
    World Scientific series in 3D printing; vol 3 | Includes bibliographical references and index.
Identifiers: LCCN 2020053053 | ISBN 9789811218354 (hardcover) |
    ISBN 9789811218934 (paperback) | ISBN 9789811218361 (ebook for institutions) |
    ISBN 9789811218378 (ebook for individuals)
Subjects: LCSH: Printed circuits--Design and construction. | Electronic apparatus and
    appliances--Design and construction. | Three-dimensional printing.
Classification: LCC TK7868.P7 C47 2021 | DDC 621.3815/31--dc23
LC record available at https://lccn.loc.gov/2020053053

British Library Cataloguing-in-Publication Data
A catalogue record for this book is available from the British Library.

For any available supplementary material, please visit
https://www.worldscientific.com/worldscibooks/10.1142/11773#t=suppl

Desk Editors: Aanand Jayaraman/Amanda Yun

Typeset by Stallion Press
Email: enquiries@stallionpress.com

Dedication

To my wife, Wendy, children, Cherie, son-in-law Darren & grandchildren Hannah, Esther and Gabriel, Clement & daughter-in-law Lynette and Cavell, whose prayer, support and motivation have made it possible for us to finish writing this book.

Chee Kai

To my husband Tee Seng and to my beautiful children, Bao Rong and Zi Kai, who are the pillars of strength for me in this journey.

Wai Yee

To my family.

Tuan

To my wife, Clarrisa, and my family for their unconditional love and support.

Hong Wei

Preface

In the age of digitalisation, electronics products are becoming increasingly ubiquitous. Over the past few decades, electronic products like computers, smartphones, televisions and gaming consoles have completely transformed the way we interact, live, work and play. In consumer products, the demands for customisation and miniaturisations are two major trends that place technical challenges in designing and manufacturing of future electronic products.

Three-dimensional (3D) printing of electronics has attracted growing interest in recent years; it offers unique advantages not achievable by traditional manufacturing of electronics. For instance, on-demand fabrication of highly-customisable electronics, direct fabrication on a wide variety of substrates and conformal surfaces and novel device designs in 3-dimensions have been demonstrated in academic research.

In general, 3D printed electronics are defined as functional electrical devices that are fabricated by additive manufacturing technologies through depositing functional inks directly onto substrates. Development and understanding of 3D printing of electronics require a combination of multiple disciplines, specifically material science of the functional inks, mechanical-electrical engineering for the digital manufacturing processes and computational engineering for the design of 3D printable electronic devices.

This textbook has been written to provide a comprehensive coverage on the principles and application in 3D electronics printing for students, researchers and engineers. This textbook begins with the introduction of conventional electronics manufacturing and an overview of 3D printing of

vii

electronics. A chapter is devoted to the printing processes of the various conventional electronics techniques for printed electronics. Four subsequent chapters are dedicated to the key components of 3D printing of electronics, which are 3D electronics printing techniques, materials and inks, substrates and processing, and sintering techniques for metallic nanoparticle inks. The book next presents a new exciting area which is gaining traction recently, namely the designs and simulations for 3D printed electronics. The last chapter is entirely devoted to the applications of 3D printed electronics, and valuable insights are provided for the existing challenges and future outlook.

A set of designed problems at the end of each chapter provide undergraduate and postgraduate students practices on key concepts covered. For tertiary-level lecturers and university professors, the topic on 3D electronics printing can be readily correlated to other subjects in material science, mechanical, electrical and industrial engineering.

Chua C. K.
Yeong W. Y.
Low H. Y.
Tran T.
Tan H. W.

About the Authors

Chee Kai Chua is the Head of Pillar for Engineering Product Development and Cheng Tsang Man Chair Professor at Singapore University of Technology and Design (SUTD). Dr Chua has extensive teaching and consulting experience in 3D Printing and Additive Manufacturing (3DP & AM). He is an active contributor to the Additive Manufacturing (AM or 3D Printing) field for over 30 years, where his re-design of AM processes for innovative devices such as tissue engineering scaffolds are highly regarded by the scientific community. He is now active in 3D printing of electronics, food, metals and polymers. He won the prestigious International Freeform and Additive Manufacturing Excellence (FAME) Award in 2018. As at 2020, he has contributed more than 400 technical papers and patents, generating more than 14,000 citations, and co-authored four books including *3D Printing and Additive Manufacturing: Principles and Applications (5th edition)* and *Bioprinting: Principles and Applications*. In addition, he is the chief editor of *Virtual and Physical Prototyping*, as well as the chief editor of *International Journal of Bioprinting*. Professor Chua can be contacted by email at *cheekai_chua@ sutd.edu.sg*.

Wai Yee Yeong is an Associate Professor at School of Mechanical and Aerospace Engineering (MAE), Nanyang Technological University, Singapore. She also serves as Associate Chair (Students) at MAE She has published more than 150 papers, generating more than 4,900 citations with a current H-index of 35, and co-authored 2 textbooks (published by World Scientific and Elsevier respectively). Her works have been featured on media such as CNA, the Straits Times and other media channels. Her portfolio also includes serving as Programme Director (Aerospace and Defence) at Singapore Centre for 3D Printing (SC3DP) and at HP-NTU Digital Manufacturing Corporate Labs. She has filed 6 patents applications and 14 know-hows. Her main research interest is in 3D printing, bioprinting and the translational of the advanced technologies for industrial applications. Her current research topics include 3D printing of new materials, hybrid electronic-mechanical structures and bioprinting for tissue engineering. She was named the winner of TCT Women in 3D Printing Innovator Award 2019. Associate Professor Yeong can be contacted by email at *wyyeong@ntu.edu.sg* and more information can be found on her website: www.yeongresearch.com.

Hong Yee Low received her PhD from the Macromolecular Science and Engineering department of Case Western Reserve University in 1998. After 2 years at Motorola Semiconductor Sector, she worked at the Institute of Materials Research and Engineering (IMRE), Agency for Science Technology and Research, Singapore for 13 years. In IMRE she spearheaded nanoimprinting research, held the position as group head of Patterning and Fabrication Capability Group and Director of Research and Innovation. She is currently an associate professor in the Engineering Product Development Pillar (EPD) at the Singapore University of Technology and Design (SUTD) and the director for the Digital Manufacturing and Design (DmanD) centre. Her primary research interest is in nanofabrication of functional surfaces. She has co-authored >150 peer reviewed publications and is a co-inventor of 30 granted patents. Associate Professor Low can be contacted by email at *hongyee_low@sutd.edu.sg*.

Tuan Tran is currently an Associate Professor of the School of Mechanical and Aerospace Engineering, Nanyang Technological University (NTU), Singapore. He also acts as the Deputy Director of the NTU hub at Singapore's National Additive Manufacturing Innovation Cluster (NAMIC) with the mission to provide financial support for translational research in 3D printing and promote adoption of 3D printing in Singapore's industrial ecosystem. His research in 3D printing encompasses a wide range of activities, from fundamental works in development of droplet- and powder-based 3D printing technologies to translational works such as standards and qualification in 3D printing and 3D printing of conventional and wearable electronics. He has authored 40 journal papers in multiphase flows, droplet-surface interactions, and 3D printing. He holds patents for embedding identifiers in 3D printed parts and *in-situ* monitoring of 3D printing processes. His work on *in-situ* monitoring of 3D printing processes has been selected by ASTM to develop into industry standards. Associate Professor Tran can be contacted by email at *ttran@ntu.edu.sg*.

Hong Wei Tan is currently a research fellow in the Engineering Product Development Pillar (EPD) at the Singapore University of Technology and Design (SUTD). He received his BEng (First-Class Honours) and PhD degrees in Mechanical Engineering from Nanyang Technological University, Singapore. He has filed one PCT patent in sintering metallic nanoparticle inks for 3D printed electronics applications. His research interests are additive manufacturing processes for 3D printed electronics and materials characterisations. Dr Tan can be contacted by email at *hongwei_tan@sutd.edu.sg*.

Acknowledgements

First, we would like to thank God for granting us His strength throughout the writing of this book. Secondly, we are especially grateful to our respective spouses, Wendy and Tee Seng Lim and our respective children, Cherie Chua, Clement Chua, Cavell Chua, son-in-law Darren (Cherie's husband) and daughter-in-law Lynette (Clement's wife), Bao Rong Lim and Zi Kai Lim for their patience, support and encouragement throughout the year it took to complete this book.

We wish to thank the valuable support from the administration of Singapore University of Technology and Design (SUTD) and Nanyang Technological University (NTU), especially to their respective departments, the Engineering Product Development (EPD) pillar and the School of Mechanical and Aerospace Engineering (MAE).

The acknowledgements would not be complete without the contributions of the following companies for supplying and helping us with the information about their products they develop, manufacture or represent:

1. BotFactory, Inc.
2. Enjet, Inc.
3. FUJIFILM Dimatix, Inc.
4. Integrated Deposition Solutions, Inc.
5. Nano Dimension Ltd.
6. Neotech AMT GmbH
7. nScrypt, Inc.
8. Optomec, Inc.

9. Sonoplot, Inc.
10. Voltera, Inc.

Your suggestions, corrections and contributions will be appreciated and reflected on the later editions of this book.

Chua C. K.
Yeong W. Y.
Low H. Y.
Tran T.
Tan H. W.

Contents

List of Abbreviations

2D	Two-Dimensional
3D	Three-Dimensional
a-ITO	Amorphous Indium–Tin–Oxide
AM	Additive Manufacturing
ASTM	American Society for Testing and Materials
C8-BTBT	2,7-Dioctyl[1]benzothieno[3,2-b][1]benzothiophene
CAD	Computer-Aided Design
CAD-CAM	Computer-Aided Design-Computer-Aided Manufacturing
CAGR	Compound Annual Growth Rate
CBM	Conduction Band Minimum
CCD	Charge-Coupled Device
CFD	Computational Fluid Dynamics
CGFR	Carrier Gas Flow Rate
CNT	Carbon Nanotube
Cu_2O	Cuprous Oxide
CuO	Cupric Oxide
CVD	Chemical Vapor Deposition
CW	Continuous-Wave
DBD	Dielectric Barrier Discharge
EC	Electrochromic
ECG	Electrocardiography
EGOFET	Electrolyte-Gated Organic Field-Effect Transistor
EGT	Electrolyte-Gated Transistor
EM	Electromagnetic
ESA	Electrically Small Antennas

FDM	Fused Deposition Modelling
FEM	Finite Element Method
FET	Field Effect Transistor
FFF	Fused Filament Fabrication
FM	Fast Marching
GO	Graphene Oxide
H_3O^+	Hydronium Ion
HASL	Hot Air Solder Levelling
HOMO	Highest Occupied Molecular Orbital
IC	Integrated Circuit
IEC	International Electrotechnical Commission
IGZO	Indium–Gallium–Zinc–Oxide
IPA	Isopropyl Alcohol
IPL	Intense Pulse Light
IR	Infrared
ITO	Indium Tin Oxide
ITO/PET	Indium Tin Oxide Coated Polyethylene Terephthalate
IZO	Indium-Zinc-Oxide
LED	Light-Emitting Diode
LHS	Latin Hypercube Sampling
LUMO	Lowest Unoccupied Molecular Orbital
MDPI	Multidisciplinary Digital Publishing Institute
MIMO	Multiple-Input Multiple-Output
MIR	Mid-Infrared
MOD	Metal-Organic Decomposition
MoS_2	Molybdenum Disulfide
MOSFET	Metal-Oxide-Semiconductor Field-Effect Transistor
MWCNT	Multi-Walled CNT
NIR	Near-Infrared
OECT	Organic Electrochemical Transistor
OFET	Organic Field-Effect Transistor
OH^-	Hydroxide Ion
OLED	Organic Light-Emitting Diode
OPV	Organic Photovoltaic
OTFT	Organic Thin-Film Transistors
P3HT	Poly(3-hexylthiophene)
P3HT:PCBM	Poly(3-hexylthiophene):[6,6]-phenyl C61 butyric acidmethyl ester
PC	Polycarbonate

PCB	Printed Circuit Board
PCS	Printed Circuit Structures
PE	Printed Electronics
PEDOT	Poly(3,4-ethylenedioxythiophene)
PEDOT:PSS	Poly(3,4-ethylenedioxythiophene)-poly(styrenesulfonate)
PEEK	Polyether Ether Ketone
PEKK	Polyetherketoneketone
PEN	Polyethylene Naphthalate
PET	Polyethylene Terephthalate
PI	Polyimide
PLA	Polylactic Acid
PLC	Programmable Logic Controller
PMMA	Poly(methyl methacrylate)
P-OLED	Polymer Organic Light-Emitting Diode
PP	Polypropylene
PPV	Polyphenylene Vinylene
PS	Polystyrene
PSS	Polystyrene Sulfonate
PTC	Positive Temperature Coefficient
PU	Polyurethane
PV	Photovoltaic
PVA	Polyvinyl Alcohol
PVC	Polyvinyl Chloride
PVDF	Polyvinylidene Fluoride
PVP	Polyvinylpyrrolidone
RF	Radiofrequency
RFID	Radio-Frequency Identification
RGB	Red, Green and Blue
rGO	reduced Graphene Oxide
RP	Rapid Prototyping
RR-P3HT	Regioregular poly(3-hexylthiophene)
SEM	Scanning Electron Microscope
ShGFR	Sheath Gas Flow Rate
SLA	Stereolithography Apparatus
SMD	Surface-Mount Device
SMT	Surface-Mount Technology
SnO	Tin Monoxide
SPR	Surface Plasmon Resonance
STL	Stereolithography

SVM	Support Vector Machine
SWCNT	Single-Walled CNT
TEM	Transmission Electron Microscopy
TFT	Thin-Film Transistor
TIPS	6,13-bis(triisopropylsilylethynyl)pentacene-pentacene
TPU	Thermoplastic Polyurethane
UHF	Ultra-High Frequency
UI	User Interface
UV	Ultraviolet
UWB	Ultra-Wide Band
VBM	Valence Band Maximum
WPT	Wireless Power Transfer
WS_2	Tungsten Disulfide
WSe_2	Tungsten Diselenide
ZnO	Zinc Oxide
ZTO	Zinc-Tin-Oxide

Chapter 1

Introduction to Conventional Electronics Manufacturing and 3D Printing of Electronics

Electronics have become one of the essential necessities of life in modern society. It is hard to imagine what the world would be like without electronics. Everything from entertainment to healthcare uses electronics in some way or another. Many electronic devices, such as smartphones, computers and gaming consoles, have also revolutionised the way we communicate, live, work and play — and with increasing convenience — over the past few decades.

1.1 Conventional Electronics

Most electrical components and circuitries in electronic devices are fabricated by conventional electronics manufacturing methods. These manufacturing methods are highly complex and comprise of a series of additive and subtractive processes [1] (for instance, photolithography [2], laser ablation [3], etching [2,4], masking [5], chemical-vapour deposition [5] and lift-off [2,4]) that take place in expensive cleanrooms [5] under well-controlled conditions.

1.1.1 *Printed Circuit Boards*

The printed circuit board (PCB) is one of the most essential components in any electronic devices or systems. A PCB is an electrical circuit board

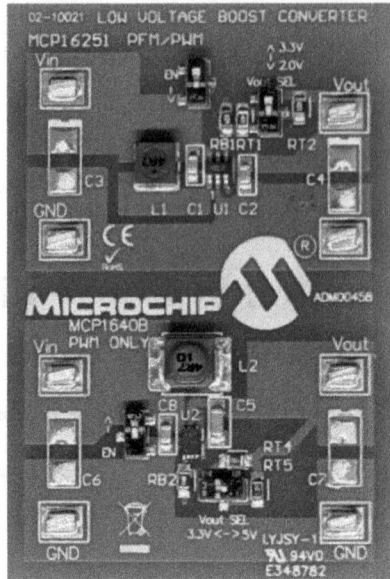

Figure 1.1.   PCB with mounted electrical components.

which is primarily used for mounting and connecting electrical components (see **Figure 1.1**) [6]. The conductive traces within the PCB are etched from thin copper sheets and are laminated onto the rigid dielectric base material substrate [7]. The FR-4 epoxy all woven glass laminate is one of the most popular and commonly used base material substrates for PCBs due to its low cost and superior mechanical, electrical and thermal properties [8]. Electrical components are usually mechanically joined and electrically connected to the PCBs by soldering [9].

There are three different types of conventional PCBs: single-layer PCBs, double-layer PCBs and multi-layer PCBs. Single-layer PCBs, also known as single-sided boards, are PCBs with only one layer of conductive material laminated on one side of the insulating substrate [10]. Single-layer PCBs are the simplest of all to design and manufacture and are commonly found in simple electronic devices. As the name suggests, the double-layer PCBs are PCBs with conductive material laminated on both sides of the insulating substrate whereas the multi-layer PCBs are circuit boards with three or more layers of circuit. Multi-layer PCBs can significantly improve the packaging density of electrical components with increased complex functionality, while at the same time, making the size of the PCBs compact [11]. However, multi-layer PCBs are more expensive and complex to design and fabricate as compared to single-layer PCBs.

The name "printed circuit board" is misleading as there is no printing process involved in the deposition of functional or conductive materials for the fabrication of PCBs [12]. The conventional PCB fabrication process is highly complex and involves a series of additive and subtractive processes. For instance, the general fabrication process of a double layer PCB can be described as the following [11]:

(1) *Board cutting*: Copper foils are first laminated on both sides of the dielectric base material substrate to form the double-sided board. The board is then cut to the desired size for the PCB fabrication [11].
(2) *Board drilling*: Holes are then drilled onto the board according to the PCB's design, in which these holes can either be positioning holes for assembling electrical components or internal electrical interconnections between two sides of the board [13].
(3) *Electroless copper plating*: Electroless copper plating is required to make the inner surfaces of the holes conductive so that it can conduct electricity from one side of the board to the other side [14]. About 40 μm thickness of copper is usually deposited onto the hole walls and copper foils during the electroless copper plating process [15].
(4) *Board pre-treatment*: The boards must first be thoroughly cleaned to remove any undesirable organic materials, dust, sulphides and oxides from their surfaces. During pre-treatment, the boards go through a series of cleaning processes which include degreasing, scrubbing, wet brushing, acid washing and washing in light quality de-ionised water [15].
(5) *Photolithography*: Photolithography is a widely used patterning process for microfabrication and nanofabrication, in which photosensitive photoresist is selectively exposed to a light source through a pattern mask [16]. The pattern mask comprises of opaque and transparent regions which allow light to pass through in a defined pattern, in which regions under the opaque mask are not exposed to light (see **Figure 1.2(a–c)**). Two different types of photoresist (positive photoresist and negative photoresist) can be used for photolithography and they react differently when exposed to photon radiation [16]. On the one hand, exposed positive photoresist becomes very soluble

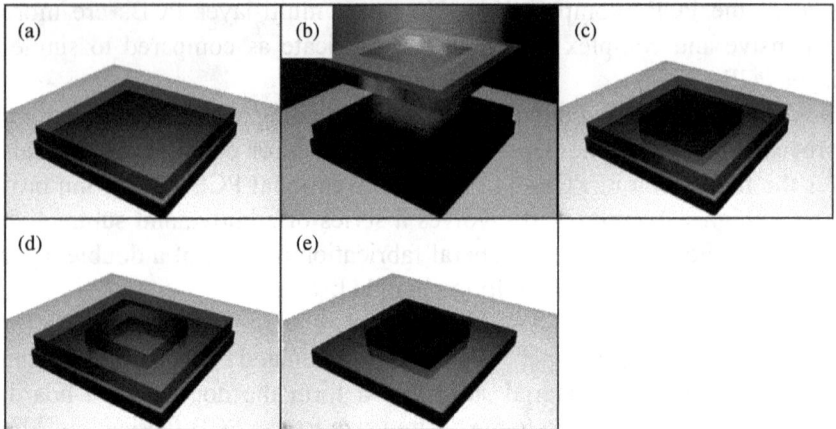

Figure 1.2.   Schematic of the photolithography process: (a) photoresist coated on a substrate, (b) photoresist exposed to light through a pattern mask, (c) exposed photoresist in the centre, (d) the unexposed regions in positive photoresist remains on substrate after washing in developing solution and (e) the exposed regions in negative photoresist remains on substrate after washing in developing solution. Reprinted with permission from Ref. [16]. Copyright (2014) from Royal Society of Chemistry.

in the developing solution and can be removed off easily (see **Figure 1.2(d)**) [16]. On the other hand, negative photoresist will crosslink when exposed to light and become insoluble in the developing solution (see **Figure 1.2(e)**). Only the unexposed negative photoresist will be removed in the developing solution [16].

The fabrication process of the double layer PCB is described in the case of using a positive photoresist. Thin layers of photoresist are coated onto both sides of the board and the pattern masks are laid on top of the photoresist layers. The pattern masks define the patterns to be patterned by photolithography. The entire panel of pattern masks, photoresist, copper foils and the dielectric substrate is exposed to intense ultraviolet light (see **Figure 1.3(a)**). The positive photoresist degrades and becomes very soluble in the developing solution when exposed to the intense ultraviolet light. The pattern masks are then removed from the panel and alkaline developing solutions are sprayed onto the panel to dissolve the highly soluble photoresist regions and expose the copper foils beneath [11] (see **Figure 1.3(b)**).

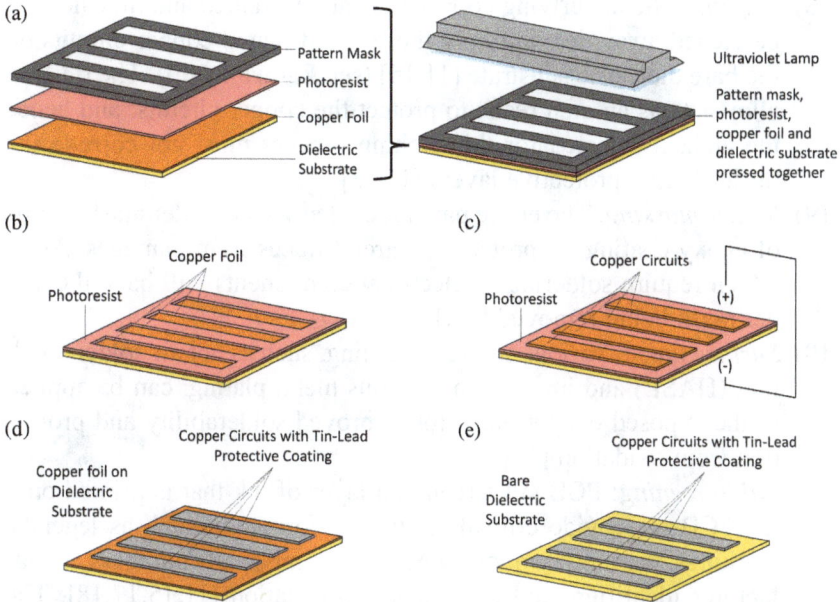

(a)

Pattern Mask
Photoresist
Copper Foil
Dielectric Substrate

Ultraviolet Lamp

Pattern mask, photoresist, copper foil and dielectric substrate pressed together

(b)

Copper Foil

Photoresist

(c)

Copper Circuits

Photoresist

(+)

(-)

(d)

Copper Circuits with Tin-Lead Protective Coating

Copper foil on Dielectric Substrate

(e)

Copper Circuits with Tin-Lead Protective Coating

Bare Dielectric Substrate

Figure 1.3.   Schematic diagrams of PCB at various stages of the fabrication process: (a) PCB during photolithography process; (b) PCB after photolithography process with exposed positive photoresist stripped off; (c) PCB after copper electroplating; (d) PCB after tin-lead plating with the remaining photoresist stripped off; and (e) PCB after etching [11].

(6) *Copper electroplating*: The entire panel next goes through a copper electroplating process. The exposed copper foils in the panel function as cathodes in the electroplating process and get electroplated with copper to form the copper circuits (see **Figure 1.3(c)**). The copper circuits typically have thicknesses ranging between 25–50 μm [11]. The regions on the panel which are covered with photoresist cannot function as cathodes and are not electroplated with copper [11,15].

(7) *Tin-lead plating*: The entire panel goes through another electroplating process again, where the tin-lead alloy is electroplated on top of the copper circuits. The tin-lead alloy serves as a protective layer for the copper circuits to resist etching in the next step and prevent oxidation of copper [11,15]. The remaining photoresist on the panel is washed away with a solvent to expose the underlying copper foils (see **Figure 1.3(d)**).

(8) *Etching*: The underlying copper foils are unwanted and they need to be etched away, leaving only the tin-lead plated copper circuits on the bare dielectric substrate [11,15] (see **Figure 1.3(e)**). The tin-lead alloy acts as an etch resist to protect the copper circuits, and hence the etchant selected for this etching process must not corrode the tin-lead alloy protective layers [11,15].

(9) *Solder masking*: The entire panel is coated with a solder mask, a type of epoxy coating, to protect the circuit traces from damages. Areas which require soldering of electrical components will have the solder mask being removed [11,15].

(10) *Surface finish*: Various surface finishing, such as hot air solder levelling (HASL) and immersion precious metal plating, can be applied to the exposed copper areas for improved solderability and protection from oxidation [15].

(11) *Silk screening*: PCB silkscreen is a layer of ink that is printed onto the PCB to provide critical circuitry information such as legends, markings, warning symbols, logos, test points, parts numbers, manufacturer identifiers and components orientation [11,15,17,18]. The silkscreen can help in better identification of the positions of the electrical components during the assembly process [18] and better identification of the circuitries for ground point, testing points and component interconnects [17]. Silk screening can be done by manual screen printing, liquid photo imaging and direct legend printing [17].

(12) *Inspection and testing*: Visual inspection and electrical testing are done to check for manufacturing defects, open circuits and short circuits on the finished PCB [11,15].

1.1.2 *Integrated Circuits*

An integrated circuit (IC) is defined by the International Electrotechnical Commission (IEC) in IEC 60748-1:2002 as a "microcircuit in which all or some of the circuit elements are inseparably associated and electrically interconnected so that it is considered to be indivisible for the purpose of construction and commerce" [19]. In other words, an integrated circuit is an integration of electrical components (passive and active devices such as resistors, capacitors, diodes and transistors) [20], circuits and base material fabricated on a piece of a semiconductor substrate (usually silicon) [11,21]. ICs are very compact and have high reliability, high processing speeds and low power requirements [11]. ICs can be used as controllers and they are commonly found in many sophisticated electronic devices such as computers, laptops, smartphones and cameras.

The IC fabrication process is highly complex and can involve hundreds of steps just to produce a single IC chip [12]. On top of that, the entire fabrication process must be done in a cleanroom where the environment is tightly controlled to maintain extremely low levels of particulates and avoid contamination by particulates [11]. For instance, the general fabrication process of ICs can be described as follows [11,12]:

(1) *Silicon cutting*: Thin slices of round silicon wafers are sliced from a huge and pure cylindrical silicon ingot [11]. The surfaces of the silicon wafers are well-polished and coated with a layer of insulating silicon dioxide to protect them from oxidisation and contamination [11].

(2) *Photolithography*: The photolithography process is critical for patterning silicon wafers in the fabrication process of ICs. The process of photolithography can be described in the steps below:

    a. *Surface preparation*: Surface preparation is required for silicon wafers to ensure that the photoresist can well adhere to their surfaces (see **Figure 1.4(a)**). Particle removal, dehydration baking, and wafer priming are the essential processes required for surface preparation.

    b. *Photoresist application*: Thin layers of photoresist are applied evenly on the wafers' surfaces by spin coating (see **Figure 1.4(b)**).

    c. *Soft baking*: Soft baking is a heating process to evaporate solvents from the photoresist and soft baking can be done through conduction, convection and radiation (see **Figure 1.4(c)**). Some of the common equipment used for soft baking include manual hot plates, in-line single-wafer hot plates, moving belt hot plates, convection ovens, moving-belt infrared ovens, microwave ovens and vacuum ovens. The presence of the solvents in the photoresist is undesirable as they can affect the photoresist adhesion to the wafers and interfere with the subsequent processing steps.

    d. *Alignment and exposure*: The photomask is aligned to the desired position of the silicon wafer. Light or other radiation sources are passed through the photomask to encode the patterns on the photoresist (see **Figure 1.4(d)**). Similar to the photolithography

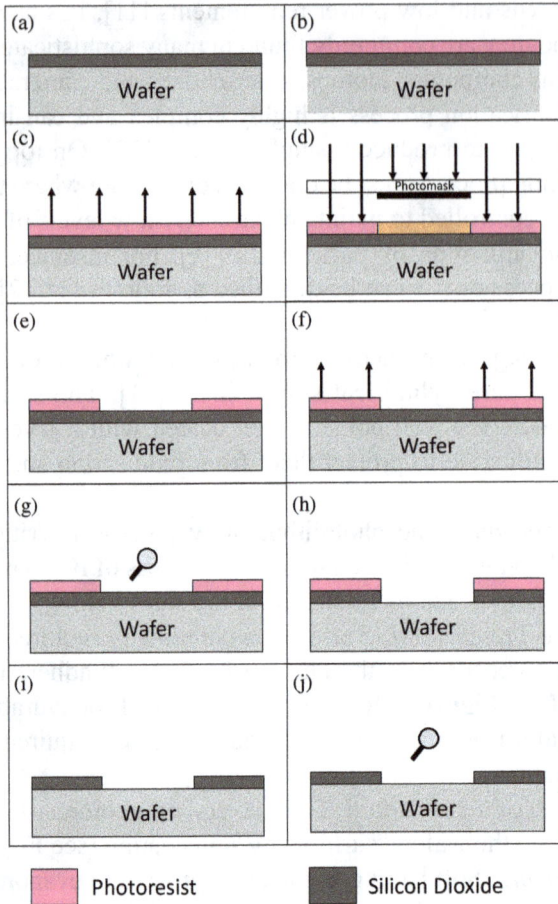

Figure 1.4.   Schematic diagrams of the steps in the photolithography process of IC fabrication: (a) surface preparation; (b) photoresist application; (c) soft baking; (d) alignment and exposure; (e) development; (f) hard baking; (g) develop inspect; (h) etching; (i) photoresist removal and (j) final inspection [12].

process in PCB fabrication, two different types of photoresist (positive photoresist and negative photoresist) can be used. For example, the negative photoresist polymerises when exposed to light as shown in **Figure 1.4(d)**.

e. *Development*: The unexposed negative photoresist that is unpolymerised by light can be dissolved easily by chemical developers (see **Figure 1.4(e)**). The exposed silicon wafer that is not covered by polymerised photoresist will be etched in the subsequent steps.

f. *Hard baking*: Hard baking dehydrates and polymerises the photoresist, which makes the photoresist more etch-resistant to etchants (see **Figure 1.4(f)**).

g. *Develop inspect*: The silicon wafers are inspected for defects and misalignments on their surfaces (see **Figure 1.4(g)**). Rejected wafers will be sent for rework, in which the photoresist layers are stripped off and the wafers go through the entire photomasking process again from the beginning.

h. *Etching*: Etching can be done by either wet etching or dry etching. The exposed top layer of the silicon wafers is etched away by etchants through the openings in the photoresist layer (see **Figure 1.4(h)**). The photoresist layer acts as an etch barrier for the underlying silicon wafers.

i. *Photoresist removal*: The photoresist layer is no longer needed after the etching process and can be stripped away by wet chemical processes (see **Figure 1.4(i)**).

j. *Final inspection*: The silicon wafers are inspected visually for defects and etch irregularities (see **Figure 1.4(j)**).

(3) *Doping*: Doping is defined as a "process that puts specific amounts of electrically active dopants in the wafer surface through openings in the surface layers" and it can be done either by thermal diffusion or ion implantation.

(4) *Making successive layers*: The photolithography and doping processes are repeated to make successive layers on the silicon wafer until the completion of the ICs. Insulating silicon dioxide can also be coated between different electrical components or layers. A layer of silicon dioxide is coated over the finalised ICs and various positions designated for contact points are etched away. Aluminium is then deposited onto the contact points to form the contact pads.

(5) *Separating individual IC*: Hundreds of ICs are fabricated on a single silicon wafer and they can be individually separated by scoring them with a fine diamond cutter.

(6) *Inspecting individual IC*: The ICs are visually inspected and electrically tested. Those ICs which fail the inspection and electrical tests are disposed off.

1.1.3 *Advantages and Disadvantages of Conventional Electronics*

This section discusses some of the advantages and disadvantages of conventional electronics. The advantages of conventional electronics include:

(1) *High resolution*: The current state-of-the-art technology for manufacturing conventional electronics can produce very high-resolution features. For instance, the ICs have feature sizes in the low-nanometre-size regime [12,22].
(2) *High integration density*: Conventional electronics have very high integration density [23]. For instance, an IC contains large numbers of electrical components in a small confined area [24].
(3) *High performance*: Conventional electronics are usually very reliable and offer high performances [25].

The disadvantages of conventional electronics include:

(1) *High capital investments*: Conventional electronics manufacturing requires very high upfront capital investments [26] for specialised machineries, equipment and facilities [27], and thereby also resulting in high barriers to entry.
(2) *Complex processing steps*: Conventional electronics manufacturing is very time-consuming and involves many complex processing steps [23,28]. Hence, conventional electronic manufacturing is not very favourable for prototyping since time bottlenecks usually arise in the prototyping phase.
(3) *Environmentally unfriendly*: Conventional electronics manufacturing is a "mixed subtractive-additive" approach [1] where excessive wastages are generated from subtractive processes. For instance, up to 90% of copper material may be etched away during manufacturing processes of PCBs [12]. In addition, the corrosive chemicals used during the manufacturing process can also cause environmental pollution [28] if they are not properly treated and disposed.
(4) *Rigid substrates*: Conventional electronics manufacturing techniques are only compatible with planar, rigid substrates (for example, FR-4 and silicon substrates) [23,25], and thereby resulting in lesser design freedom.

1.2 Printed Electronics

Printed electronics, as the name suggests, is a set of electrical devices fabricated by conventional printing techniques on a variety of substrates [29,30]. In other words, the printed electronics technology integrates the matured conventional print media printing technology with electronics manufacturing to allow direct deposition of electrically functional inks onto substrates for the fabrication of functional electronic devices [1], while achieving high cost-effectiveness and high-throughputs at the same time. Some of the conventional printing techniques used include screen printing, flexographic printing, gravure printing, gravure offset printing, pad printing and offset lithography.

The fundamental motivation for printed electronics is to fabricate electrically functional structures and devices at a lower cost, higher speed and lesser production complexity as compared to conventional electronics [5,30] when stringent electrical properties and high performances are not required [31]. In addition, printed electronics can also be flexible, thin, light-weight, large-scale and environmental friendly [10]. With all these advantages, the printed electronics technology is increasingly drawing significant interests from the industries as a viable manufacturing method for functional electronic devices [14,30]. A research market study done by BCC Research [32,33], forecasts that the global printed electronics market is expected to grow at a compound annual growth rate (CAGR) of 13.6%, from $14.0 billion US dollars in 2017 to $26.6 billion US dollars by 2022. Experts have also predicted that printed electronics also have the potential to revolutionise product innovations and generating new applications.

To many people's surprise, the printed electronics technology is not new. The first attempts at using conventional printing equipment to fabricate functional printed electronics can be traced back as early as the 1950s [10]. Japanese researchers from Nippon Telegraph and Telephone first started utilising the gravure printing technique for the fabrication of printed wiring boards due to its ability to achieve fine pitch accuracy [10]. Cheek *et al.* [34] demonstrated screen printing of silver contacts on silicon solar cells in 1984 [33]. Bao *et al.* [35], in 1997, also demonstrated the fabrication of high-mobility transistors on indium tin oxide (ITO)-coated poly(ethylene terephthalate) film entirely by screen printing. Until the recent decades, there are tremendous advancements in printed electronics technologies. This can be highly attributed to the better printing capabilities and the new inks developed over the years [10].

Different layers of functional materials are printed and overlapped on top of each other to fabricate printed electronics devices [12]. Various liquid forms of these functional materials, such as dispersions, suspensions and solutions, are usually used in conventional printing techniques for fabricating printed electronics. These functional materials can either be organic or inorganic materials, and function as conductors, insulators, dielectric or semi-conductors [36]. The interfacial interactions between different material layers, such as adhesion, wetting and solubility, can also affect the printability and electrical functionality. Hence, compatibility of different inks must be considered especially for multi-layer, multi-material printing. Post-processing of the functional inks, such as drying, curing or sintering, is usually required after the printing process. Many commercial printed electronics products can also be readily found in the markets for applications such as flexible electronics, flexible displays, wearable electronics, radio-frequency identification (RFID) tags and smart labels [30].

1.2.1 *Advantages and Disadvantages of Printed Electronics*

This section discusses some of the advantages of printed electronics over conventional electronics.

The advantages of printed electronics include:

(1) *Cost Benefits*: Printed electronics can be low-cost alternatives to conventional electronics components, especially when stringent electrical properties and high performances are not required [31]. Printed electronics may also be more cost-effective for low-volume and high-customisation productions [37]. Printed electronics can typically enjoy cost savings in these areas:

    a. *Lower capital expenditure*: The printing equipment used for fabricating printed electronics generally requires lower capital expenditures, as compared to the traditional lithography equipment used for fabricating conventional electronics components [38].

    b. *Lower overall process complexity*: Printing is an additive process which only involves the direct material deposition of functional inks onto the substrates and is technically less complex as compared to the traditional lithography process [38].

    c. *Lower substrate costs*: Low-cost substrates such as polymer films and paper can be used for printed electronics applications. Hence, pushing down further the costs for fabricating printed electronics further [38].

(2) *High-volume fabrication*: Conventional roll-to-roll (R2R) printing techniques, such as flexographic printing, gravure printing, gravure-offset printing and offset lithography, typically have high print speeds and high throughputs characteristics [1]. Hence, these printing techniques can be utilised for high-volume fabrications of printed electronics for better cost-effectiveness.

(3) *Wide range of substrates can be used*: The printing technologies in printed electronics can deposit functional inks directly onto the substrates. Hence, a wide range of rigid and flexible substrates, such as polymer films, pressure-sensitive foils, glass, paper and textiles, can be used for various printed electronics applications [1].

(4) *Flexible electronics*: One of the major benefits of printed electronics is the ability to fabricate electronic devices on flexible substrates. This unique type of electronics, also commonly known as flexible electronics, can be rolled, bent, folded and even stretched to some extent without losing their functionality [14]. These features are not found in rigid conventional electronic devices.

(5) *Lesser material wastage and more environmentally friendly*: The printed electronics processes are additive and can bring about lesser material wastages [39]. The printed electronics technologies are also more environmentally friendly as no corrosive and toxic etchants are required for the printing process.

(6) *Large-area processing*: Functional inks can be deposited directly onto large-area substrates that span over several metres in length with printed electronics technologies, which is an added advantage over conventional electronics manufacturing [12,40]. These large-area electronics usually used in applications such as solar panels or information displays [12].

(7) *Wide range of potential applications*: Printed electronics can be used in a wide range of potential applications, such as RFID tags, passive electrical components, displays, sensors, photovoltaics, lighting and many more [31].

The disadvantages of printed electronics include:

(1) *Low printing resolution*: Printed electronics have much low printing resolution as compared with conventional electronics. Printed

electronics can produce feature sizes in the micrometre-size regime at best, whereas conventional electronics have feature size in low-nanometre-size regime [12,22].

(2) *Low alignment and overlay accuracy*: Conventional printing techniques used for fabricating printed electronics have lower alignment and overlay accuracy [12] as compared to conventional electronics manufacturing techniques, such as photolithography. Alignment and overlay accuracy can affect the overall uniformity, electrical performances and functionality, especially when layers of different materials are printed on top of each other to fabricate the printed electronics devices.

(3) *High surface roughness*: The surface geometry of printed electronics, such as surface roughness and thickness, can significantly affect their performances [41]. Functional materials deposited by conventional printing techniques usually have high surface roughness, which can cause problems like charge leakage or electric breakdown at the interfaces when layers of different materials are printed on top of each other [12].

(4) *Post-processing required*: Post-processing of the functional inks, such as drying, curing or sintering, are usually required after the printing process.

The comparisons between conventional electronics and printed electronics are also summarised in Table 1.1.

Table 1.1.   Comparison between printed electronics and conventional electronics [5].

Parameters	Conventional Electronics	Printed Electronics
Process	Batch	Continuous (R2R printing)
Capital expenditure	Extremely high	Low to moderate
Production speed	Slow	Fast (R2R printing with high throughput is possible)
Accuracy and resolution	High	Low to moderate
Cost	Moderate in high volume	Low to moderate
Substrates	Rigid substrates	Wide range of substrates
Environmental friendliness	Less environmentally friendly	More environmentally friendly

1.3 Additive Manufacturing

Additive manufacturing (AM) [42–48], also commonly known as three-dimensional (3D) printing or formerly known as rapid prototyping (RP), is defined by the American Society for Testing and Materials (ASTM) in ISO/ASTM 52900:2015 as a "process of joining materials to make parts from 3D model data, usually layer upon layer, as opposed to subtractive manufacturing and formative manufacturing methodologies" [49]. The emerging additive manufacturing technology is revolutionising the entire manufacturing industry with its added advantages over traditional manufacturing methods, such as fabricating complex parts from digital data and rapid production of prototypes [50–63]. The global market for additive manufacturing is forecasted to grow at a rate of 15% (CAGR, 2015–2025) and expected to exceed $10 billion US dollars by 2021 [2].

1.3.1 *Fundamentals of Additive Manufacturing*

The additive manufacturing technology fabricates a physical 3D object by successive addition of material [49], joined together layer by layer. Most additive manufacturing processes usually have a similar process chain. A generalised process chain for additive manufacturing can be described in five main steps: 3D modelling, data conversion and transmission, checking and preparing, building and post-processing [1].

(1) *3D modelling*: A 3D model is necessary for any additive manufacturing process to build the desired physical object. The 3D model can either be modelled as a surface representation or 3D solid on a computer-aided design/computer-aided manufacturing (CAD/CAM) system or 3D scanned as a digital CAD file with reverse engineering equipment [64].
(2) *Data conversion and transmission*: The digital CAD file of the 3D model is usually converted to the .STL file format [42], as it is the de facto file format that most AM systems and CAD systems can accept and output respectively [64]. The .STL file format is defined in ISO/ASTM 52915:2016 as a "file format for model data describing the surface geometry of an object as a tessellation of triangles used to communicate 3D geometries to machines to build physical parts". The STL file is then transferred to the AM systems for checking and preparing before building.

(3) *Checking and preparing*: The STL file must be first checked for errors (such as missing facets or gaps, degenerate facets, overlapping facets and non-manifold conditions) to prevent potential printing failures. The AM system's computer then analyses the error-free STL file and slices the model into thin cross-sections [42]. The model can also be manipulated by software to change its building orientation, position and size, and generate the necessary support structures [42,64]. The AM system must also be prepared and set up with the desired building parameters and slice parameters prior to the build process [64].

(4) *Building*: The part building is a fully automated process for most AM systems, in which the systems can build the parts without much supervision [42]. However, superficial monitoring, to prevent errors like insufficient feedstocks and software glitches, may be required to ensure minimal disruptions during the building process [64].

(5) *Post-processing*: Post-processing is the last step of the additive manufacturing process chain. This step is manual and highly dependent on the operator's skills, and hence it often carries the high risks of damaging the printed parts [42]. Post-processing can include tasks like cleaning, post-curing, sintering and finishing. For instance, some AM parts may require cleaning to remove the excess materials or supporting structures, and additional finishing to improve their aesthetic appearances and surface finishing. Post-curing, in particular, is required for curing any unreacted photosensitive resin in stereolithography apparatus (SLA) parts.

The additive manufacturing wheel in **Figure 1.5** can be used to describe the four primary aspects of the development of additive manufacturing: input, material, method and applications [42].

(1) *Input*: The input can either be a computer model or a physical object. The computer model can be created by a CAD/CAM system, whereas the 3D model of the physical object can be scanned with 3D scanners or laser digitisers.

(2) *Material*: The materials used for additive manufacturing can come in solid, powder and liquid forms [42]. Solid materials can also be further classified as laminates, pellets or wire. The range of materials includes, but not limited to, ceramics, metals, paper, polymers, resins and wax.

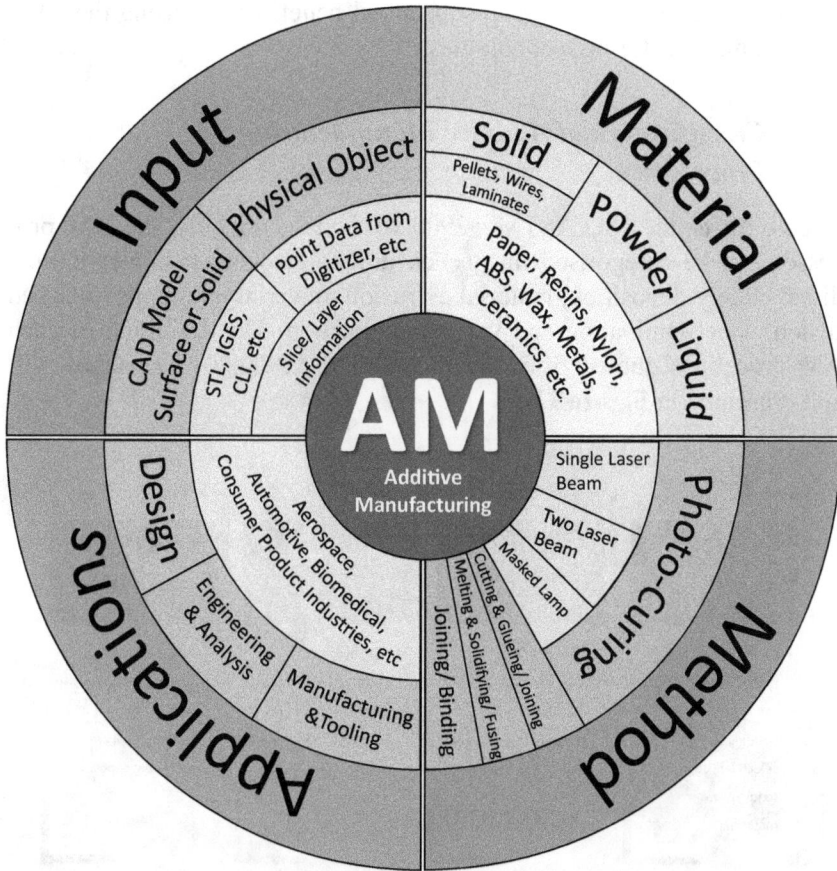

Figure 1.5.   The additive manufacturing wheel depicting the four key aspects of AM [42].

(3) *Method*: The different methods of additive manufacturing [42] can be generally categorised as: joining or binding; melting and solidifying or fusing; cutting and glueing or joining; and photo-curing. The photo-curing method can be further classified into a single laser beam, double laser beam and masked lamp.

(4) *Applications*: The applications of additive manufacturing [42] can be generally categorised as: design; engineering and analysis; and manufacturing and tooling. Various industries such as, but not limited to, automotive, aerospace, biomedical, electronics, and consumer

product industries can reap substantial benefits from using the additive manufacturing technologies.

1.3.2 *Classification of Additive Manufacturing Processes*

According to the ISO/ASTM 52900:2015 standard [49], the AM processes can be categorised into seven unique categories: binder jetting, direct energy deposition, material extrusion, material jetting, powder bed fusion, sheet lamination and vat polymerisation. The definition of each AM process is quoted from the ISO/ASTM 52900:2015 standard [49] and presented in **Figure 1.6**.

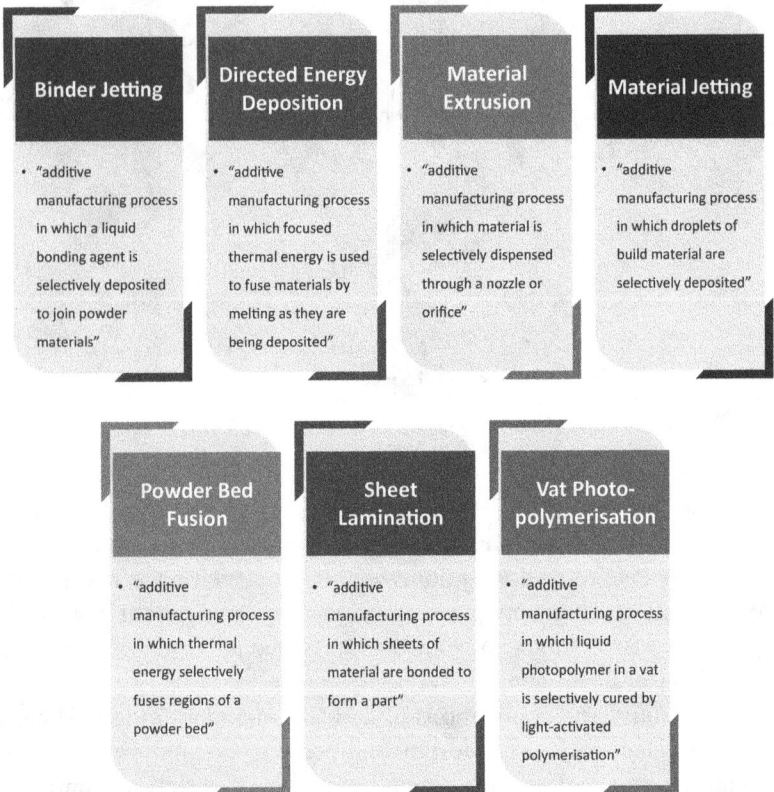

Figure 1.6.   Classification of additive manufacturing processes [49].

1.3.3 *Advantages of Additive Manufacturing*

With continuing research efforts in the additive manufacturing sector, there are increasing uses of this new technology for more novel applications [65–71]. This section will discuss some of the advantages of additive manufacturing over conventional manufacturing techniques. The advantages of additive manufacturing include:

(1) *Rapid prototyping*: Additive manufacturing can accelerate prototyping and reduce product development costs. Hence, allowing companies to have competitive leverage over their competitors by the significant reduction in their time-to-market for new product innovations [42,72].

(2) *Ease of modifications and redesigns*: Additive manufacturing can allow product designs to be changed easily for fabrication to facilitate modifications and redesigns easily, without incurring any additional tooling costs or preparations [73].

(3) *On-demand manufacturing*: Additive manufacturing can allow on-demand manufacturing of parts. Hence, allowing production capacity to change according to the market demand with minimal impact on the manufacturing facilities [42]. In addition, benefits like lower inventory costs, shorter supply chain and product lifecycle leverage can be enjoyed too [72].

(4) *Mass customisation*: Additive manufacturing can allow mass customisation of products at a lower cost as compared to conventional manufacturing techniques [72], in which different product designs can be changed easily for fabrication without incurring any additional costs or tooling preparations [73].

(5) *Small volume manufacturing*: Additive manufacturing is more cost-effective for small volume manufacturing as it can eliminate the need for expensive moulds and tooling [42,72].

(6) *Complex geometries and structures*: Complex geometries, such as lattice structures, can be fabricated easily with additive manufacturing at low cost with short lead time and minimal waste [42,72,73]. It is challenging to machine complex geometries with traditional manufacturing techniques due to time-consuming and complex tooling processes [73].

(7) *Reduce part counts*: Part counts can be reduced significantly with the additive manufacturing technology, by combining several part

features into single-piece parts [42]. This attribute is previously not achievable by traditional manufacturing techniques due to poor tool accessibility or the need to minimise waste and machining. For instance, GE Aviation had successfully reduced 900 separate components of a helicopter engine to 16 parts [74] with the help of the additive manufacturing technology, and also enjoyed significant reductions in weight and production cost.

1.4 3D Printing of Electronics

3D printing of electronics is an emerging interdisciplinary field with the integration of engineering principles and materials science, which allow functional inks to be deposited digitally and precisely onto substrates by additive manufacturing technologies to fabricate functional electronic devices [75]. The functional electronic devices fabricated by additive manufacturing technologies are defined as 3D printed electronics.

3D printed electronics has become one of the world's fastest growing technologies in recent years. The 3D printed electronics sector in the AM industry had generated $681 million US dollars of revenue in 2015, which was equivalent to 13% of the larger AM industry. The 3D printed electronics' market size is also expected to grow and exceed $1 billion US dollars by 2025 [4–5]. AM experts have also predicted that 3D printed electronics have the potentials of revolutionising product innovations, generating innovative applications, and paving new opportunities for markets [76].

3D printed electronics has attracted tremendous interests over the recent years for on-demand fabrication of highly customisable electronics, by depositing functional materials directly onto conformal surfaces and wide varieties of substrates [75] (including transparent, flexible, stretchable and wearable substrates). Apart from that, 3D printed electronics with multiple functionalities can also be fabricated by embedding functional materials within complex 3D structures [76] by multi-material printing platforms [77].

The aerospace and electronics industries express rising interest in fabrications of functional antennas and sensors directly onto conformal surfaces through AM technologies [1,42,78]. The aerospace industry anticipates the use of AM technologies for more significant weight reduction and space utilisation to seek better fuel efficiency and aircraft performances, whereas the electronics industry aims to reduce footprints of electronic devices with increased functionalities [78].

1.4.1 *Key Processes in 3D Printing of Electronics*

There are two key processes in 3D printing of electronics: the printing process and sintering process (see **Figure 1.7**). The printing process is a process in which the electronics printer directly deposits functional inks onto the substrates digitally according to the computer-aided design (CAD) files tool paths. The functional inks that are commonly used in 3D printing of electronics include metallic nanoparticle inks, carbon nanomaterials inks, metal-organic decomposition (MOD) inks, semiconductor inks, conductive polymer and dielectric inks.

The printed patterns obtained at the end of the printing process are usually wet and non-functional. Thus, the printed patterns must go through an additional post-processing process, also known as the sintering process, to turn them into functional printed electronics. The sintering process is a process in which the sintering equipment introduces energy to the printed patterns through either thermal, photonic, microwave, electrical, plasma or chemical means.

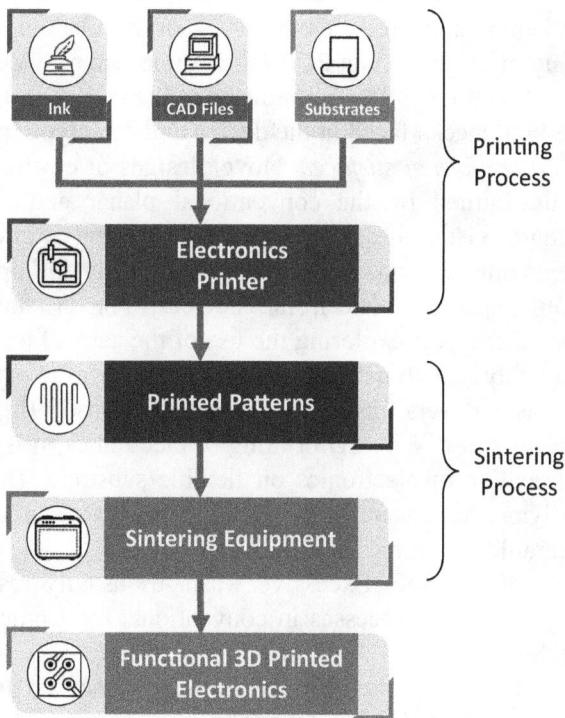

Figure 1.7.   Key processes in 3D printing of electronics.

1.4.2 *Advantages of 3D Printing of Electronics*

With the added advantages over conventional electronics manufacturing techniques, additive manufacturing could revolutionise the electronics industry in the near future. This section discusses some of the advantages of 3D printing of electronics.

The advantages of 3D printing of electronics include:

(1) *In-house prototyping*: PCBs are usually outsourced to external vendors for prototyping, but companies may face intellectual property infringement concerns when outsourcing. 3D printing of electronics can allow companies to keep the design, prototyping and fabrication capabilities in-house, and hence safeguarding their PCB intellectual property [1].

(2) *Eliminating time-bottlenecks in prototyping*: Conventional fabrication methods may take weeks to fabricate a prototype PCB before its actual testing can be done [1]. Further amendments on the current PCB's design may require weeks of fabrication again for the new board. Hence, causing time bottlenecks in the prototyping process and lengthening the product's time-to-market. On the other hand, 3D printing of electronics in PCB fabrications can provide users with additional flexibility and ease of editing of the PCB's design to eliminate time-bottlenecks faced in the design and prototyping phase.

(3) *Printing on conformal surfaces*: Novel designs of electronic devices are usually limited by the conventional planar and rigid PCBs. 3D printing of electronics can allow direct printing of circuitries and components onto conformal surfaces which favour full optimisations of available spaces in electronic devices. For instance, Lite-On Mobile Mechanical is exploring the use of the aerosol jet technology to directly fabricate functional antennas and sensors onto pre-cast devices' plastic covers for slimmer product designs [79].

(4) *Flexible substrates*: With 3D printing of electronics, it is possible to allow fabrication of electronics on flexible substrates (for instance, paper, polymer films and textiles) for innovative applications like smart wearable electronics [1].

(5) *Low material wastage*: Excessive wastages usually result from a series of subtractive processes in conventional electronics manufacturing techniques [1,2,4]. 3D printing of electronics, on the other hand, only involves material deposition processes on the required areas to minimise material wastage [1].

(6) *Streamline product development*: Fully functional electronics and structures can now be fabricated together in a single manufacturing process with additive manufacturing technologies to streamline product development. 3D printing of electronics can allow customisation and integration of electronics into product design to reap potential benefits like space optimisations, weight reduction and material saving.

(7) *Mass customisation*: 3D printed electronics can offer mass-customisation and greater versatility at a lower cost as compared to conventional electronics, in which product designs can be changed accordingly to individual customers' preferences and requirements without incurring any additional costs or tooling preparations.

(8) *Ease of changing designs*: The use of CAD can allow designers to make numerous modifications and design iterations easily, without incurring any additional tooling costs or preparations. Early modifications during the design phase help to add more agility to the product development cycle. The new designs can then be sent to 3D printers to have their prototypes fabricated immediately for functional testing.

References

[1] Tan, H. W., Tran, T. and Chua, C. K. (2016). A review of printed passive electronic components through fully additive manufacturing methods, *Virtual Phys. Prototyp.*, 11, pp. 271–288.

[2] Chang, J., Zhang, X., Ge, T. and Zhou, J. (2014). Fully printed electronics on flexible substrates: High gain amplifiers and DAC, *Org. Electron.*, 15, pp. 701–710.

[3] Haglund, R. F. and Itoh, N. (1994). *Laser Ablation. Springer Series in Materials Science, vol. 28*, eds. John C. Miller, Chapter 2: Electronic processes in laser ablation of semiconductors and insulators (Springer, Berlin, Heidelberg) pp. 11–52.

[4] Zhang, X., Ge, T. and Chang, J. S. (2015). Fully-additive printed electronics: Transistor model, process variation and fundamental circuit designs, *Org. Electron.*, 26, pp. 371–379.

[5] Hrehorova, E. (2007). *Materials and Processes for Printed Electronics: Evaluation of Gravure Printing in Electronics Manufacture* (Doctoral dissertation, Western Michigan University).

[6] Robertson, C. T. (2004). *Printed Circuit Board Designer's Reference: Basics* (Prentice Hall Professional, USA).

[7]   Awasthi, A. K. and Zeng, X. (2019). *Waste Electrical and Electronic Equipment (WEEE) Handbook* 2nd edn, eds. Vannessa Goodship, Ab Stevels and Jaco Huisman, Chapter 11: Recycling printed circuit boards (Woodhead Publishing, USA) pp. 311–325.

[8]   Urey, H., Holmstrom, S. and Yalcinkaya, A. D. (2008). Electromagnetically actuated FR4 scanners, *IEEE Photonics Technology Letters*, 20, pp. 30–32.

[9]   Tiwari, J. N. (2019). *A to Z Computer Acronyms with Explanations* (Shri Balaji, India).

[10] Suganuma, K. (2014). *Introduction to Printed Electronics* (Springer Science+Business Media, New York).

[11] Lim, K.-S. (1996). *How Products Are Made: An Illustrated Guide to Product Manufacturing, Volume 2* (Gale Research, Detroit, Michigan).

[12] Cui, Z. (2016). *Printed electronics: Materials, technologies and applications* (John Wiley & Sons, Singapore).

[13] Circuits, A.-T. Introduction about PCB drill holes. Retrieved from https://www.atechcircuit.com/pcb-news-resource/pcb-knowledge/90-introduction-about-pcb-drill-holes.

[14] Cruz, S. M. F., Rocha, L. A. and Viana, J. C. (2018). *Flexible Electronics*, Chapter 2: Printing technologies on flexible substrates for printed electronics (IntechOpen).

[15] Khandpur, R. S. (2005). *Printed Circuit Boards: Design, Fabrication, Assembly and Testing* (Tata McGraw-Hill Education, New Delhi).

[16] Falcaro, P., Ricco, R., Doherty, C. M., Liang, K., Hill, A. J. and Styles, M. J. (2014). MOF positioning technology and device fabrication, *Chem. Soc. Rev.*, 43, pp. 5513–5560.

[17] WellPCB. How To Get Satisfied PCB Silkscreen — A Step-by-Step Guide. Retrieved from https://www.wellpcb.com/pcb-silkscreen.html.

[18] Willis, B. (2004). Printed circuit board basics — An introduction to the PCB industry, *Circuit World*, 30(3).

[19] International Electrotechnical Commission (2002). IEC 60748-1:2002 *Semiconductor Devices — Integrated Circuits — Part 1: General*. Retrieved from https://webstore.iec.ch/publication/3291.

[20] Saxena, A. N. (2009). *Invention of Integrated Circuits: Untold Important Facts*, Chapter 1: Introduction (World Scientific Publishing Singapore) pp. 1–30.

[21] Judy Lynne, S. and Christopher, S. Integrated circuit. Retrieved from https://www.britannica.com/technology/integrated-circuit.

[22] Murr, L. E. (2015). *Handbook of Materials Structures, Properties, Processing and Performance*, Photolithography applied to integrated circuit (IC) microfabrication (Springer Cham) pp. 607–612.

[23] Bandyopadhyay, A. and Bose, S. (2019). *Additive Manufacturing*, 2nd edn (CRC Press, USA).

[24] Moore, G. E. (1965). Cramming more components onto integrated circuits, *Electronics*, 38, pp. 114–117.

[25] Khan, S. (2016). *Towards Merging of Microfabrication and Printing of Si μ-Wires for Flexible Electronics* (Doctoral dissertation, University of Trento).

[26] King, B. and Renn, M. (2009). Aerosol Jet direct write printing for mil-aero electronic applications, *presented at the Lockheed Martin Palo Alto Colloquia*, Palo Alto, CA.

[27] Cole, P., Turner, L., Hu, Z. and Ranasinghe, D. (2011). *Unique Radio Innovation for the 21st Century*, eds. D. Ranasinghe, Q. Sheng and S. Zeadally, The Next Generation of RFID Technology (Springer, Berlin, Heidelberg).

[28] Patil, B. H. (2015). *Formulation and Evaluation of Soy Polymer Based, Gravure Printed Resistive Inks for Applications in Printed Electronics* (Masters thesis, Western Michigan University).

[29] Wu, W. (2017). Inorganic nanomaterials for printed electronics: A review, *Nanoscale*, 9, pp. 7342–7372.

[30] Aijazi, A. T. (2014). *Printing Functional Electronic Circuits and Components* (Doctoral dissertation, Western Michigan University).

[31] Gregor-Svetec, D. (2018). *Nanomaterials for Food Packaging*, 1st edn, eds. Jose Maria Lagaron Miguel Ângelo Parente Ribeiro Cerqueira, Lorenzo Miguel Pastrana Castro and António Augusto Martins de Oliveira Soares Vicente, Chapter 8: Intelligent packaging, pp. 203–247.

[32] Sullivan, M. (2018). *Printed Electronics: Global Markets to 2022*.

[33] Huang, Q. and Zhu, Y. (2019). Printing conductive nanomaterials for flexible and stretchable electronics: A review of materials, processes, and applications, *Adv. Mater. Technol.*, 4, p. 1800546.

[34] Cheek, G. C., Mertens, R. P., Overstraeten, R. V. and Frisson, L. (1984). Thick-film metallization for solar cell applications, *IEEE Trans. Electron Devices*, 31, pp. 602–609.

[35] Bao, Z., Feng, Y., Dodabalapur, A., Raju, V. R. and Lovinger, A. J. (1997). High-performance plastic transistors fabricated by printing techniques, *Chem. Mater.*, 9, pp. 1299–1301.

[36] Bao, Z. (2000). Materials and fabrication needs for low-cost organic transistor circuits, *Adv. Mater.*, 12, pp. 227–230.

[37] Unander, T. (2011). *System Integration of Electronic Functionality in Packaging Application* (Doctoral dissertation, Mid Sweden University).

[38] Subramanian, V., Chang, J. B., Vornbrock, A. D. L. F., Huang, D. C., Jagannathan, L., Liao, F., Mattis, B., Molesa, S., Redinger, D. R., Soltman, D., Volkman, S. K. and Qintao, Z. (2008). Printed electronics for low-cost electronic systems: Technology status and application development, *presented at the ESSCIRC 2008 — 34th European Solid-State Circuits Conference*, Edinburgh, UK.

[39] Kunnari, E., Valkama, J., Keskinen, M. and Mansikkamäki, P. (2009). Environmental evaluation of new technology: Printed electronics case study, *J. Cleaner Prod.*, 17, pp. 791–799.

[40] Khan, S., Lorenzelli, L. and Dahiya, R. S. (2015). Technologies for printing sensors and electronics over large flexible substrates: A review, *IEEE Sens. J.*, 15, pp. 3164–3185.

[41] Nguyen, H.-A.-D., Nguyen, H., Shin, K. and Lee, S. (2011). Improvement of surface roughness and conductivity by calendering process for printed electronics, *presented at the 8th International Conference on Ubiquitous Robots and Ambient Intelligence (URAI)*, Incheon, South Korea.

[42] Chua, C. K. and Leong, K. F. (2017). *3D Printing and Additive Manufacturing — Principles and Applications*, 5th edn (World Scientific Publishing, Singapore).

[43] Choong, Y. Y. C., Maleksaeedi, S., Eng, H., Su, P.-C. and Wei, J. (2017). Curing characteristics of shape memory polymers in 3D projection and laser stereolithography, *Virtual Phys. Prototyp.*, 12, pp. 77–84.

[44] Choong, Y. Y. C., Maleksaeedi, S., Eng, H., Wei, J. and Su, P.-C. (2017). 4D printing of high performance shape memory polymer using stereolithography, *Mater. Des.*, 126, pp. 219–225.

[45] Choong, Y. Y. C., Maleksaeedi, S., Eng, H., Yu, S., Wei, J. and Su, P.-C. (2020). High speed 4D printing of shape memory polymers with nanosilica, *Appl. Mater. Today*, 18, p. 100515.

[46] Choong, Y. Y. C., Tan, H. W., Patel, D. C., Choong, W. T. N., Chen, C.-H., Low, H. Y., Tan, M. J., Patel, C. D. and Chua, C. K. (2020). The global rise of 3D printing during the COVID-19 pandemic, *Nature Reviews Materials*, 5, pp. 637–639.

[47] Eng, H., Maleksaeedi, S., Yu, S., Choong, Y. Y. C., Wiria, F. E., Kheng, R. E., Wei, J., Su, P.-C. and Tham, H. P. (2017). Development of CNTs-filled photopolymer for projection stereolithography, *Rapid Prototyping J.*, 23, pp. 129–136.

[48] Eng, H., Maleksaeedi, S., Yu, S., Choong, Y. Y. C., Wiria, F. E., Tan, C. L. C., Su, P. C. and Wei, J. (2017). 3D stereolithography of polymer composites reinforced with orientated nanoclay, *Procedia Engineering*, 216, pp. 1–7.

[49] ASTM International (2015). ISO/ASTM 52900-15, *Standard Terminology for Additive Manufacturing–General Principles–Terminology*. Retrieved from https://www.astm.org/Standards/ISOASTM52900.htm.

[50] Khoo, Z. X., Teoh, J. E. M., Liu, Y., Chua, C. K., Yang, S., An, J., Leong, K. F. and Yeong, W. Y. (2015). 3D printing of smart materials: A review on recent progresses in 4D printing, *Virtual Phys. Prototyp.*, 10, pp. 103–122.

[51] Bai, Y. L., Srikanth, N., Chua, C. K. and Zhou, K. (2019). Density functional theory study of M(n+1)AX(n) phases: A Review, *Crit Rev. Solid State Mater. Sci.*, 44, pp. 56–107.

[52] An, J., Teoh, J. E. M., Suntornnond, R. and Chua, C. K. (2015). Design and 3D printing of scaffolds and tissues, *Engineering*, 1, pp. 261–268.

[53] Lee, J.-Y., An, J. and Chua, C. K. (2017). Fundamentals and applications of 3D printing for novel materials, *Appl. Mater. Today*, 7, pp. 120–133.

[54] Tan, X., Kok, Y., Tan, Y. J., Descoins, M., Mangelinck, D., Tor, S. B., Leong, K. F. and Chua, C. K. (2015). Graded microstructure and mechanical properties of additive manufactured Ti–6Al–4V via electron beam melting, *Acta Mater.*, 97, pp. 1–16.

[55] Yu, W., Sing, S. L., Chua, C. K. and Tian, X. (2019). Influence of remelting on surface roughness and porosity of AlSi10Mg parts fabricated by selective laser melting, *J. Alloys Compd.*, 792, pp. 574–581.

[56] Kuo, C., Chua, C., Peng, P., Chen, Y., Sing, S., Huang, S. and Su, Y. (2020). Microstructure evolution and mechanical property response via 3D printing parameter development of Al–Sc alloy, *Virtual Phys. Prototyp.*, 15, pp. 120–129.

[57] Li, Y., Zhou, K., Tan, P., Tor, S. B., Chua, C. K. and Leong, K. F. (2018). Modeling temperature and residual stress fields in selective laser melting, *Int. J. Mech. Sci.*, 136, pp. 24–35.

[58] Loh, L.-E., Chua, C.-K., Yeong, W.-Y., Song, J., Mapar, M., Sing, S.-L., Liu, Z.-H. and Zhang, D.-Q. (2015). Numerical investigation and an effective modelling on the selective laser melting (SLM) process with aluminium alloy 6061, *Int. J. Heat Mass Transfer*, 80, pp. 288–300.

[59] Yu, W., Sing, S., Chua, C., Kuo, C. and Tian, X. (2019). Particle-reinforced metal matrix nanocomposites fabricated by selective laser melting: A state of the art review, *Prog. Mater. Sci.*, 104, pp. 330–379.

[60] Yuan, S., Shen, F., Chua, C. K. and Zhou, K. (2019). Polymeric composites for powder-based additive manufacturing: Materials and applications, *Prog. Poly. Sci.*, 91, pp. 141–168.

[61] Ng, W. L., Chua, C. K. and Shen, Y.-F. (2019). Print me an organ! Why we are not there yet, *Prog. Poly. Sci.*, 97, p. 101145.

[62] Yap, C. Y., Chua, C. K., Dong, Z. L., Liu, Z. H., Zhang, D. Q., Loh, L. E. and Sing, S. L. (2015). Review of selective laser melting: Materials and applications, *Appl. Phys. Rev.*, 2, p. 041101.

[63] Sun, Z., Tan, X., Tor, S. B. and Chua, C. K. (2018). Simultaneously enhanced strength and ductility for 3D-printed stainless steel 316L by selective laser melting, *NPG Asia Mater.*, 10, pp. 127–136.

[64] Gibson, I., Rosen, D. W. and Stucker, B. (2010). *Additive Manufacturing Technologies: Rapid Prototyping to Direct Digital Manufacturing*, Chapter 3: Generalized additive manufacturing process chain (Springer, New York) pp. 59–77.

[65] Chua, C.-K., Yeong, W.-Y. and Leong, K.-F. (2005). Rapid prototyping in tissue engineering: A state-of-the-art report, *presented at the 2nd Int. Conf.*

on *Advanced Research in Virtual and Rapid Prototyping*, Leiden, Netherlands.

[66] Lee, J. M., Ng, W. L. and Yeong, W. Y. (2019). Resolution and shape in bioprinting: Strategizing towards complex tissue and organ printing, *Appl. Phys. Rev.*, 6, p. 011307.

[67] Lee, J. M., Sing, S. L., Tan, E. Y. S. and Yeong, W. Y. (2016). Bioprinting in cardiovascular tissue engineering: A review, *Int. J. Bioprint.*, 2, pp. 27–36.

[68] Ng, W. L., Goh, M. H., Yeong, W. Y. and Naing, M. W. (2018). Applying macromolecular crowding to 3D bioprinting: Fabrication of 3D hier-archical porous collagen-based hydrogel constructs, *Biomater. Sci.*, 6, pp. 562–574.

[69] Ng, W. L., Yeong, W. Y. and Naing, M. W. (2016). Development of polye-lectrolyte chitosan-gelatin hydrogels for skin bioprinting, *Procedia Cirp*, 49, pp. 105–112.

[70] Ng, W. L., Yeong, W. Y. and Naing, M. W. (2017). Polyvinylpyrrolidone-based bio-ink improves cell viability and homogeneity during drop-on-demand printing, *Materials*, 10, p. 190.

[71] Suntornnond, R., An, J., Yeong, W. Y. and Chua, C. K. (2015). Biodegradable polymeric films and membranes processing and forming for tissue engi-neering, *Macromol. Mater. Eng.*, 300, pp. 858–877.

[72] Attaran, M. (2017). The rise of 3-D printing: The advantages of addi-tive manufacturing over traditional manufacturing, *Bus. Horiz.*, 60, pp. 677–688.

[73] Ngo, T. D., Kashani, A., Imbalzano, G., Nguyen, K. T. Q. and Hui, D. (2018). Additive manufacturing (3D printing): A review of materials, methods, applications and challenges, *Compos. Part B: Eng.*, 143, pp. 172–196.

[74] Kellner, T. (2017). An epiphany of disruption: GE additive chief explains how 3D printing will upend manufacturing. Retrieved from https://www. ge.com/reports/epiphany-disruption-ge-additive-chief-explains-3d-printing-will-upend-manufacturing/.

[75] Tan, H. W., An, J., Chua, C. K. and Tran, T. (2019). Metallic nanoparticle inks for 3D printing of electronics, *Adv. Electron. Mater.*, 5, p. 1800831.

[76] Lu, B., Lan, H. and Liu, H. (2018). Additive manufacturing frontier: 3D printing electronics, *Opto-Electron. Adv.*, 1, p. 170004.

[77] Lewis, J. A. and Ahn, B. Y. (2015). Three-dimensional printed electronics, *Nature*, 518, pp. 42–43.

[78] Paulsen, J. A., Renn, M., Christenson, K. and Plourde, R. (2012). Printing conformal electronics on 3D structures with Aerosol Jet technology, *pre-sented at the 2012 Future of Instrumentation International Workshop (FIIW)* Gatlinburg, Tennessee, USA.

[79] Kira. (2016). LITE-ON using Optomec 3D printing to mass-produce consumer electronics. Retrieved from https://www.3ders.org/articles/20160324-lite-on-using-optomec-3d-printing-to-mass-produce-consumer-electronics.html.

Problems

1. Compare and discuss the advantages and disadvantages of printed electronics over conventional electronics.
2. Describe briefly the general fabrication process of a double layer PCB and state the disadvantages of this process.
3. The fabrication processes of conventional electronics are complex and not environmentally friendly. Is it fair to make this statement? Discuss.
4. Define the term "additive manufacturing" according to the ISO/ASTM 52900:2015 standard terminology.
5. List the seven categories of additive manufacturing processes.
6. Describe three advantages of additive manufacturing.
7. Describe the key processes in 3D printing of electronics.
8. Describe three advantages of 3D printing of electronics.
9. What are the implications of 3D printing of electronics for different industries?

Chapter 2

Conventional Contact Printing Techniques for Printed Electronics

As the name suggests, contact printing techniques require the contact of a printing plate or an image carrier with the substrates for ink transfer [1]. The images or patterns can be printed on the substrates, either directly or indirectly, from a master plate. The master plate carries the image information in the form of engravings, but it cannot be amended freely once fabricated. Hence, the contact printing techniques are only cost-effective for high volume printing [2].

Screen printing, flexographic printing, gravure printing, offset gravure printing, pad printing and offset lithography are conventional contact printing techniques that have been around for many decades. These printing techniques are commonly used for mass printing of print media, fabrics and packaging in various industries due to their low cost and high throughputs characteristics [3]. By replacing the inks with functional materials (for instance, conductive metallic nanoparticle inks or conductive polymers), it is possible to fabricate functional electrical devices and circuitries on the desired substrates for printed electronics applications with these contact printing techniques.

This chapter will cover some of the most commonly used contact printing techniques used for printed electronics applications, namely screen printing, flexographic printing, gravure printing, offset gravure printing, pad printing and offset lithography. This chapter focuses on each printing technique's working principle, printing process, its strengths and weaknesses, and its applications in printed electronics.

2.1 Screen Printing

Screen printing is one of the simplest and oldest contact printing techniques dating back to ancient China [4]. Screen printing is commonly used to print patterns on textiles, T-shirts and plastic bags packaging [3].

2.1.1 *Working Principle and Printing Process*

The screen consists of a combination of mesh, stencil and screen frame as shown in **Figure 2.1**, in which the mesh can be made from silk fibres, plastic fibres or metal wires [3]. The printed patterns are defined by the open mesh areas that are not covered by the stencil, while the stencil blocks out the non-image areas on the substrate from getting in contact with the ink.

A rubber-edged squeegee blade spreads the ink across the screen during the screen printing process and hence, forces the ink through the open mesh areas. The screen only contacts the substrate at the edge of the squeegee blade at any point in time, and the ink is deposited onto the substrate as the screen peels off from the substrate after contact. High viscosity inks with thixotropic properties are specially chosen for screen printing applications as these inks can prevent ink leakage from the screen to substrates when not in use.

The print properties of screen printing can be largely affected by the attributes of the mesh such as mesh material, mesh count, wire diameter, wire bias and mesh tension [3,5]. Plastic fibres meshes are commonly used for decorative screen printing, whereas stainless steel wire meshes are preferred for printed electronics applications. The stainless steel meshes can better retain their dimensional stability after many repeated

Figure 2.1.   Schematic diagram of the screen.

prints and give the printed electronics the required high geometric tolerances [5]. The mesh count is defined as the number of fibres per inch and can be used to control the print thickness directly, in which higher mesh count generally gives smaller print thickness [5]. In addition, the print resolution is also inversely affected by the wire diameter, where higher print resolution is achievable with smaller wire diameter meshes [3]. Nevertheless, the screen frame [2], stencil material, stencil thickness [3] and ink viscosity are some of the factors that will also affect the printing properties of screen printing.

Manual screen printing is relatively slow, and it is not suited for roll-to-roll (R2R) processing of printed electronics. The automated flatbed screen printing and rotary screen printing techniques are some of the screen printing techniques that are commonly used for roll-to-roll (R2R) applications [6]. The automated flatbed screen printing technique (see **Figure 2.2**) is a stepwise process, in which the screen is raised after one print and a fresh roll of the substrate is rolled in for the next print [6]. The schematic diagram of the rotary screen printing process is shown in **Figure 2.3**. The screen used in rotary screen printing is in the form of a cylinder and it rotates at the same speed as the impression cylinder [6]. The ink is dispensed within the cylindrical screen and is forced through the open mesh areas of the screen by the stationary squeegee. The impression cylinder then pushes the substrate against the rotating screen for ink deposition. In comparison, the rotary screen printing technique can give better print resolution and higher printing speed than the flatbed screen printing technique [6], and hence allowing high throughput printing [2].

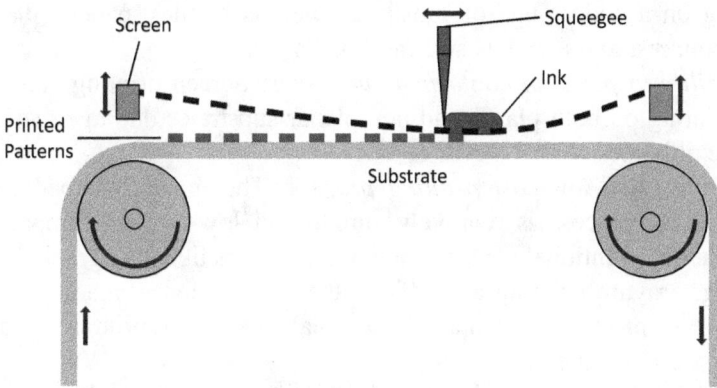

Figure 2.2.   Schematic diagram of the automatic flatbed screen printing process.

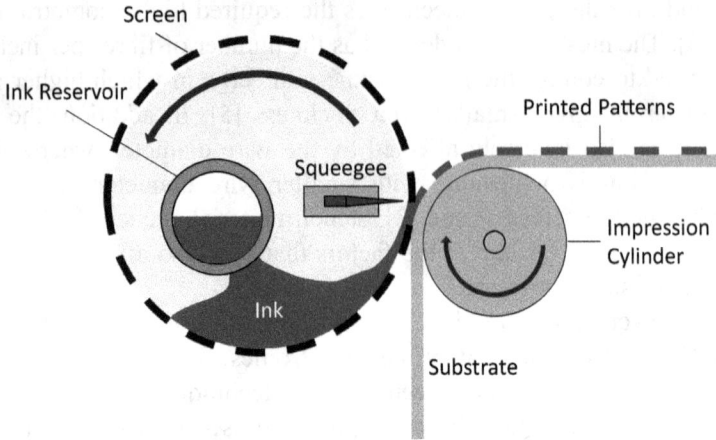

Figure 2.3.   Schematic diagram of the rotary screen printing process.

The rotary screen printing technique can achieve printing speeds over 100 m/min whereas the flatbed screen printing technique can only achieve printing speeds ranging from 0–35 m/min [6]. Moreover, rotary screen printing systems are considerably more costly than flatbed printing systems [5].

2.1.2 *Strengths and Weaknesses*

The strengths of screen printing include:

(1) *Adaptable for various substrates*: Screen printing is suitable for printing on a wide range of substrates such as textiles, papers, plastics, ceramics, glass, metals and cardboards [7].
(2) *Ability to print on conformal substrates*: Screen printing can allow printing on both planar and non-planar substrates due to its soft and flexible mesh [2].
(3) *Simple and low-cost printing process*: The manual flatbed screen printing process is relatively simple and low-cost as compared to other conventional contact printing techniques like flexographic printing, gravure printing and offset lithography. Hence, manual flatbed screen printing is suitable for small-scale lab printing and for cost-effectiveness.
(4) *Thick layers are possible*: Screen printing process can allow printing of thick layers, which are typically between 10–500 μm thick [6].

(5) *Scalable for large-area processing*: Screen printing is scalable for large-area processing at relatively low cost [8].

The weaknesses of screen printing include:

(1) *Only highly viscous inks or paste-like inks can be used*: Low viscosity inks tend to cause undesirable spreading and bleed out [9], and hence render them unsuitable for screen printing processes. Low viscosity inks also tend to leak out from the screen to substrates when not in use due to gravity [6]. Highly viscous inks are used to prevent these problems; addition of polymers binders to control ink viscosity has been found to degrade the electrical properties and functionalities of the inks [7]. Hence, preventing wider adoption of the screen printing techniques for printed electronics applications [1].
(2) *High wastage of materials*: Large amount of inks are wasted during the screen printing process.
(3) *High surface roughness*: Highly viscous inks are used in the screen printing process and hence, leading to printed patterns with high surface roughness [2,10].
(4) *Slow printing speed*: As compared to other conventional contact printing techniques like flexographic printing, gravure printing and offset lithography, the manual flatbed screen printing process is significantly slower [6]. However, rotary screen printing can achieve printing speeds over 100 m/min and hence, allowing applications in R2R processes [6].

2.1.3 *Applications*

Screen printing technique has been used in the following applications for printed electronics:

(1) *Conductive traces and electrodes*: Screen printing is widely used for printing conductive traces and electrodes [6] on different types of substrates, such as textiles [11–13], silicone [14] and thermoplastic polyurethane (TPU) [15] (see **Figure 2.4**).
(2) *Passive devices*: Screen printing has also be used for fabricating passive devices such as resistors, capacitors and inductors (see **Figure 2.5**) [16]. Screen printing is advantageous for fabricating passive devices due to its ability to produce printed patterns with large

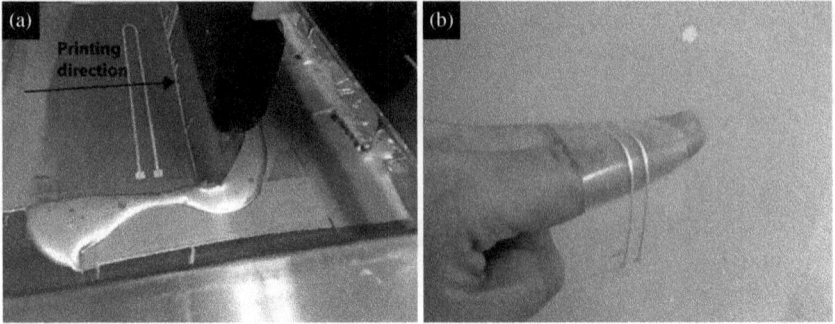

Figure 2.4.   (a) Screen printing process of strain test patterns on TPU substrate and (b) screen printed strain test patterns on flexible TPU substrate for wearable electronics applications. Reprinted with permission from Ref. [15]. Copyright (2016) from Springer Nature.

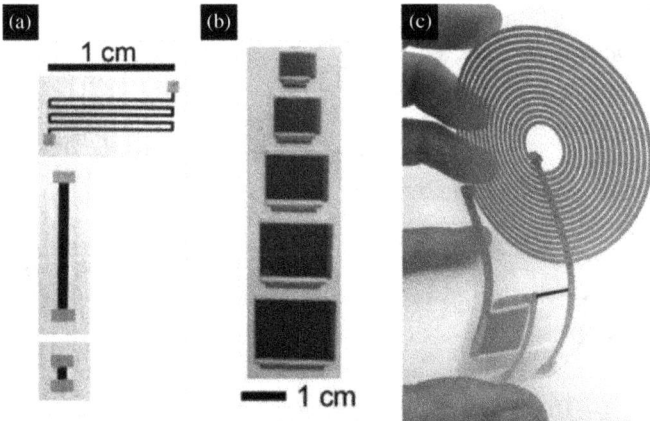

Figure 2.5.   Screen printed (a) resistors, (b) capacitors, and (c) inductor on flexible plastic substrates. Reproduce with permission from Ref. [16]. Copyright (2015) from Springer Nature.

film thickness and hence reducing the electrical resistance of these printed passive devices [16].

(3) *Sensors*: Screen printing has also be used to fabricate flexible sensors for pressure sensing [17] or large area sensing [18]. Khan *et al.* [17] fabricated foldable pressure sensors arrays with screen printing technique for low-cost electronic skin applications, and they also demonstrated that screen printing technique is also suitable for fabricating

multilayered electronic devices. Chang *et al.* screen printed flexible electronics sensor on polyimide substrates with thixotropy sol-gel materials for low-cost large area sensing applications [18].

(4) *Organic light-emitting devices (OLEDs)*: Pardo *et al.* [19] demonstrated the feasibility of using the screen printing technique for fabricating OLEDs, and this printing technique can be extended to logos and low information-content displays fabrications.

Other printed electronics applications by screen printing reported in literature also include active-matrix electrochromic displays [20], thin-film transistors [8], solar cells [21,22] and carbon-fibre supercapacitors for wearable electronics applications [23].

2.2 Flexographic Printing

Flexographic printing, also known as flexography, is another conventional contact printing technique used for printing labels on packaging materials, carriers, bags and labels [3]. Soft printing plate cylinder, fast drying and low viscosity inks are some of the prominent characteristics of the flexographic printing technique [3].

2.2.1 *Working Principle and Printing Process*

A typical flexographic printing process is shown in **Figure 2.6**. The fountain roller partially submerges in the ink reservoir and transports the ink up to the *anilox* roller [1].

The *anilox* roller is an engraved cylinder that comprises of many finely engraved cells. The size and frequency of these engraved cells can directly control the volume of ink that the *anilox* roller can hold, which in return affects the layer thickness of the printed patterns. For instance, an *anilox* roller with a lower frequency of engraved cells, or lower screen ruling, can increase the *anilox* volume. The doctor blade skims off excess ink from the non-engraved surfaces of the *anilox* roller, leaving only ink in the cavities. The *anilox* roller is then brought into contact with the soft printing plate cylinder to start the ink transfer process from the *anilox* roller to the raised relief areas of the printing patterns through surface tension. The printing patterns are embossed on the printing plate cylinder, which is usually made of soft materials such as rubber or polymers. The

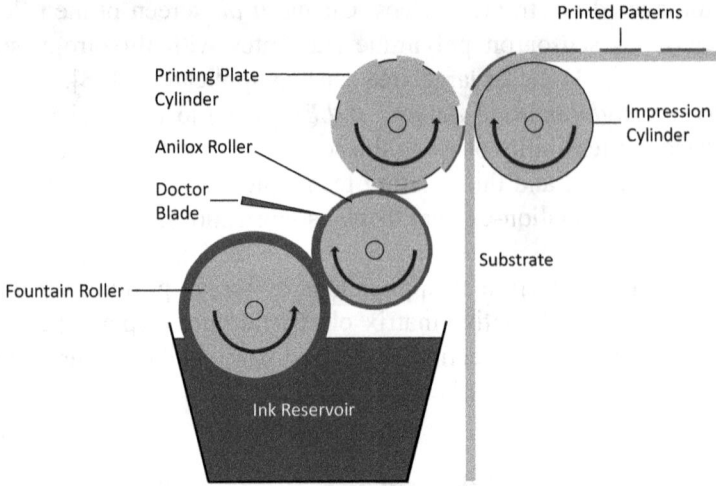

Figure 2.6.   Schematic diagram of the flexographic printing process.

impression cylinder then presses the substrates onto the printing plate cylinder for printing, which functions like a stamping process.

2.2.2 *Strengths and Weaknesses*

The strengths of flexographic printing include:

(1) *Low printing pressure*: The soft printing plate cylinder in flexographic printing can minimise damage on fragile substrates due to its low printing pressure on the substrates. The low printing pressure of flexographic printing is also ideal for multilayer printing [24], as damages to the pre-printed patterns can be minimised.

(2) *Suitable for a wide variety of substrates*: The flexographic printing technique can allow printing on a wide range of substrates, including compressible substrates [15], rigid substrates [2], flexible substrates, pressure-sensitive films and foils and metalised films [1,7]. Some of the substrates that can be used for flexographic printing include glass, textiles, polymer films, laminates, papers, paperboards and corrugated cardboards [15,25].

(3) *High print speed*: Flexographic printing can achieve high print speed (ranging from 100–500 m/min [7,17]), while still maintaining good

print resolution in the range of 50–100 μm [10,26] at the same time. The fast-drying inks used in the flexographic printing process allow the printing at high speed [25].

(4) *Wide variety of inks can be used*: Both water-based and oil-based inks can be used for flexographic printing, and the viscosity of these inks can range between 50–500 cP [7].

(5) *Thin and uniform layers*: Flexographic printing can achieve thin and uniform layers [27]. The anilox roller can directly control the volume of ink that is transferred to the printing plate cylinder, which in return affects the layer thickness of the printed patterns. The achievable printed layer thickness by flexographic printing is up to 1 μm [3].

The weaknesses of flexographic printing include:

(1) *Planar printing process*: The flexographic printing process applies only to planar substrates, and hence limiting its applications.

(2) *High initial cost and complex preparations prior to printing*: The flexographic printing process has a high setup cost and is more complex as compared to the screen printing process [1].

(3) *Distortion*: The soft printing plate cylinder can deform during the printing process and cause undesirable distortions in the printed patterns [2] along the printing direction. Hence, these distortions have to be accounted for when designing the printing plate.

2.2.3 *Applications*

Flexographic printing technique has been used in the following applications for printed electronics:

(1) *Conductive traces, patterns and electrodes*: Flexographic printing is used for printing conductive traces [28], patterns [24,29] and electrodes [27,30] on flexible substrates for printed electronics applications.

(2) *Strain gauges*: Strain gauges have been printed on paper substrates by the flexographic printing technique with conductive silver ink [31].

(3) *Chipless Radio-frequency identification (RFID) Tags*: Vena *et al.* [32] utilised the flexographic printing technique for the fabrication of low-cost chipless RFID tags on paper substrates that were operational within the ultra-wideband (UWB) frequency. The cost of fabricating

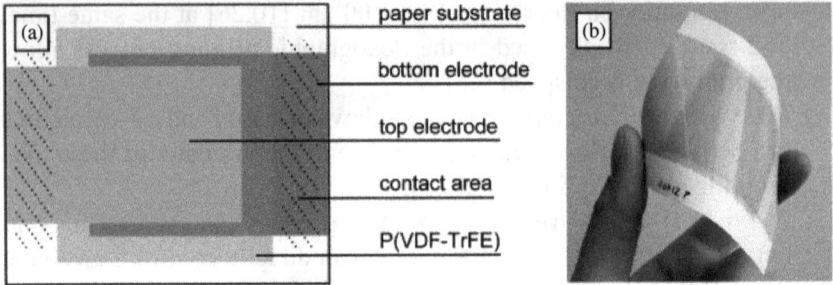

Figure 2.7.   (a) Schematic of the flexographic printed paper loudspeaker, and (b) flexible flexographic printed paper loudspeaker. Reprinted with permission from Ref. [36]. Copyright (2012) from Elsevier.

of these chipless RFID tags was comparable to optical barcodes, and hence demonstrated the feasibility of using the chipless RFID technology for low-cost identification applications.

(4) *Thin-film transistors* (*TFTs*): Higuchi *et al.* [33] fabricated flexible, high mobility carbon nanotubes TFTs on poly(ethylene naphthalate) (PEN) substrates by flexographic printing and their TFTs were able to achieve high mobility of 157 cm^2 V^{-1} s^{-1} with an ON/OFF ratio of 104.

Other printed electronics applications by flexographic printing reported in literature also include solar cells [30,34], complementary ring oscillators [35] and large-area piezoelectric loudspeakers on paper [36] (see **Figure 2.7**).

2.3 Gravure Printing

Gravure printing, also known as rotogravure printing, is commonly used for printing large volumes of coloured graphics on rolls of papers, textiles and plastic films (for instance, gift wrappers, wallpapers, packaging etc.) [3].

2.3.1 *Working Principle and Printing Process*

Unlike flexographic printing which transfer inks from a relief [6], the gravure printing process directly transfers the ink from the engraved cavities on the gravure cylinder to the substrate upon contact [10]. The steel gravure cylinder is usually electroplated with a layer of base copper and has a thin layer of engravable copper applied on top of it [3]. Prior to printing, the image areas are first engraved on the surface of the gravure cylinder to form the printing patterns master. The engraved gravure cylinder is then electroplated again with chrome to protect it from wear and tear during the printing process [10].

A typical gravure printing process is shown in **Figure 2.8**. The gravure cylinder partially submerges in the ink reservoir and rotates at constant velocity to ensure a consistent layer of ink coating around the entire cylinder. The doctor blade scrapes the excess ink off the non-image areas and only leaves the ink in the cavities of the gravure cylinder. The impression cylinder then pushes the roll of substrates onto the rotating gravure cylinder to absorb the ink from the cavities and thus, forming the printed patterns [1].

Low viscosity inks, in the range of 10–1100 cP [10], are typically used for gravure printing. The line definition of gravure printing can be improved with a faster printing speed of low viscosity inks [37], as slower printing speed may allow low viscosity inks to bleed out from the cavities before getting transferred onto the substrates. Ink receptivity, ink viscosity, ink wettability, ink drying rate, solvent evaporation rate, impression cylinder pressure, doctor blade angle, doctor blade pressure and printing

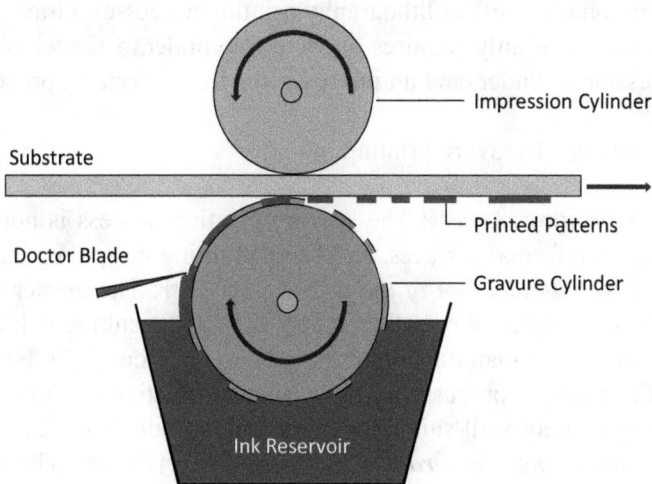

Figure 2.8. Schematic diagram of the gravure printing process.

speed are some of the factors that can also affect the printing properties of gravure printing [10].

2.3.2 *Strengths and Weaknesses*

The strengths of gravure printing include:

(1) *High print speed*: Gravure printing can achieve high print speed, up to 15 m/s [6].
(2) *Capable of printing different layer thickness in a single print*: Gravure printing is capable of printing different layer thickness in a single print, in which this unique feature is not available in other printing techniques [1]. The engraved cavities in the gravure cylinder can have different depths to give different desired printed layer thickness, with printed layer thickness ranging from 50 nm–5 μm [7]. Deeper cavities will produce printed patterns with a larger layer thickness.
(3) *Durable gravure cylinder and long working life*: The metallic surface of the gravure cylinder is usually electroplated with chrome to protect the cylinder from wear and tear during the printing process [10] and extend its working life. This makes the cylinder very durable and allowing it for long print runs while maintaining high-quality prints.
(4) *Mechanically straightforward printing process*: The gravure printing process is mechanically straightforward [3,7,38] as compared to flexographic and offset lithography printing processes. Gravure printing essentially only requires a gravure cylinder, a doctor blade, an impression cylinder, and an ink reservoir for its printing process.

The weaknesses of gravure printing include:

(1) *Planar printing process*: The gravure printing process is not able to print on conformal surfaces, and hence limiting its applications.
(2) *Expensive manufacture of the gravure cylinder*: The master patterns are directly engraved onto the gravure cylinder, resulting in high costs incurred for the manufacture of the gravure cylinder [3,7]. Hence, the gravure printing process can be only cost-effective for high volume printing and not well-suited for small batch productions [2].
(3) *Time-consuming preparations for gravure cylinder*: The gravure cylinder has to be engraved and electroplated prior to printing, and

these preparation processes are very time-consuming [39]. In addition, very limited alterations can be made to the gravure cylinder if any revisions are needed due to mistakes in the engraving or design processes.

(4) *Inability to print on fragile substrates*: The gravure printing technique is not ideal for printing directly on hard and fragile substrates [2], as the hard gravure cylinder may cause damages to these substrates during the printing process. However, Hrehorova *et al.* successfully demonstrated gravure printing of conductive patterns with silver inks on glass substrates [38].

(5) *Inability to produce sharp edges*: Gravure printing is unable to produce sharp edges and tends to produce serrated edges [2,10]. Hence, gravure printing may not be suitable for fabricating electronic devices which require well-defined edges and high resolution.

2.3.3 *Applications*

The gravure printing technique has been used in the following applications for printed electronics:

(1) *Conductive traces, patterns and electrodes*: Gravure printing can be used for printing conductive traces [40], patterns [41,42] and electrodes [43–45] on flexible substrates.

(2) *Organic Photovoltaic (OPV) Modules*: Yang *et al.* [37] had successfully gravure printed both hole transport and light-emitting layers with poly(3,4-ethylenedioxythiophene)-poly(styrenesulfonate) (PEDOT:PSS) and poly(3-hexylthiophene):[6,6]-phenyl C61 butyric acidmethyl ester (P3HT:PCBM) respectively on indium tin oxide coated polyethylene terephthalate (ITO/PET) substrates for fabricating large-area, flexible OPV modules; opens up the potentials for large-area, high throughputs printing of flexible OPV using gravure printing.

(3) *Organic field-effect transistors (OFETs)*: Hambsch *et al.* [46] fabricated OFETs completely by gravure printing on flexible PET substrates. They fabricated more than 50,000 OFETs and achieved approximately 75% yield rate of functional OFETs.

(4) *Humidity sensors*: Reddy *et al.* [47] fully gravure printed a humidity sensor on PET substrate. Interdigitated capacitors (IDCs) patterns were first printed onto PET substrate with silver nanoparticle ink, and

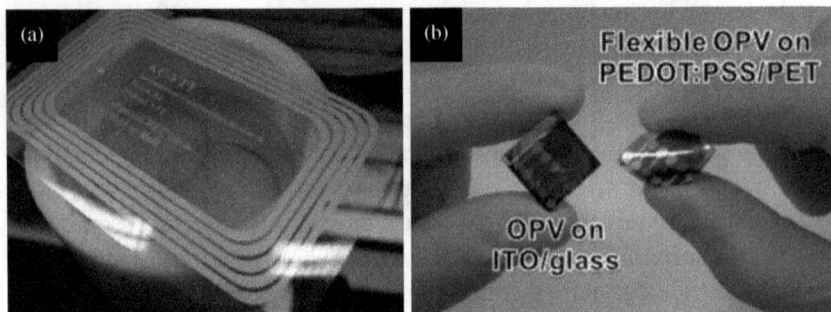

Figure 2.9.   (a) Gravure printed radiofrequency (RF) resonant tag. Reprinted with permission from Ref. [53]. Copyright (2005) from Elsevier; (b) Gravure printed PEDOT:PSS electrode on flexible PET substrate for flexible organic solar cells application. Reprinted with permission from [54]. Copyright (2011) from Elsevier.

then a layer of humidity sensitive polymer [poly (2-hydroxyethyl methacrylate) (pHEMA)] were deposited over the IDC patterns.

(5) *Rectenna*: Park *et al.* [48] fully fabricated a rectenna with R2R gravure printing process on PET substrate. The printed rectenna reportedly achieved more than 90% device yield and supply at least 0.3 W of power from a standard 13.56 MHz power transmitter.

Other printed electronics applications by gravure printing reported in the literature also include electrochemical biosensors [49], flexible antennas on transparent nanopaper [50], organic diode rectifiers [51], radio frequency identification (RFID) tags (see **Figure 2.9(a)**) [41,52,53], flexible organic solar cells (see **Figure 2.9(b)**) [54–56], and thin film transistors (TFTs) [57,58].

2.4 Gravure Offset Printing

The gravure offset printing technique is very similar to gravure printing, except that the former printing technique has an additional soft blanket cylinder to help transfer the ink from the hard gravure cylinder onto the substrates [2,10]. The soft blanket cylinder can help to enhance ink transfer and reduce printing pressures [59]. Hence, the gravure offset printing technique can allow printing directly onto fragile substrates, not achievable in gravure printing.

2.4.1 *Working Principle and Printing Process*

A typical gravure offset printing process is shown in **Figure 2.10**. Similar to the gravure printing process, the gravure cylinder is also partially submerged in an ink reservoir and rotates at constant velocity to ensure a consistent layer of ink coating around the entire cylinder. The doctor blade scrapes the excess ink off the non-image areas and only leaves the ink in the cavities of the gravure cylinder. The soft blanket cylinder, usually made up of silicone or rubber [60], picks up the ink from the cavities of the gravure cylinder and transfers the printed patterns onto the substrate as the impression cylinder presses the substrate onto the blanket cylinder [2,10].

The adhesive force between the gravure and ink, the adhesive force between blanket and ink, the cohesive force within the ink, and the adhesive force between substrate and ink [10,61] are involved in ink transfer in the gravure offset printing process. These interfacial forces and ink interactions should be optimised prior to printing to improve the ink transfer process. 100% transfer of inks is highly desirable during the gravure offset printing process as any ink that is not transferred from the blanket cylinder to the substrates may result in broken patterns and incomplete printed features [10]. Non-uniform printing pressure and high print speed may affect the print quality of gravure offset printing [10], causing issues in the printed patterns like wavy-edges and blank areas.

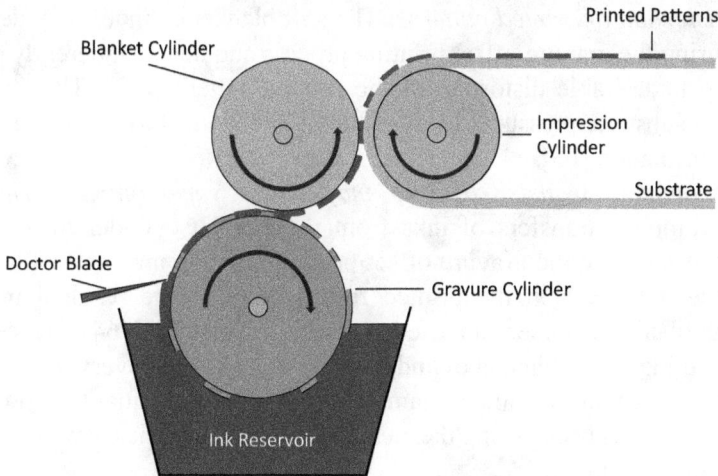

Figure 2.10. Schematic diagram of the gravure offset printing process.

2.4.2 *Strengths and Weaknesses*

The strengths of gravure offset printing include:

(1) *Low printing pressure*: The soft blanket cylinder in gravure offset printing technique can minimise damage on fragile substrates due to its low printing pressure on the substrates and also allow printing on hard [62], flexible and fragile [2] substrates.
(2) *Multilayer printing*: The low printing pressure of the gravure offset printing technique is also ideal for multilayer printing, as damages to the pre-printed patterns can be minimised [59].

The weaknesses of gravure offset printing include:

(1) *Expensive manufacture of the gravure cylinder*: The master patterns are directly engraved onto the gravure cylinder, resulting in high costs incurred for the manufacture of the gravure cylinder [3,7]. Hence, the gravure printing process can be only cost-effective for high volume printing and not well-suited for small batch productions [2].
(2) *Time-consuming preparations for gravure cylinder*: The gravure cylinder has to be engraved and electroplated prior to printing, and the preparation processes are very time-consuming [39]. In addition, once fabricated very limited alterations can be made to the gravure cylinder.
(3) *Distortion of printed patterns*: The soft blanket cylinder may deform during the gravure offset printing process and hence, inherently causing undesirable distortion of the printed patterns [62]. The gravure and substrate pressures ideally should be identical to prevent blanket deformation [63].
(4) *Incomplete transfer of inks may cause variations in printing*: Incomplete transfers of inks from the blanket cylinder to the substrates during the gravure offset printing process may cause undesirable variations in the printed results [60], as the residual ink on the blanket cylinder can affect subsequent print jobs [64]. Hence, the cleaning of the blanket cylinder must be done when every incomplete transfer of inks occurs to maintain the desired print quality. However, the solvents used during the cleaning process can quickly wear out the blanket cylinder.

2.4.3 *Applications*

The gravure offset printing technique can be used in the following applications for printed electronics:

(1) *Conductive traces, patterns and electrodes*: The gravure offset printing technique can deposit functional materials on various substrates to print conductive traces and patterns [60] (see **Figure 2.11**), and electrodes [65].

Figure 2.11. Gravure offset printed patterns with hydrocarbon resin-based inks with (a) 300 μm and (b) 150 μm line widths respectively; Dried gravure offset printed 300 μm and 20 μm wide lines with hydrocarbon resin-based inks: (c) (light through) and (d) (light from top respectively); Dried gravure offset printed wide lines with hydrocarbon resin-based inks of line widths: (e) 75 μm and 150 μm and (f) 37.5 μm and 50 μm. Reprinted with permission from Ref. [60]. Copyright (2004) from Elsevier.

(2) *Ultra-high frequency (UHF) RFID tag antennas*: UHF RFID tags have been printed with the gravure offset printing technique, and Choi *et al.* [66] demonstrated gravure offset printed UHF RFID tags achieved 60% of commercially available copper-etched antennas' identification range.

2.5 Pad Printing

Pad printing, also known as pad transfer printing [2] or tampography, is another form of indirect gravure printing technique. This unique printing technique allows the transfer of a two-dimensional image onto a three-dimensional substrate [3], and hence it is widely used for printing on non-planar surfaces in various industries (e.g. letters on calculators and computer keyboards, pens, tennis balls, golf balls, bowls, glasses, mugs, badges etc.).

2.5.1 *Working Principle and Printing Process*

A simple pad printing machine comprises of a flat gravure plate (also known as *gravure cliché*), a silicone printing pad, a doctor blade and an object holding device [3]. Unlike the gravure printing and gravure offset printing techniques, the pad printing technique usually has a flat gravure plate as the image carrier [3], usually made from plastics, steel or even ceramics [67]. Cavities are engraved on the surface of the flat gravure plate to form the image areas. The depths of the cavities are typically 15–30 μm deep [68], and can directly affect the layer thickness of the printed patterns.

The printing pads are made up of silicone rubber and they come in different hardness, qualities, forms and sizes [67]. The hardness, shape and size of the printing pads are selected depending on the printing patterns, and the shape and type of the substrates [68]. Generally, as a rule of thumb, the printing pads should be chosen to have the highest hardness, largest pad volume and steepest curvature if the printing conditions can allow [67]. The amount of silicone oil in the silicone rubber formulation can determine the hardness of the printing pads, in which lesser silicone oil in the formulation can result in higher hardness [69]. The addition of silicone rubber can also decrease the swelling resistance and tensile strength of the silicone rubber [69]. Hard printing pads can offer better print definitions and longer working lifespans, but they are not suitable for printing on fragile substrates [69]. Hard printing pads are also capable of printing into the surface dimples of uneven substrates' surfaces to prevent slur [69]. Larger pads deform relatively lesser and hence help to minimise image distortions during printing [68]. Printing pads with steeper profiles are also preferred for printing on flat substrates, as the shape can help to reduce air entrapment between the pad and substrate during the ink pickup and ink transfer processes. The steep profile can also allow the printing pad to roll the ink onto the substrates with minimal distortion [69].

The typical pad printing process can be described in six steps: ink flooding, pad wetting, pick up, head stroke, ink deposit and pad release [2] (see **Figure 2.12**).

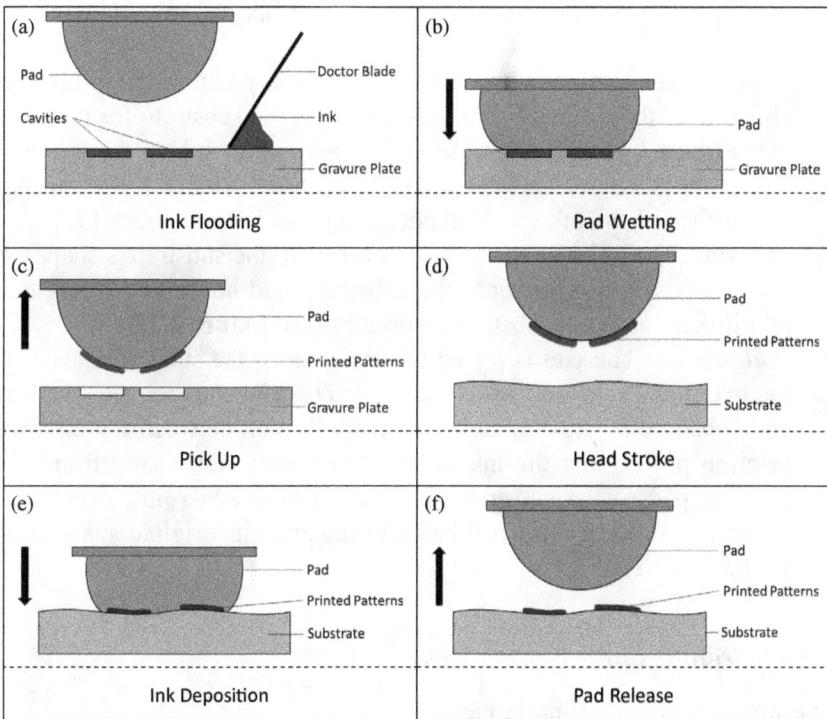

Figure 2.12.   Schematic diagram of the pad printing process: (a) ink flooding, (b) pad wetting, (c) pick up, (d) head stroke, (e) ink deposit and (f) pad release.

(1) *Ink flooding*: The entire gravure plate is first flooded with ink and the doctor blade then scrapes the excess ink off from the non-image areas, leaving only the ink in the cavities of the gravure plate (see **Figure 2.12(a)**). The surface chemistry of the ink changes as the volatile solvents evaporated [70], causing the surface of the ink that is exposed to air to become tacky [71]. The tacky ink improves the inks' ability to adhere onto the printing pad. Solvent-based inks are typically preferred for pad printing for their relatively quick drying [69]. Also, the inks' solvent contents are tailorable to optimize the drying process according to the printing speed.

(2) *Pad wetting*: The printing pad is positioned directly over the image areas and is lowered down onto the gravure plate. The soft printing pad deforms and covers the printed areas entirely (see **Figure 2.12(b)**). The tacky ink adheres onto the printing pad.

(3) *Pick up*: The pad is then raised vertically, picking up the ink from the gravure pad (see **Figure 2.12(c)**). The ink has now adhered to the pad.

(4) *Head stroke*: The pad now moves laterally to position itself directly above over the desired printing location on the substrate for printing (see **Figure 2.12(d)**). The surface chemistry of the ink on the printing pad changes again as more volatile solvents evaporated [70], causing the surface of the ink that is exposed to air to become tacky [71].

(5) *Ink deposition*: The printing pad conforms to the substrate's shape as it lowers itself to press onto the substrate, and hence depositing the ink directly onto the substrates' surfaces (see **Figure 2.12(e)**).

(6) *Pad release*: The pad is raised vertically from the substrate, leaving the ink on the substrate's surfaces [71]. The adhesion between the ink and substrate must be greater than the adhesion between the ink and printing pad so that the ink can be completely transferred from the printing pad to the substrate. The ink is now adhering to the substrate's surfaces and the pad has also regained its original shape (see **Figure 2.12(f)**).

2.5.2 *Strengths and Weaknesses*

The strengths of pad printing include:

(1) *Suitable for various substrates*: Pad printing is suitable for printing on a wide range of substrates such as plastics, ceramics, glass and metals.

(2) *Printing on three-dimensional surfaces*: Pad printing can allow printing on any shape or surface (for instance, flat, spherical, cylindrical, convex, concave surfaces [72] or surfaces with multiple curvatures [70]) due to its soft and deformable silicone printing pad [73]. This feature is not achievable in other conventional contact printing techniques such as flexographic printing, gravure printing and offset lithography.

(3) *Suitable for smaller runs*: Pad printing is more cost-effective than gravure printing for smaller runs [71], which is more suitable for small-scale lab printing.

The weaknesses of pad printing include:

(1) *Size of printed areas*: Pad printing is not suited for printing large areas.

(2) *Slow printing speed*: As compared to other conventional contact printing techniques like flexographic printing, gravure printing and offset lithography, the pad printing technique is significantly slower.

(3) *Thin layer thickness*: The layer thickness approximately range between 4–10 μm and can also be affected by factors such as the atmospheric conditions, ink type, ink viscosity and printing pad material [67]. The thickness of the printed patterns can be increased by printing multiple passes, but it may affect the printing resolution [68].

(4) *Image distortion*: Image distortion is one common problem in pad printing, and it is due to the flexibility of the soft printing pad [74]. The gravure plate must be designed carefully with consideration of the printing pad's deformation, especially when printing on complex surfaces, to prevent any undesirable image distortions [2]. Larger printing pads have lower degrees of deformation and can be used to minimise image distortions [68].

2.5.3 *Applications*

The pad printing technique has been used in the following applications for printed electronics:

(1) *Conductive traces, patterns and electrodes*: The pad printing technique can deposit functional materials on various substrates to print conductive traces [68], patterns and electrodes [75]. Laine-Ma *et al.* [68]

demonstrated pad printing of conductive traces on thermoplastic foils with three layers of silver ink (thickness ranging from 5.7–8.5 μm) and achieved sheet resistances between 20–110 mΩ/sq.

(2) *RFID tags*: Merilampi *et al.* [71] pad printed UHF RFID tags on both flat and convex surfaces with silver inks, and demonstrated its versatile printing ability on non-planar substrates for printed electronics applications.

(3) *Antenna*: Pad printing is advantageous for antenna fabrications [76] as it offers more geometric design freedom and flexibility, in which the antenna patterns can be directly printed on the desired three-dimensional device structure (see **Figure 2.13(a)–(c)**) [72,76].

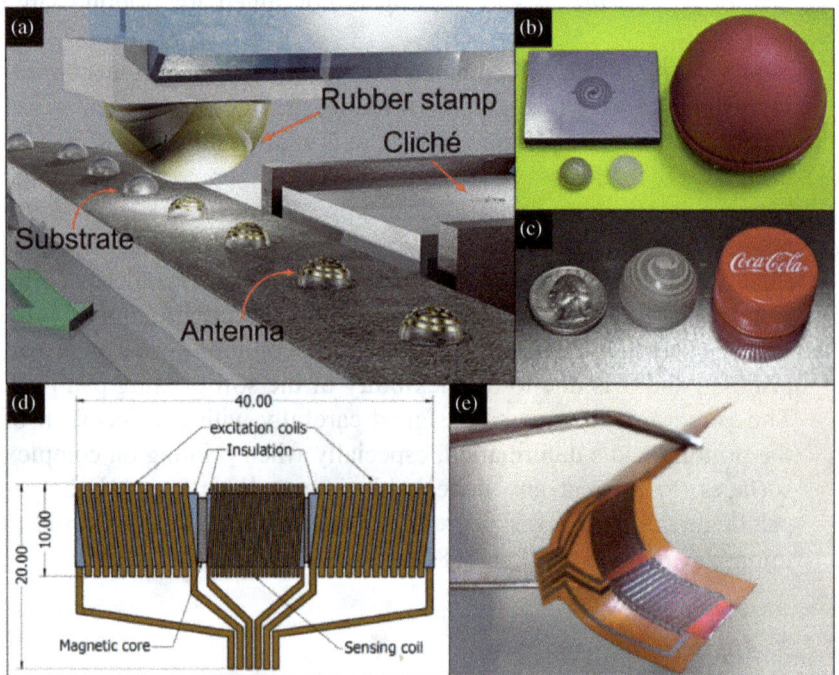

Figure 2.13.   (a) Pad-printing process of electrically small antennas (ESAs) on conformal substrates; (b) images of gravure plate, printing pad, poly(methyl methacrylate) (PMMA) substrate and printed ESA (from clockwise); (c) pad printed double helix ESA. Reprinted with permission from Ref. [76]; (d) schematic illustration of the fluxgate sensor; and (e) flexible pad printed fluxgate sensor on a polyimide substrate. Reprinted with permission from Ref. [77].

Other printed electronics applications by pad printing reported in literature include solar cells [69,70], transducer [74], disposable electrodes [75] and fluxgate sensor (see **Figure 2.13(d)–(e)**) [77].

2.6 Offset Lithography

The offset lithography technique is one of the most widely used printing techniques for mass printing of print media such as newspapers, magazines, posters and books [3].

2.6.1 *Working Principle and Printing Process*

A typical offset lithography printing process [3,7,78] is shown in **Figure 2.14**. The plate cylinder consists of both image and non-image areas, in which they are treated with either oleophilic or hydrophilic coatings respectively. The water rollers first damp the hydrophilic non-image areas of the plate cylinder with a mixture of water and other chemicals to prevent inks from staining these areas. The ink rollers then wet the oleophilic image areas on the plate cylinder with ink. The plate cylinder next transfers the inked patterns onto the blanket cylinder after the entire plate

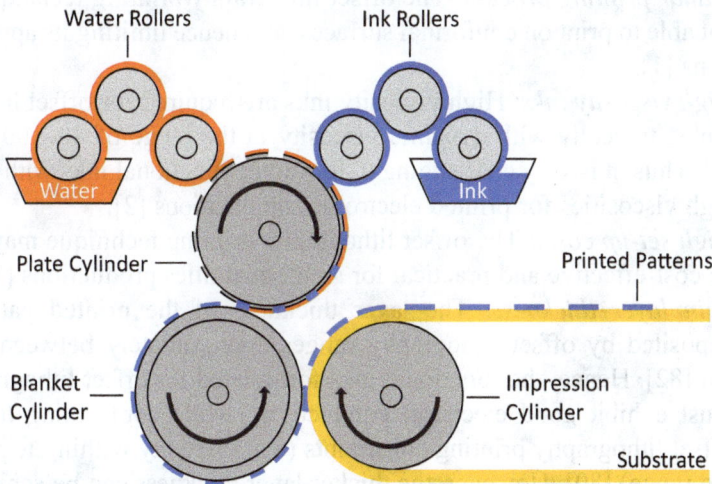

Figure 2.14.   Schematic diagram of the offset lithography printing process.

cylinder is coated with ink. The impression cylinder lastly presses the substrate onto the blanket cylinder as the substrate runs in between these two cylinders, and the printed patterns are transferred from the blanket cylinder onto the substrate.

The blanket cylinder is usually made of flexible rubber, and it can help to improve printing quality and consistency in the offset lithography printing process [79]. The intermediary blanket cylinder also helps the plate cylinder to avoid direct contact with the substrates for wear reduction, and hence prolonging the plate cylinder's working lifespan.

2.6.2 *Strengths and Weaknesses*

The strengths of offset lithography include:

(1) *High print speed and high throughput*: A basic offset lithography printing machine can achieve 6,000–10,000 impressions per hour [80], and hence it is very suitable for mass production [81].
(2) *Good resolution and dimensional control*: The offset lithography printing technique can print 80–100 micron traces with 60 micron gap [80].

The weaknesses of offset lithography include:

(1) *Planar printing process*: The offset lithography printing technique is not able to print on conformal surfaces, and hence limiting its applications [1].
(2) *High viscosity inks*: High viscosity inks are required for offset lithography, typically with dynamic viscosity in the range of 40–100 Pa·s [7]. Thus, it is very challenging to formulate functional inks with such high viscosities for printed electronics applications [2].
(3) *High set-up costs*: The offset lithography printing technique may not be cost-effective and practical for small quantities productions [1,7].
(4) *Thin layer thickness*: The layer thickness of the printed patterns deposited by offset lithography ranges approximately between 3–5 μm [82]. Hence, the functional inks formulated for offset lithography must exhibit good electrical conductivity while conforming to the offset lithography printing constraints (e.g. viscosity within the printing range) [80]. However, the thicker layer thickness can be achieved by over-printing [83].

2.6.3 *Applications*

The offset lithography printing technique can be used in the following applications for printed electronics:

(1) *Conductive traces, patterns and electrodes*: Offset lithography print-ing of conductive traces and circuits on paper-like substrates were first demonstrated by Ramsey *et al.* [84]. They proposed these con-ductive lithographic films as cheaper and more environmentally-friendly alternatives to traditional copper printed circuit boards (PCBs) [84,85]. Offset lithography was also used for printing elec-trodes for light-emitting diodes (LEDs) [86], voltaic cells [87] and organic field-effect transistors (OFETs) [88].

(2) *Interdigitated capacitors*: Harrey *et al.* demonstrated interdigitated capacitors of 100 μm linewidth and 150 μm separation gaps can be fabricated by offset lithography [89].

Other printed electronics applications by offset lithography reported in literature also include capacitive-type humidity sensors [82], analogue filter structures [90], microwave integrated circuits [83] and RF circula-tors [91].

2.7 Comparison of Conventional Contact Printing Techniques

The main printing characteristics of each conventional contact printing techniques are tabulated in Table 2.1. Each contact printing technique's strengths, weaknesses and applications are also tabulated in Table 2.2. These printing techniques each have their own set of strengths and weak-nesses, and the suitable choice of printing technique is highly dependent on the required printing characteristics and the intended applications [78].

Screen printing, flexographic printing and gravure are direct image transfer printing techniques, whereas gravure offset printing, pad printing and offset lithography are indirect image transfer printing techniques. The gravure printing technique is the fastest printing technique among all. On the other end, flatbed screen printing and pad printing have very slow printing speed. Although the offset lithography printing technique can achieve very good printing resolution, it requires very viscous inks for printing. Hence, limiting its applications in printed electronics due to the

Table 2.1.   Comparison of various conventional contact printing techniques.

Parameters	Conventional Contact Printing Techniques					
	Screen Printing	Flexo-graphic Printing	Gravure Printing	Gravure Offset Printing	Pad Printing	Offset Litho-graphy
Image transfer	Direct	Direct	Direct	Indirect	Indirect	Indirect
Ink viscosity (Pa·S)	0.5–50	0.05–0.5	0.05–0.2	2–5	2040	40–100
Printing speed (m/min)	0–35 (flatbed) >100 (rotary)	100–500	100–1000	0.6–15	Not available	200–800
Smallest line width (μm)	50 150	20–50	10–50	20–50	~30	10–15
Print thickness (μm)	3–30	1 2.5	<0.1–5	0.6–2	410	0.5–1.5
R2R compatibility	Yes	Yes	Yes	Yes	No	Yes
Printing on three-dimensional surfaces	Yes	No	No	No	Yes	No
Refs.	[6,78]	[78]	[78]	[10]	[67,69,92]	[78]

Table 2.2. Advantages, disadvantages and applications of the various contact printing techniques.

Printing Techniques	Strengths	Weaknesses	Applications
Screen printing	• Suitable for various substrates [7] • Ability to print on conformal substrates [2] • Simple and low-cost printing process • Thick layers are possible [6] • Scalable for large-area processing [8]	• Only highly viscous inks or paste-like inks can be used [1,6,7,9] • High wastage of materials • High surface roughness [2,10] • Slow printing speed [6]	• Conductive traces and electrodes [6,11–15] • Passive devices [16] • Sensors [17,18] • OLEDs [19] • Active matrix electrochromic displays [20] • Thin-film transistors [8] • Solar cells [21,22] • Carbon-fibre supercapacitors [23]
Flexographic printing	• Low printing pressure [24] • Suitable for wide variety of substrates [1,2,7,15,25] • High print speed [7,17] • Wide variety of inks can be used [7] • Thin and uniform layers [3,27]	• Planar printing process • High initial cost and complex preparations prior printing [1] • Distortion [2]	• Conductive traces, patterns and electrodes [24,27–30] • Strain gauges [31] • Chipless RFID tags [32] • Thin-film transistors [33] • Solar cells [30,34] • Complementary ring oscillators [35] • Large-area piezoelectric loudspeakers [36]
Gravure printing	• High print speed [6] • Capable of printing different layer thickness in a single print [1,7] • Durable gravure cylinder and long working life [10] • Mechanically straightforward printing process [3,7,38]	• Planar printing process • Expensive manufacture of the gravure cylinder [2,3,7] • Time-consuming preparations for gravure cylinder [39] • Inability to print on fragile substrates [2,38] • Inability to produce sharp edges [2,10]	• Conductive traces, patterns and electrodes [40–45] • OPV modules [37] • OFETs [46] • Humidity sensors [47] • Rectenna [48] • Electrochemical biosensors [49] • Flexible antennas [50] • Organic diode rectifiers [51] • RFID tags [41,52,53]

(Continued)

Table 2.2.   (*Continued*)

Printing Techniques	Strengths	Weaknesses	Applications
Gravure offset printing	• Low printing pressure [2,62] • Multilayer printing [59]	• Expensive manufacture of the gravure cylinder [2,3,7] • Time-consuming preparations for gravure cylinder [39] • Distortion of printed patterns • Incomplete transfer of inks may cause variations in printing [60,64]	• Flexible organic solar cells [54–56] • TFTs [57,58] • Conductive traces, patterns and electrodes [60,65] • Ultra-high frequency (UHF) RFID tag antennas [66]
Pad printing	• Suitable for various substrates • Printing on three-dimensional surfaces [70,72,73] • Suitable for smaller runs [71]	• Small printing area • Slow printing speed • Thin layer thickness [67,68] • Image distortion [2,68,74]	• Conductive traces, patterns and electrodes [68,75] • RFID tags [71] • Electro-luminescence devices [73] • Antenna [72,76] • Solar cells [69,70] • Transducers [74] • Disposable electrodes [75] • Fluxgate sensors [77]
Offset lithography	• High print speed [80] • High throughput [81] • Good resolution and dimensional control [80]	• Planar printing process [1] • High viscosity inks [7] • High set-up costs [1,7] • Thin layer thickness [80,82,83]	• Conductive traces, patterns and electrodes [84–88] • Interdigitated capacitors [89] • Capacitive-type humidity sensors [82] • Analogue filter structures [90] • Microwave integrated circuits [83] • RF circulators [91]

difficulties in formulating functional inks with good electrical properties. Some of these conventional contact printing techniques, such as rotary screen printing, flexographic printing, gravure printing, gravure offset printing and offset lithography, are R2R compatible. They are very suitable for mass production of printed electronics due to their high printing speeds, high throughputs and their high cost-efficiency. Both screen printing and pad printing can allow printing on three-dimensional surfaces, and this unique feature cannot be found on other printing techniques.

References

[1] Tan, H. W., Tran, T. and Chua, C. K. (2016). A review of printed passive electronic components through fully additive manufacturing methods, *Virtual Phys. Prototyp.*, 11, pp. 271–288.

[2] Cui, Z. (2016). *Printed Electronics: Materials, Technologies and Applications*, eds. Zheng Cui and Jian Lin, Chapter 4: Printing processes and equipments, pp. 106–144.

[3] Kipphan, H. (2001). *Handbook of Print Media -Technologies and Production Methods* (Springer Science & Business Media, New York).

[4] Riemer, D. E. (1989). The theoretical fundamentals of the screen printing process, *Microelectron. Int*, 6, pp. 8–17.

[5] Bacher, R. J. (2001). *Encyclopedia of Materials: Science and Technology*, eds. K. H. Jürgen Buschow, Robert W. Cahn, Merton C. Flemings, Bernhard Ilschner, Edward J. Kramer, Subhash Mahajan and Patrick Veyssière, Screen Printing (Elsevier, Oxford) pp. 8281–8283.

[6] Roth, B., Søndergaard, R. R. and Krebs, F. C. (2015). *Handbook of Flexible Organic Electronics* (Woodhead Publishing, Oxford).

[7] Rosa, P. (2015). *Minimal Computation Structures for Visual Information Applications based on Printed Electronics* (Doctoral Dissertation, New University of Lisbon).

[8] Cao, X., Chen, H., Gu, X., Liu, B., Wang, W., Cao, Y., Wu, F. and Zhou, C. (2014). Screen printing as a scalable and low-cost approach for rigid and flexible thin-film transistors using separated carbon nanotubes, *ACS Nano*, 8, pp. 12769–12776.

[9] Cruz, S. M. F., Rocha, L. A. and Viana, J. C. (2018). *Flexible Electronics*, Chapter 2: Printing technologies on flexible substrates for printed electronics (IntechOpen).

[10] Khan, S., Lorenzelli, L. and Dahiya, R. S. (2015). Technologies for printing sensors and electronics over large flexible substrates: A review, *IEEE Sens. J.*, 15, pp. 3164–3185.

[11]   Paul, G., Torah, R., Beeby, S. and Tudor, J. (2014). The development of screen printed conductive networks on textiles for biopotential monitoring applications, *Sens. Actuators, A*, 206, pp. 35–41.

[12]   Paul, G., Torah, R., Yang, K., Beeby, S. and Tudor, J. (2014). An investigation into the durability of screen-printed conductive tracks on textiles, *Meas. Sci. Technol.*, 25, p. 025006.

[13]   Yang, K., Torah, R., Wei, Y., Beeby, S. and Tudor, J. (2013). Waterproof and durable screen printed silver conductive tracks on textiles, *Tex. Res. J.*, 83, pp. 2023–2031.

[14]   Ho, X., Cheng, C. K., Tan, R. L. S. and Wei, J. (2015). Screen printing of stretchable electrodes for large area LED matrix, *J. Mater. Res.*, 30, pp. 2271–2278.

[15]   Suikkola, J., Björninen, T., Mosallaei, M., Kankkunen, T., Iso-Ketola, P., Ukkonen, L., Vanhala, J. and Mäntysalo, M. (2016). Screen-printing fabrication and characterization of stretchable electronics, *Sci. Rep.*, 6, p. 25784.

[16]   Ostfeld, A. E., Deckman, I., Gaikwad, A. M., Lochner, C. M. and Arias, A. C. (2015). Screen printed passive components for flexible power electronics, *Sci. Rep.*, 5, p. 15959.

[17]   Khan, S., Tinku, S., Lorenzelli, L. and Dahiya, R. S. (2015). Flexible tactile sensors using screen-printed P(VDF-TrFE) and MWCNT/PDMS composites, *IEEE Sens. J.*, 15, pp. 3146–3155.

[18]   Chang, W., Fang, T., Lin, H., Shen, Y. and Lin, Y. (2009). A large area flexible array sensors using screen printing technology, *J. Disp. Technol.*, 5, pp. 178–183.

[19]   Pardo, D. A., Jabbour, G. E. and Peyghambarian, N. (2000). Application of screen printing in the fabrication of organic light-emitting devices, *Adv. Mater.*, 12, pp. 1249–1252.

[20]   Cao, X., Lau, C., Liu, Y., Wu, F., Gui, H., Liu, Q., Ma, Y., Wan, H., Amer, M. R. and Zhou, C. (2016). Fully screen-printed, large-area, and flexible active-matrix electrochromic displays using carbon nanotube thin-film transistors, *ACS Nano*, 10, pp. 9816–9822.

[21]   Gemeiner, P., Kuliček, J., Syrový, T., Ház, A., Khunová, V., Hatala, M., Mikula, M., Hvojnik, M., Gál, L., Jablonský, M. and Omastová, M. (2019). Screen-printed PEDOT:PSS/halloysite counter electrodes for dye-sensitized solar cells, *Synth. Met.*, 256, p. 116148.

[22]   Zhou, Z., Wang, Y., Xu, D. and Zhang, Y. (2010). Fabrication of Cu2ZnSnS4 screen printed layers for solar cells, *Sol. Energy Mater. Sol. Cells*, 94, pp. 2042–2045.

[23]   Jost, K., Stenger, D., Perez, C. R., McDonough, J. K., Lian, K., Gogotsi, Y. and Dion, G. (2013). Knitted and screen printed carbon-fiber supercapacitors for applications in wearable electronics, *Energy Environ. Sci.*, 6, pp. 2698–2705.

[24] Krzyżkowski, J., Dąbrowa, T., Hamerliński, J. and Śleboda, P. (2015). Investigation on lineworks printed with different types of flexographic printing forms for purposes of printed electronics, *Chall. Mod. Technol.*, 6, pp. 3–9.

[25] Izdebska, J. (2016). *Printing on Polymers*, eds. Joanna Izdebska and Sabu Thomas, Chapter 11: Flexographic printing (William Andrew Publishing) pp. 179–197.

[26] Vena, A., Perret, E., Tedjini, S., Tourtollet, G. E. P., Delattre, A., Garet, F. and Boutant, Y. (2013). Design of chipless RFID tags printed on paper by flexography, *IEEE Trans. Antennas Propag.*, 61, pp. 5868–5877.

[27] Lo, C., Huttunen, O., Hiitola-Keinanen, J., Petaja, J., Fujita, H. and Toshiyoshi, H. (2010). MEMS-controlled paper-like transmissive flexible display, *J. Microelectromech. Syst.*, 19, pp. 410–418.

[28] Deganello, D., Cherry, J. A., Gethin, D. T. and Claypole, T. C. (2010). Patterning of micro-scale conductive networks using reel-to-reel flexographic printing, *Thin Solid Films*, 518, pp. 6113–6116.

[29] Faddoul, R., Reverdy-Bruas, N., Blayo, A., Haas, T. and Zeilmann, C. (2012). Optimisation of silver paste for flexography printing on LTCC substrate, *Microelectron. Reliab.*, 52, pp. 1483–1491.

[30] Hösel, M., Søndergaard, R. R., Angmo, D. and Krebs, F. C. (2013). Comparison of fast roll-to-roll flexographic, inkjet, flatbed, and rotary screen printing of metal back electrodes for polymer solar cells, *Adv. Eng. Mater.*, 15, pp. 995–1001.

[31] Maddipatla, D., Narakathu, B. B., Avuthu, S. G. R., Emamian, S., Eshkeiti, A., Chlaihawi, A. A., Bazuin, B. J., Joyce, M. K., Barrett, C. W. and Atashbar, M. Z. (2015). A novel flexographic printed strain gauge on paper platform, *presented at the 2015 IEEE Sens.*, Busan, South Korea.

[32] Vena, A., Perret, E., Tedjini, S., Tourtollet, G. E. P., Delattre, A., Garet, F. and Boutant, Y. (2013). Design of chipless RFID tags printed on paper by flexography, *IEEE Trans. Antennas Propag.*, 61, pp. 5868–5877.

[33] Higuchi, K., Kishimoto, S., Nakajima, Y., Tomura, T., Takesue, M., Hata, K., Kauppinen, E. I. and Ohno, Y. (2013). High-mobility, flexible carbon nano-tube thin-film transistors fabricated by transfer and high-speed flexographic printing techniques, *Appl. Phys. Express*, 6, p. 085101.

[34] Thibert, S., Chaussy, D., Beneventi, D., Reverdy-Bruas, N., Jourdan, J., Bechevet, B. and Mialon, S. (2012). Silver ink experiments for silicon solar cell metallization by flexographic process, *presented at the 38th IEEE Photovoltaic Specialists Conference*, Austin, TX, USA.

[35] Kempa, H., Hambsch, M., Reuter, K., Stanel, M., Schmidt, G. C., Meier, B. and Hubler, A. C. (2011). Complementary ring oscillator exclusively prepared by means of gravure and flexographic printing, *IEEE Trans. Electron. Devices*, 58, pp. 2765–2769.

[36] Hübler, A. C., Bellmann, M., Schmidt, G. C., Zimmermann, S., Gerlach, A. and Haentjes, C. (2012). Fully mass printed loudspeakers on paper, *Org. Electron.*, 13, pp. 2290–2295.

[37] Yang, J., Vak, D., Clark, N., Subbiah, J., Wong, W. W. H., Jones, D. J., Watkins, S. E. and Wilson, G. (2013). Organic photovoltaic modules fabricated by an industrial gravure printing proofer, *Sol. Energy Mater. Sol. Cells*, 109, pp. 47–55.

[38] Hrehorova, E., Rebros, M., Pekarovicova, A., Bazuin, B., Ranganathan, A., Garner, S., Merz, G., Tosch, J. and Boudreau, R. (2011). Gravure printing of conductive inks on glass substrates for applications in printed electronics, *J. Disp. Technol.*, 7, pp. 318–324.

[39] Robertson, G. L. (2005). *Food Packaging: Principles and Practice* (CRC press, USA).

[40] Sung, D., Vornbrock, A. D. L. F. and Subramanian, V. (2010). Scaling and optimization of gravure-printed silver nanoparticle lines for printed electronics, *IEEE Trans. Compon. Packag. Technol.*, 33, pp. 105–114.

[41] Jung, M., Kim, J., Noh, J., Lim, N., Lim, C., Lee, G., Kim, J., Kang, H., Jung, K., Leonard, A. D., Tour, J. M. and Cho, G. (2010). All-printed and roll-to-roll-printable 13.56-MHz-operated 1-bit RF tag on plastic foils, *IEEE Trans. Electron Devices*, 57, pp. 571–580.

[42] Pudas, M., Halonen, N., Granat, P. and Vähäkangas, J. (2005). Gravure printing of conductive particulate polymer inks on flexible substrates, *Prog. Org. Coat.*, 54, pp. 310–316.

[43] Noh, J., Yeom, D., Lim, C., Cha, H., Han, J., Kim, J., Park, Y., Subramanian, V. and Cho, G. (2010). Scalability of Roll-to-Roll Gravure-Printed Electrodes on Plastic Foils, *IEEE Trans. Electron. Packag. Manuf.*, 33, pp. 275–283.

[44] Puetz, J. and Aegerter, M. A. (2008). Direct gravure printing of indium tin oxide nanoparticle patterns on polymer foils, *Thin Solid Films*, 516, pp. 4495–4501.

[45] Alsaid, D. A., Rebrosova, E., Joyce, M., Rebros, M., Atashbar, M. and Bazuin, B. (2012). Gravure printing of ITO transparent electrodes for applications in flexible electronics, *J. Disp. Technol.*, 8, pp. 391–396.

[46] Hambsch, M., Reuter, K., Stanel, M., Schmidt, G., Kempa, H., Fügmann, U., Hahn, U. and Hübler, A. C. (2010). Uniformity of fully gravure printed organic field-effect transistors, *Mater. Sci. Eng. B*, 170, pp. 93–98.

[47] Reddy, A. S. G., Narakathu, B. B., Atashbar, M. Z., Rebros, M., Rebrosova, E. and Joyce, M. K. (2011). Fully printed flexible humidity sensor, *Procedia Eng.*, 25, pp. 120–123.

[48] Park, H., Kang, H., Lee, Y., Park, Y., Noh, J. and Cho, G. (2012). Fully roll-to-roll gravure printed rectenna on plastic foils for wireless power transmission at 13.56 MHz, *Nanotechnology*, 23, p. 344006.

[49] Reddy, A. S. G., Narakathu, B. B., Atashbar, M. Z., Rebros, M., Rebrosova, E. and Joyce, M. K. (2011). Gravure printed electrochemical biosensor, *Procedia Eng.*, 25, pp. 956–959.

[50] Zhu, H., Narakathu, B. B., Fang, Z., Tausif Aijazi, A., Joyce, M., Atashbar, M. and Hu, L. (2014). A gravure printed antenna on shape-stable transparent nanopaper, *Nanoscale*, 6, pp. 9110–9115.

[51] Lilja, K. E., Bäcklund, T. G., Lupo, D., Hassinen, T. and Joutsenoja, T. (2009). Gravure printed organic rectifying diodes operating at high frequencies, *Org. Electron.*, 10, pp. 1011–1014.

[52] Subramanian, V., Frechet, J. M. J., Chang, P. C., Huang, D. C., Lee, J. B., Molesa, S. E., Murphy, A. R., Redinger, D. R. and Volkman, S. K. (2005). Progress toward development of all-printed RFID tags: Materials, processes, and devices, *Proceedings of the IEEE*, 93, pp. 1330–1338.

[53] Allen, M., Lee, C., Ahn, B., Kololuoma, T., Shin, K. and Ko, S. (2011). R2R gravure and inkjet printed RF resonant tag, *Microelectron. Eng.*, 88, pp. 3293–3299.

[54] Cho, C.-K., Hwang, W.-J., Eun, K., Choa, S.-H., Na, S.-I. and Kim, H.-K. (2011). Mechanical flexibility of transparent PEDOT:PSS electrodes prepared by gravure printing for flexible organic solar cells, *Sol. Energy Mater. Sol. Cells*, 95, pp. 3269–3275.

[55] Kopola, P., Aernouts, T., Guillerez, S., Jin, H., Tuomikoski, M., Maaninen, A. and Hast, J. (2010). High efficient plastic solar cells fabricated with a high-throughput gravure printing method, *Sol. Energy Mater. Sol. Cells*, 94, pp. 1673–1680.

[56] Voigt, M. M., Mackenzie, R. C. I., Yau, C. P., Atienzar, P., Dane, J., Keivanidis, P. E., Bradley, D. D. C. and Nelson, J. (2011). Gravure printing for three subsequent solar cell layers of inverted structures on flexible substrates, *Sol. Energy Mater. Sol. Cells*, 95, pp. 731–734.

[57] Yi, M., Yeom, D., Lee, W., Jang, S. and Cho, G. (2013). Scalability on roll-to-roll gravure printed dielectric layers for printed thin film transistors, *J. Nanosci. Nanotechnol.*, 13, pp. 5360–5364.

[58] Kang, H., Kitsomboonloha, R., Jang, J. and Subramanian, V. (2012). High-performance printed transistors realized using femtoliter gravure-printed sub-10 μm metallic nanoparticle patterns and highly uniform polymer dielectric and semiconductor layers, *Adv. Mater.*, 24, pp. 3065–3069.

[59] Kim, K., Kim, C. H., Kim, H.-Y. and Kim, D.-S. (2010). Effects of blanket roller deformation on printing qualities in gravure-offset printing method, *Jpn. J. Appl. Phys.*, 49, p. 05EC04.

[60] Pudas, M., Hagberg, J. and Leppävuori, S. (2004). Printing parameters and ink components affecting ultra-fine-line gravure-offset printing for electronics applications, *J. Eur. Ceram. Soc.*, 24, pp. 2943–2950.

[61] Lee, T.-M., Noh, J.-H., Kim, I., Kim, D.-S. and Chun, S. (2010). Reliability of gravure offset printing under various printing conditions, *J. Appl. Phys.*, 108, p. 102802.

[62] Kim, C. H., Jo, J. and Lee, S.-H. (2012). Design of roll-to-roll printing equipment with multiple printing methods for multi-layer printing, *Rev. Sci. Instrum.*, 83, p. 065001.

[63] Pudas, M., Hagberg, J. and Leppävuori, S. (2004). Gravure offset printing of polymer inks for conductors, *Prog. Org. Coat.*, 49, pp. 324–335.

[64] Lee, T.-M., Lee, S.-H., Noh, J.-H., Kim, D.-S. and Chun, S. (2010). The effect of shear force on ink transfer in gravure offset printing, *J. Micromech. Microeng.*, 20, p. 125026.

[65] Lee, T.-M., Noh, J.-H., Kim, C. H., Jo, J. and Kim, D.-S. (2010). Development of a gravure offset printing system for the printing electrodes of flat panel display, *Thin Solid Films*, 518, pp. 3355–3359.

[66] Choi, B.-O., Kim, C. H. and Kim, D. S. (2010). Manufacturing ultra-high-frequency radio frequency identification tag antennas by multilayer printings, *Proc. Inst. Mech. Eng., Part C: J. Mech. Eng. Sci.*, 224, pp. 149–156.

[67] Teca-Print. The pad printing overview: Pad printing, printing pads, plates, pad printing inks, adhesion und antistatic. Retrieved from https://teca-print.com/pdf_eng/Tampondruckverfahren/The_pad_printing_process_706-000-465.pdf.

[68] Laine-Ma, T., Ruuskanen, P., Pasanen, S. and Karttunen, M. (2016). Pad printing of polymeric silver ink conductors on thermoplastic foils, *Circuit World*, 42, pp. 170–177.

[69] Hahne, P., Hirth, E., Reis, I. E., Schwichtenberg, K., Richtering, W., Horn, F. M. and Eggenweiler, U. (2001). Progress in thick-film pad printing technique for solar cells, *Sol. Energy Mater. Sol. Cells*, 65, pp. 399–407.

[70] Krebs, F. C. (2009). Pad printing as a film forming technique for polymer solar cells, *Sol. Energy Mater. Sol. Cells*, 93, pp. 484–490.

[71] Merilampi, S., Björninen, T., Ukkonen, L., Ruuskanen, P. and Sydänheimo, L. (2010). Characterization of UHF RFID tags fabricated directly on convex surfaces by pad printing, *Int. J. Adv. Manuf. Technol.*, 53, pp. 577–591.

[72] Zengchao, Q., Ye, X., Zhan, L., Yongfa, F. and Yong, Y. (2010). The pad printing technology evaluation in mobile phone antenna manufacture, *presented at the 9th International Symposium on Antennas, Propagation and EM Theory*, Guangzhou, China.

[73] Lee, T.-M., Hur, S., Kim, J.-H. and Choi, H.-C. (2009). EL device pad-printed on a curved surface, *J. Micromech. Microeng.*, 20, p. 015016.

[74] Filoux, E., Lou-Moeller, R., Callé, S., Lethiecq, M. and Levassort, F. (2013). Optimised properties of high frequency transducers based on

curved piezoelectric thick films obtained by pad printing process, *Adv. Appl. Ceram.*, 112, pp. 75–78.

[75] Mooring, L., Karousos, N. G., Livingstone, C., Davis, J., Wildgoose, G. G., Wilkins, S. J. and Compton, R. G. (2005). Evaluation of a novel pad printing technique for the fabrication of disposable electrode assemblies, *Sens. Actuators, B*, 107, pp. 491–496.

[76] Wu, H., Chiang, S. W., Yang, C., Lin, Z., Liu, J., Moon, K.-S., Kang, F., Li, B. and Wong, C. P. (2015). Conformal pad-printing electrically conductive composites onto thermoplastic hemispheres: Toward sustainable fabrication of 3-cents volumetric electrically small antennas, *PLoS ONE*, 10, pp. e0136939–e0136939.

[77] Schoinas, S., Guamra, A.-M. E., Moreillon, F. and Passeraub, P. (2017). A flexible pad-printed fluxgate sensor, *presented at the Eurosensors 2017*, Paris, France.

[78] Rosa, P., Câmara, A. and Gouveia, C. (2015). The potential of printed electronics and personal fabrication in driving the internet of things, *Open J. Internet Things*, 1, pp. 16–36.

[79] Neuvo, Y. and Ylönen, S. (2010). Bit Bang: Rays to the future.

[80] Evans, P. S. A. (2001). Lithographic film circuits — A review, *Circuit World*, 27, pp. 31–35.

[81] Southee, D., Hay, G. I., Evans, P. S. A. and Harrison, D. J. (2006). Development and characterisation of lithographically printed voltaic cells, *presented at the Electronics Systemintegration Technology Conference*, Dresden, Germany.

[82] Harrey, P. M., Ramsey, B. J., Evans, P. S. A. and Harrison, D. J. (2002). Capacitive-type humidity sensors fabricated using the offset lithographic printing process, *Sens. Actuators, B*, 87, pp. 226–232.

[83] Shepherd, P. R., Evans, P. S. A., Ramsey, B. J. and Harrison, D. J. (1997). Lithographic technology for microwave integrated circuits, *Electron. Lett.*, 33, pp. 483–484.

[84] Ramsey, B. J., Evans, P. S. A. and Harrison, D. (1997). A novel circuit fabrication technique using offset lithography, *J. Electron. Manuf.*, 07, pp. 63–67.

[85] Ramsey, B. J., Evans, P. and Harrison, D. (1997). Conductive lithographic films, *presented at the IEEE International Symposium on Electronics and the Environment*, San Francisco, CA, USA.

[86] Lochun, D., Kilitziraki, M., Jarrett, E., Green, S., Ramsey, B., Samuel, I. and Harrison, D. (1999). Post-processing of conductive lithographic films for multilayer device fabrication, *presented at the Twenty Fourth IEEE/CPMT International Electronics Manufacturing Technology Symposium*, Austin, TX, USA, USA.

[87] Southee, D., Hay, G. I., Evans, P. S. A. and Harrison, D. J. (2006). Development and characterisation of lithographically printed voltaic cells,

presented at the 1st Electronic Systemintegration Technology Conference, Dresden, Germany.

[88] Zielke, D., Hübler, A. C., Hahn, U., Brandt, N., Bartzsch, M., Fügmann, U., Fischer, T., Veres, J. and Ogier, S. (2005). Polymer-based organic field-effect transistor using offset printed source/drain structures, *Appl. Phys. Lett.*, 87, p. 123508.

[89] Harrey, P. M., Evans, P. S. A., Ramsey, B. J. and Harrison, D. J. (2000). Interdigitated capacitors by offset lithography, *J. Electron. Manuf.*, 10, pp. 69–77.

[90] Evans, P., Ramsey, B. J., Harrey, P. M. and Harrison, D. (1999). Printed analogue filter structures, *Electron. Lett.*, 35, pp. 306–308.

[91] Evans, P. S. A., Harrey, P. M., Ramsey, B. J. and Harrison, D. J. (1999). RF circulator structures via offset lithography, *Electron. Lett.*, 35, pp. 1634–1636.

[92] Lahti, M., Leppävuori, S. and Lantto, V. (1999). Gravure-offset-printing technique for the fabrication of solid films, *Appl. Surf. Sci.*, 142, pp. 367–370.

Problems

1. List the advantages and disadvantages of the screen printing process.
2. What are some of the critical factors that influence the printing properties of screen printing?
3. Using a sketch to illustrate your answer, describe the flexographic printing process.
4. List the advantages and disadvantages of the flexographic printing process.
5. Using a sketch to illustrate your answer, describe the gravure printing process.
6. List the advantages and disadvantages of the gravure printing process.
7. Describe the differences between the gravure printing process and the gravure offset printing process.
8. Using a sketch to illustrate your answer, describe the pad printing process.
9. Discuss the processes, strengths and limitations of the following processes:
   a. Screen printing
   b. Pad printing
   c. Offset Lithography

Chapter 3

3D Electronics Printing Techniques

In recent years, 3D electronics printing has been emerging as a technology capable of harnessing the potentials of additive manufacturing in fabricating electronics and structural electronics [1]. Electronics fabricated by additive manufacturing techniques, i.e. those involve materials deposition processes [2–4], are normally termed as "3D printed electronics" or "additive manufactured electronics (AMEs)".

Unlike the conventional contact printing techniques previously discussed in Chapter 2, 3D electronics printing includes digital and direct-writable printing technologies, both of which do not require master plates to transfer images, but instead deposit functional materials directly onto the substrates. Hence, these 3D electronics printing techniques are more cost-effective for low volume and on-demand manufacturing. In addition, some of the available 3D electronics printing techniques are even capable of depositing materials directly onto conformal surfaces or embedding conductive traces and components within complex 3D structures [1], thereby increasing designs freedom and flexibility.

This chapter focuses on commercially available printing systems and their working principles used to deposit functional materials and fabricate 3D printed electronics. These printing systems are categorised into several major technologies, including inkjet, aerosol-based, extrusion-based, electrohydrodynamic and other micro-dispensing mechanisms. A wide range of functional materials can be used with varying capabilities such as printing speed and printing resolution. This chapter also discusses the strengths and weaknesses, and practical applications of these commercially available printing systems.

3.1 Inkjet Printing

Inkjet printing has been widely utilised in printing applications for many decades. It is most commonly associated with desktop printers used in offices and modern homes to print digital images on paper substrates. Inkjet printing is a fully digital, non-contact printing technique [2], involving the generation and precise deposition of small ink droplets on substrates [5]. It also allows a wide range of materials to be deposited, including polymers, biological materials, metallic particle suspensions and ceramics [5].

Figure 3.1 shows a classification of inkjet printing technologies based on the droplet generation methods: continuous inkjet printing (CIJ) and drop-on-demand (DoD) inkjet printing [6]. Each method can be further classified into several other minor categories. The CIJ and DoD methods differ in the nature of flow through the nozzle [5]: the CIJ method ejects a continuous jet of ink droplets from the nozzle, whereas the DoD method only ejects an ink droplet upon demand.

The CIJ's drop generation process works as follows: a jet of pressurised ink exits the nozzle and is broken down into a stream of ink droplets

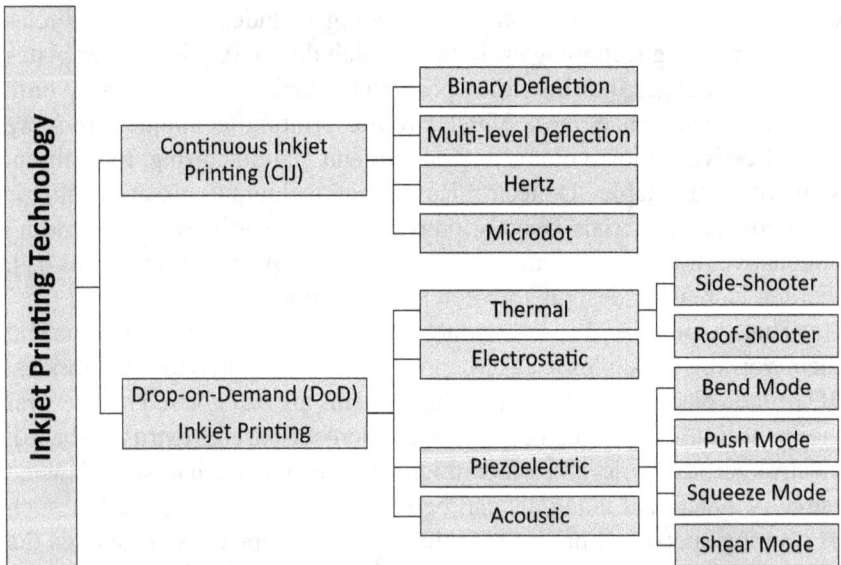

Figure 3.1.   Classification of inkjet printing technologies [6].

with well-controlled spacing and uniform droplet size by the disturbance imposed by a piezoelectric transducer vibrating at high frequency [5–7]. As the droplets pass through a charging electrode, each ink droplet is selectively induced with an electric charge. These ink droplets subsequently pass through another electric field generated by the deflection plates. Uncharged ink droplets, unaffected by the electric field, fall straight into the gutter. The collected ink is recirculated back into the CIJ system to minimise ink wastages. Charged ink droplets are directed by the deflection plates to their deposition locations on the substrate [6]. CIJ systems have the characteristics of high-speed printing, whereby ink droplets have rapid droplet formation rate and exit the nozzle at high velocity [5,6]. However, CIJ systems are generally more complex than DoD systems. CIJ systems require sophisticated mechanisms to remove, collect and reuse the unrequired ink droplets for printing from the continuous jet of droplets [5].

For the DoD inkjet printing method, several different types of actuation mechanisms are used in the printheads: thermal, electrostatic, piezoelectric and acoustic [6] (see **Figure 3.1**). These methods have similar working principles: the pressure pulse generated by the various types of actuation mechanisms forces the ink through the nozzle and causes an ejection of ink droplet in the due process [5,6]. As additional droplet selection or deflection devices are not required, DoD inkjet printheads generally have smaller distance from the nozzles to the substrates during printing, thereby enhance the printing accuracy and resolution [5].

Parameters such as viscosity, surface tension and fluid inertia of the ink that are critical to the formation of the ink droplet and the jetting behaviour in DoD inkjet printing process [8] can be related by the dimensionless Ohnesorge number, *Oh* [5], which measures the ratio of internal viscosity dissipation against surface energy in the formation of droplets [9]:

$$Oh = \frac{\eta}{\sqrt{\sigma \rho d}}, \tag{3.1}$$

where η is the viscosity of the ink, σ is the surface tension of the ink, ρ is the density of the ink and d is the characteristic length scale which is typically the droplet diameter or nozzle diameter [5].

The Ohnesorge number can also be expressed in terms of the Reynolds number, *Re*, and Weber number, *We*:

$$Oh = \frac{\sqrt{We}}{Re} \tag{3.2}$$

The dimensionless Reynolds number, *Re*, which measures the ratio of inertia against viscous forces in a moving fluid [5] and can be expressed as:

$$Re = \frac{\rho dV}{\eta}, \tag{3.3}$$

where *V* is the velocity of the moving fluid. The Weber number, *We* is defined as the inertia forces to the forces resulting from surface tension [10] and can be expressed as:

$$We = \frac{\rho dV^2}{\sigma} \tag{3.4}$$

The Ohnesorge number closely relates the physical properties of the fluid and the scale of the ejected droplet to their behaviours [5]. However, it does not consider the driving conditions of the inkjet printhead. **Figure 3.2** shows the schematic diagram of the various regimes in the drop-on-demand inkjet printing process, with Reynolds number and Ohnesorge number on the x-axis and y-axis respectively [11]. It can be observed that fluids whose Ohnesorge number ranges between 0.1 and 1 are inkjet printable. In addition, the Ohnesorge number must also lies within two boundary lines (the insufficient energy for drop formation line and the onset of splashing line) for optimised printing performances [5]. This is to ensure sufficient energy is given to eject the ink from the nozzle and to prevent splashing of the ink droplet upon impact on the substrate [5]. Fluids with Ohnesorge number exceeding 1 indicate that the fluids are too viscous to be printed by inkjet as the viscous forces impede drop separation. Many undesired satellite droplets will be formed during the printing process if the fluids have Ohnesorge number less than 0.1 [9].

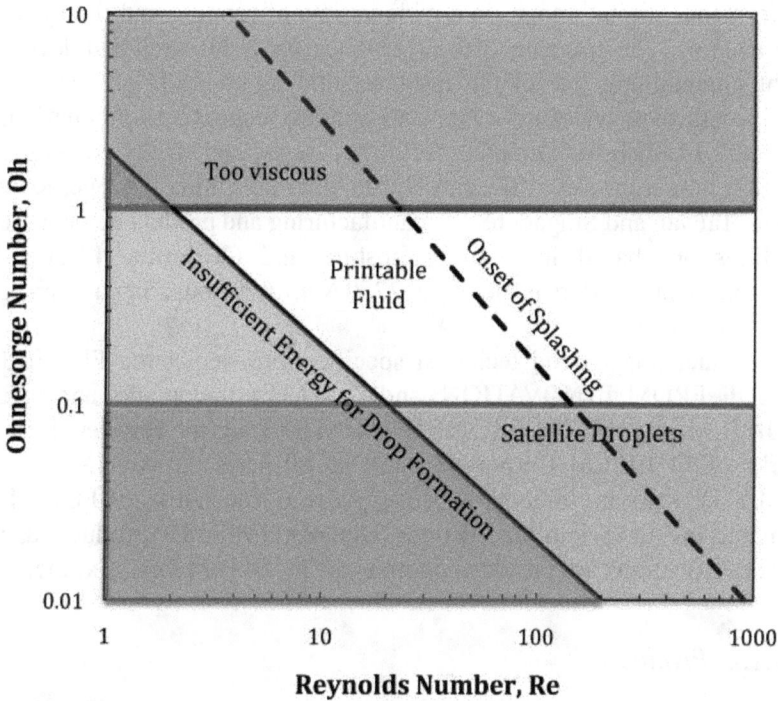

Figure 3.2. Schematic diagram of various regimes in the drop-on-demand inkjet printing process. Reprinted from Ref. [11], Copyright (2011), with the permission of AIP Publishing.

3.1.1 *FUJIFILM Dimatix, Inc.*

3.1.1.1 *Company*

A wholly-owned subsidiary of FUJIFILM Corporation, FUJIFILM Dimatix, Inc., is a leading global manufacturer of piezoelectric drop-on-demand inkjet products used for industrial and commercial applications. The company was initially founded in 1984 as Spectra, Inc., then renamed as Dimatix in 2005, and finally renamed as FUJIFILM Dimatix, Inc. after an acquisition by FUJIFILM Corporation in 2006. FUJIFILM Dimatix has developed significant intellectual property and multiple generations of proprietary drop-on-demand inkjet printheads capable of producing high performance digital images in a wide variety of printing and fluid jetting applications. The company also invests heavily in research and development, maintaining one of the most capable inkjet R&D groups in the world with over one-third of its staff actively engaged in product engineering. FUJIFILM Dimatix facilitates sales and technical support through numerous key offices in North America, China, Europe, Korea, Japan, Taiwan and Singapore. Its manufacturing and product development facilities are based in New Hampshire and California. FUJIFILM Dimatix's headquarters is located at 2250 Martin Avenue, Santa Clara, CA 95050, USA.

Product images and technical specifications depict the FUJIFILM, VALUE FROM INNOVATION, and DIMATIX trademarks and logos. FUJIFILM and VALUE FROM INNOVATION are the registered trademarks of FUJIFILM Corporation and its affiliates. DIMATIX and the DIMATIX logo are trademarks and registered trademarks of FUJIFILM Dimatix, Inc. in various jurisdictions. Dimatix DMP-2850 product images and specifications are provided courtesy of FUJIFILM Dimatix, Inc.

3.1.1.2 *Products*

(a) *Dimatix Materials Printer DMP-2850*

The Dimatix Materials Printer DMP-2850 (see **Figure 3.3**) is a benchtop precision materials deposition system which utilises the piezoelectric inkjet technology for precise deposition a variety of functional

Figure 3.3.   Dimatix Materials Printer DMP-2850 precision materials deposition system. (Courtesy of FUJIFILM Dimatix, Inc.).

fluidic materials. Users can fill their own fluidic materials for their applications in the single-use, disposable piezoelectric inkjet print cartridges. The print cartridges are easily interchangeable and can bring much convenience in changing different materials during the printing process.

The DMP-2850 is also an ideal printer for applications such as fluid and substrate interactions evaluation, material and fluid development and evaluation, product development, prototype and sample generation, optimization and evaluation of digital patterns, and deposition of biological fluids.

The DMP-2850 can support print resolution up to 5080 dots per inch (DPI). The cartridge mounting angle can be adjusted to change the effective spacing of the nozzles, which affects the drop spacing and print resolution. For instance, the print resolution can be increased from 100 dpi to 5080 dpi by decreasing the cartridge mounting angle from 90° to 1.1°, which decreases the drop spacing from 254 μm to 5 μm [12].

The DMP-2850 has a maximum printing area of 210 × 315 mm (for substrate thickness less than 0.5 mm) and the maximum allowable substrate thickness is 25 mm. The heated vacuum platen is used to secure the substrates in place and can be heated up to 60°C with ± 2°C temperature variability. Note that the DMP-2850 can only print on flat substrates and the printhead is moveable in the z-axis. Some of the substrates that can be used include paper products, thin polymer films, glass, silicon, ceramics and as well as membranes.

The DMP-2850 also has a built-in waveform editor and drop jetting observation system, which can allow optimisations of the ejected droplets from the nozzles by manipulating the input of the electronic pulse to the piezoelectric jetting devices. The in-built fiducial camera in DMP-2850 also helps to align substrates with reference marks and position the print origin to match substrate placement. In addition, the fiducial camera can provide the measurement of locations and features, inspection and image capture of the printed patterns, cartridge alignment when using multiple cartridges and drop placement matching to the previously patterned substrates. The technical specifications of the Dimatix Materials Printer DMP-2850 are tabulated in Table 3.1.

(b) *Dimatix Materials Printer DMP-2850 print cartridge*
The Dimatix Materials Printer DMP-2850 print cartridge (see **Figure 3.4**) is a single-use, disposable, interchangeable cartridge-based piezoelectric inkjet printhead with integrated reservoir and heater, specially designed to

Table 3.1.   Specifications of the Dimatix Materials Printer DMP-2850. (Courtesy of FUJIFILM Dimatix, Inc.).

Machine model	Dimatix Materials Printer DMP-2850
System description	• Flat substrate, xyz stage, inkjet deposition system
	• Low cost, user-fillable piezo-based inkjet print cartridges
	• Built-in drop jetting observation system
	• Fiducial camera for substrate alignment and measurement
	• Variable jetting resolution and pattern creation PC-controlled with
	• Graphical User Interface (GUI) application software
	• Capable of jetting a wide range of fluids
	• Heated vacuum platen
	• Cartridge cleaning station
	• Includes software
Printable area	< 0.5 mm thickness: 210 mm × 315 mm
	0.5–25 mm thickness: 210 mm × 260 mm
Repeatability	± 25 μm
Substrate holder	Vacuum platen
	Temperature adjustable; ambient to 60°C
System footprint	673 mm × 584 mm × 419 mm
Weight	approximately 50.7 kg
Power	100–120/200–240 VAC
	50/60 Hz
	375 W maximum
Operating range	15–40°C at 5–80% RH non-condensing
Altitude	up to 2000 m
Safety compliance	NRTL Certified to EN 61010-1, UL 61010-1, CSA 22.2 No. 61010-1
EMC compliance	EN61326-1 Class A, FCC Part 15 Class A
Replaceable items	• Print cartridge with one-time user-fillable reservoir
	• Cleaning station nozzle blotting pad
	• Drop watcherfluid absorbing pad

be used with the Dimatix Materials Printer DMP-2850 for high resolution, non-contact jetting of functional fluids in a broad range of applications.

The print cartridge has two main components: a jetting module and a fluid module (see **Figure 3.5**). The jetting module comprises of the

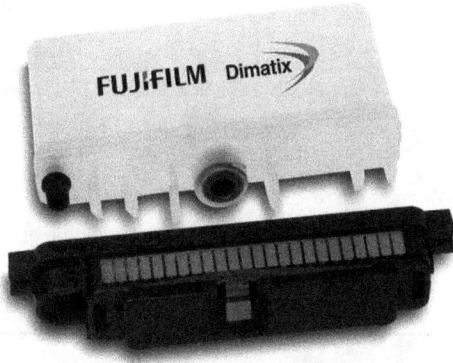

Figure 3.4. Dimatix Materials Printer DMP-2850 print cartridge. (Courtesy of FUJIFILM Dimatix, Inc.).

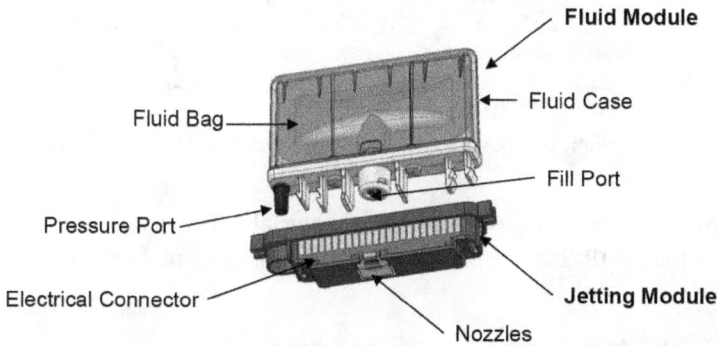

Figure 3.5. Components of the DMP-2850 print cartridge. (Courtesy of FUJIFILM Dimatix, Inc.).

nozzles, thermistor, heater, fluid connection and electrical connectors, whereas the fluid module comprises of a pressure port, fill port, fluid bag and fluid case. Each inkjet printhead has a single row of 16 nozzles intended for high-resolution printing, and the print cartridge is available in either 1 or 10 picolitre nominal drop volumes. The 10 picolitres and 1 picolitre droplet can produce spot size ranging from 40–50 μm and 20–30 μm respectively.

Users can fill their own fluidic materials for their applications in these piezoelectric inkjet print cartridges. Each cartridge has a maximum capacity of 1.5 ml and only requires a minimum volume of 0.2 ml, which aids

Table 3.2.   Specifications of Dimatix Materials Printer DMP-2850 print cartridge. (Courtesy of FUJIFILM Dimatix, Inc.).

Cartridge model	Dimatix Materials Printer DMP-2850 Print Cartridge
Type	Piezo-driven jetting device with an integrated reservoir and heater
Usable ink capacity	Up to 1.5 ml (user-fillable)
Material compatibility	Many water-based, solvent, acidic or basic fluids
Nozzles	16 nozzles, single row, 100 dpi
Drop volume	1 (DMC-11601) and 10 (DMC-11610) picolitre nominal

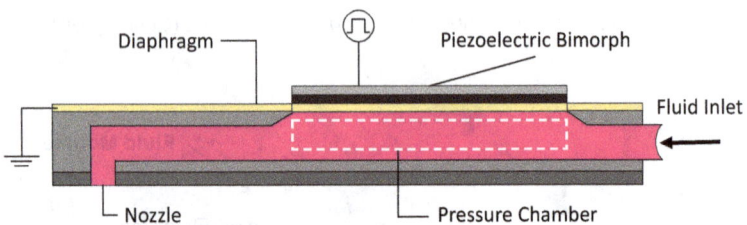

Figure 3.6.   Schematic diagram of a bend-mode piezoelectric DoD inkjet printhead [13].

in minimising wastage of expensive materials. The technical specifications of the DMP-2850 print cartridge are tabulated in Table 3.2.

3.1.1.3 *Working principles and process*

The Dimatix Materials Printer DMP-2850 utilises the bend-mode piezoelectric drop-on-demand (DoD) inkjet printing technology for material deposition. **Figure 3.6** shows a bend-mode piezoelectric DoD inkjet printhead, which comprises of piezoelectric bimorph actuator, diaphragm, pressure chamber, ink inlet and nozzle [13]. The piezoelectric bimorph actuator is bonded to the diaphragm and deforms when voltage is applied to it. The droplet generation process in bend-mode inkjet printing highly dependent on the fluid properties, waveform and nozzle structure [14].

The waveform is also known as the electrical signal that triggers and affects drop ejection. The waveform editor in the DMP Drop Manager software allows parameters within the waveform segments to be changed (such as the voltage amplitude, duration and slew rate), to optimise the drop ejection for different fluidic materials. The amplitude of the applied voltage can affect the volume of the pressure chamber directly, whereas a larger change in the applied voltage results in bigger volume change. The slew rate determines the rate of change of the pressure chamber's volume. As a rule of thumb, higher viscosity fluids require longer pulse duration and vice versa.

To better understand the jetting process, the waveform is analysed together with the printhead. In general, a basic waveform is typically segmented into four main phases (see **Figure 3.7**).

(a) *Phase 0: Standby phase*
At the standby position, the piezoelectric bimorph element slightly contracts when a bias voltage is applied prior to the jetting pulse (see **Figure 3.7(a)**). The bias voltage is the threshold voltage when the fluid is just about to eject from the nozzle. The pressure chamber is now slightly depressed by the diaphragm.

(b) *Phase 1: Fluid drawing phase*
The piezoelectric bimorph element returns to its neutral straight position (see **Figure 3.7(b)**) when the voltage is decreased to zero, and thereby causing the pressure chamber to return to its maximum volume.

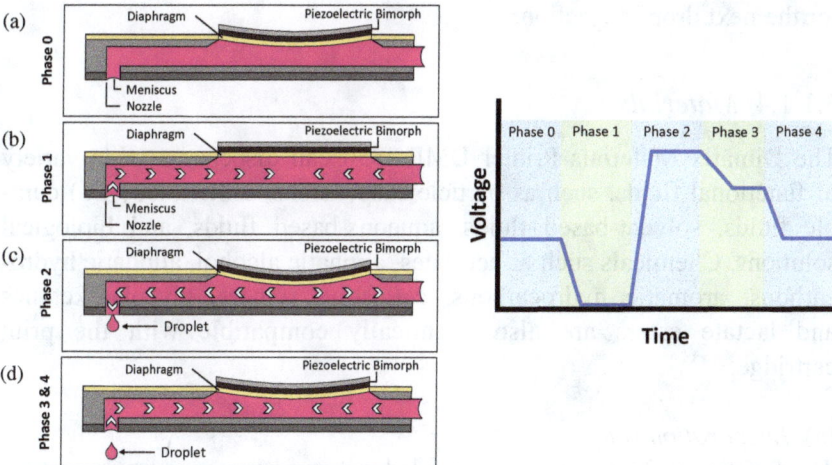

Figure 3.7.   Schematic diagram of the bend-mode piezoelectric DoD inkjet printhead at various phases: (a) phase 0; (b) phase 1; (c) phase 2; and (d) phase 3&4 [15].

This increase of volume results in more fluid being drawn into the chamber through the inlet and forming a meniscus at the nozzle.

(c) *Phase 2: Drop ejection phase*
The piezoelectric bimorph element contracts when jetting voltage is applied, and hence flexing the diaphragm inwards into the pressure chamber (see **Figure 3.7(c)**). The sudden decrease in chamber volume generates pressure on the non-compressible fluid, which forces fluid towards the nozzle [13]. A droplet will begin to form at the nozzle and get ejected from the nozzle.

(d) *Phase 3: Breakoff phase*
The piezoelectric bimorph element relaxes as the jetting voltage is brought back down slightly (see **Figure 3.7(d)**). The pressure chamber decompresses partially and thereby, initiate a pullback effect on the ejected droplet that allows quick breakoff from the bulk-fluid. This phase is also essential to prevent air to be sucked back into the printhead.

(e) *Phase 4: Refill phase*
The voltage is lastly brought back down to the bias voltage level (same voltage level as phase 0). The pressure chamber decompresses again and refills the fluid in its chamber to prepare for another jetting sequence. Phase 4 is essentially the same as phase 0, as the printhead is on standby for the next droplet ejection.

3.1.1.4 *Materials*

The Dimatix Materials Printer DMP-2850 can dispense a wide variety of functional fluids, such as particle suspensions, ultraviolet (UV) curable fluids, solvent-based fluids, aqueous-based fluids, and biological solutions. Chemicals such as acrylates, aliphatic alcohol, aliphatic hydrocarbons, aromatic hydrocarbons, cellosolve, ethers, glycols, ketones and lactate esters are also chemically compatible with the print cartridge.

(a) *Evaporation rate*
The functional fluids are recommended to have a low evaporation rate to avoid drying at the nozzle-air interface. Humectants, such as glycol, can be added to aqueous-based fluids to lower evaporation rate and improve

the jetting performance. Solvents with high boiling point and low evaporation rate are also recommended to be formulated for solvent-based
fluids.

(b) *Viscosity and surface tension*
For optimal jetting performances, the recommended viscosity and surface
tension of the fluidic materials should range between 10–12 cps and
28–33 dynes/cm respectively. The printhead can also be warmed up to
60°C to reduce the working viscosity as long as the viscous fluidic material can withstand the heating temperature. Surfactants can also be added
to the fluidic materials to tailor their surface tension to lie within the
optimal range.

(c) *Particle size*
The 10 picolitre and 1 picolitre nozzles have an effective nozzle size of
21 μm and 9 μm respectively. The particles and aggregates in the fluids
should be smaller than 1% of the nozzle size (0.2 μm and 0.09 μm for the
10 picolitres and 1 picolitre nozzles respectively) to prevent clogging.
Particle suspensions should also be thermally stable without aggregation
and particle settling for optimal jetting performances.

(d) *pH Level*
To prevent damage and corrosion to the print head, only fluids with
pH level within the 4–9 range are recommended for use in the print
cartridge.

3.1.1.5 *Strengths and weaknesses*

The key strengths of FUJIFILM Dimatix Materials Printer DMP-2850 are:

(a) *Use of own customised materials*: For users who wish to experiment
    their own customised materials, they also can fill their materials in the
    empty print cartridges.
(b) *Minimising wastage of expensive materials*: Each print cartridge only
    requires a minimum volume of 0.2 ml for printing. Hence, only a
    small amount of expensive materials is required for conducting
    experiments and wastage can be also minimised.
(c) *Wide variety of substrates can be used*: The DMP-2850 can deposit
    materials on any flat planar substrates.

(d) *Easy material change*: The interchangeable print cartridge can facilitate fast and easy changing of materials. Print cartridges filled with different materials can be changed quickly and easily during the printing process and may aid towards multi-material printing. Besides, the single-use and disposable print cartridges also eliminate the need for washing the printheads and prevents cross-contamination between different materials.

The key weaknesses of FUJIFILM Dimatix Materials Printer DMP-2850 are:

(a) *Limited printable materials*: Inkjet printing generally can only dispense fluidic materials within a specific viscosity and surface tension range, which may highly limit the choice of printable materials. For optimal jetting in DMP-2850, the recommended viscosity and surface tension of the printable materials should range between 10–12 cps and 28–33 dynes/cm respectively. Hence, high viscosity materials cannot be dispensed and materials with very low surface tension may leak out from the nozzles.
(b) *Printing on flat planar substrates only*: The DMP-2850 is unable to print on conformal substrates and hence, may limit its applications for 3D printed electronics.

3.1.1.6 *Applications*

The FUJIFILM Dimatix Materials Printer can be used in the following applications for printed electronics:

(a) *Printing conductive traces*: Conductive traces and patterns can be printed directly on a wide variety of substrates.
(b) *Flexible electronics*: Fujifilm Dimatix Materials Printer can deposit functional inks directly onto flexible substrates for flexible electronics applications. Castro *et al.* [16] printed low-pass filters with adjustable cutoff frequency directly onto flexible polyethylene naphthalate (PEN) substrates with FUJIFILM Dimatix Materials Printers. **Figure 3.8(a)** shows a first-order low-pass filter that comprises a capacitor and resistor, whereas **Figure 3.8(b)** shows a second-order low-pass filter that comprises a capacitor and inductor. To achieve a second-order low-pass filter with adjustable cutoff frequency, an

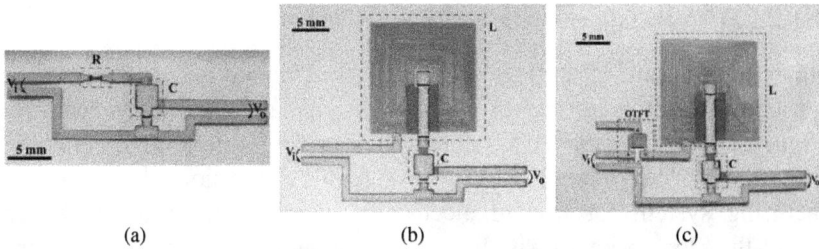

(a)   (b)   (c)

Figure 3.8.   (a) An inkjet-printed first-order low-pass filter; (b) an inkjet-printed second-order low-pass filter; and (c) an inkjet-printed second-order low-pass filter with OTFT. Adapted from Ref. [16]. Copyright (2016), with permission from Elsevier.

organic thin-film transistor (OTFT) can be integrated into the second-order low-pass filter (see **Figure 3.8(c)**).

(c) *Temperature sensors*: Ali *et al.* [17] fabricated a human body temperature sensor that exhibits good sensitivity properties. It has an operating temperature range that ranges from 28–50°C and it requires 4s response time and 8.5s recovery time. Silver nanoparticle ink was deposited by the Dimatix DMP-2850 directly onto the polyimide substrate to fabricate the interdigital electrodes (IDEs). Carbon ink was then deposited over the IDEs with a doctor blade machine to form the carbon sensing film.

3.1.2 *Nano Dimension Ltd.*

3.1.2.1 *Company*

Nano Dimension Ltd., established since 2012, is one of the world's leading providers for additive electronics with a focus on disrupting, reshaping, and defining the future fabrication of electronic devices. Nano Dimension develops and manufactures their products in-house in Israel, which include advanced additive manufacturing technology specifically developed for multilevel and simultaneous multilateral fabrication of electronic systems and nanotechnology-based conductive and dielectric inks. Their products integrate various technologies from nanomaterials, inkjet and software. Nano Dimension has three offices globally, and they are situated in US (Sunrise, FL), Israel and Hong Kong. Nano Dimension's headquarters is located at 13798 NW 4th St., Suite 315, Sunrise, FL 33325-6227.

3.1.2.2 *Products*

Nano Dimension's latest flagship system, the DragonFly LDM™ (Lights-out Digital Manufacturing) precision additive manufacturing system (see **Figure 3.9**) is an upgraded version of their previous model (DragonFly™ Pro system). The DragonFly LDM™ system is a precision additive manufacturing system for printed electronics which features uninterrupted round-the-clock printing with little or no human intervention. Some of the printed electronic components that can be printed include multi-layer AME circuit boards, with integrated antennas, capacitors, inductors and sensors.

The DragonFly LDM™ system is equipped with new advanced software management algorithms and automatic-self-cleaning function for printheads. The previous DragonFly™ Pro system required manual cleaning of the printheads every few hours to ensure good print quality. With the new technology upgrade, the printheads in DragonFly LDM™ system

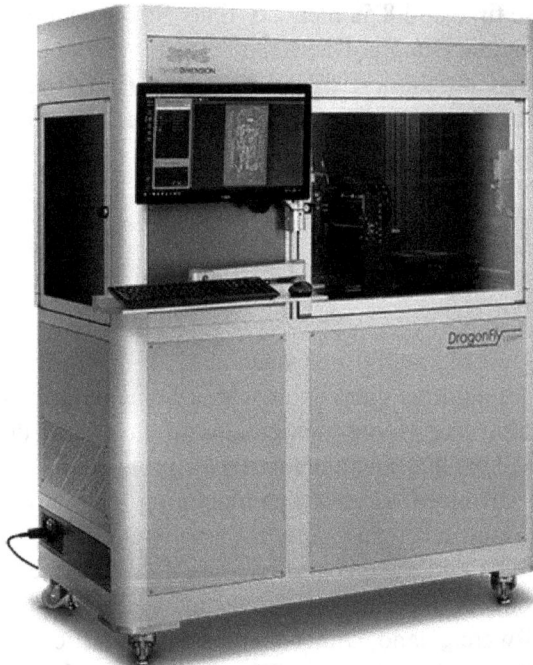

Figure 3.9.   Nano Dimension DragonFly LDM™. (Courtesy of Nano Dimension — Electrifying Additive Manufacturing).

are cleaned automatically every few hours and hence reduces the need for tedious human interventions and allows for around the clock printing. The DragonFly LDM™ also comes with real-time automatic material monitoring capabilities which can help improve workflow and production processes, maximising runtime and optimising the overall equipment effectiveness. The duration and frequency of scheduled downtime are significantly reduced to one weekly maintenance operation typically.

The DragonFly LDM™ utilises the piezoelectric drop-on-demand inkjet technology for ink deposition. It has two printheads which enable the printer to print both conductive and dielectric inks. Each printhead has 512 nozzles arranged in two rows and each row is divided into three phases. Parameters such as the voltage, pulses and temperature can affect the droplet size, droplet structure and ejection speed.

The DragonFly LDM™ can also move in three degrees of freedom: x, y and z axes. The printing x-axis contains the heated tray, linear motor and encoder. The group assembly lies in the y- and z-axes comprises of the printheads, ultraviolet (UV) lamp, camera for calibration, and fume suction, ink and cooling tubes. The z-axis is used for height build up during the printing process. The heated tray can be heated up to 175°C and is used for evaporating solvents present in the conductive ink. The heated tray also has vacuum tunnels on its surface to hold down the substrates. A stationary infrared (IR) lamp is used to dry and sinter the silver ink.

The DragonFly LDM™ has mechanical accuracy of <1 μm, and its minimum achievable conductive trace layer thickness and dielectric layer thickness are 17 μm and 35 μm respectively. The minimum trace and spacing achievable is 108 μm each. The minimum via and drill sizes are 200 and 400 μm respectively. Its build volume is 160 mm × 160 mm × 3 mm. The minimum and maximum allowable AME thickness are 0.7 mm and 3 mm respectively.

The proprietary optimised software package which comes with the DragonFly LDM™ system, helps to prepare the print jobs. SWITCH™ software can support numerous conventional file formats that are currently used in the electronics industry (for instance, Gerber, VIA and DRILL files). Various print parameters, such as layer order, layer thickness, conductor width, shape outlines, punching and rotation options, can be adjusted with the SWITCH™ software. The SWITCH™ software is available as a licensed stand alone for installation on engineers' design stations. The technical specifications of the DragonFly LDM™ are summarised in Table 3.3 respectively.

Table 3.3.   Specifications* of Nano Dimension DragonFly LDM™. (Courtesy of Nano Dimension — Electrifying Additive Manufacturing).

Machine model	DragonFly LDM™
Deposition technology	Piezo drop on demand inkjet
Number of printheads	2 printheads One printhead for each type of ink
Minimum trace layer thickness	17 microns
Minimum dielectric layer thickness	35 microns
Inks	Nano Dimension's optimised conductive silver nanoparticle ink and dielectric ink
Trace conductivity relative to copper	5–30 % (process dependent)
Dielectric constant	From 2.9 @ 200 MHz to 2.69 at 20 GHz
Build volume	160 mm × 160 mm × 3 mm
Mechanical accuracy	1 micron
Software	Proprietary Dragonfly and SWITCH
External file compatibility	Gerber, ODB++
Operating system	Windows
Network connectivity	Ethernet TCP/IP 10/100/1000
Availability	> 85%
Dimensions	1400 mm × 800 mm × 1800 mm
Weight	520 kg
Power supply	230V AC, 20A, 50–60 Hz
Operational temperature	18–22°C
Operational humidity	35–55 %, non-condensing
Regulatory compliance	UL, CE, FCC, EAC

Subject to change

3.1.2.3 *Working principles and process*

(a) *Working principles*

Nano Dimension DragonFly LDM™ utilises the shear-mode piezoelectric DoD inkjet printing technology for material deposition, in which ink can be ejected by deforming small ink channels made of piezoelectric material with an applied voltage [18]. **Figure 3.10(a)** shows the cross-sectional area of the shear-mode piezoelectric DoD inkjet printhead parallel to the ink channel, whereas **Figure 3.10(b)** shows the bottom view of the printhead at the nozzles' surface.

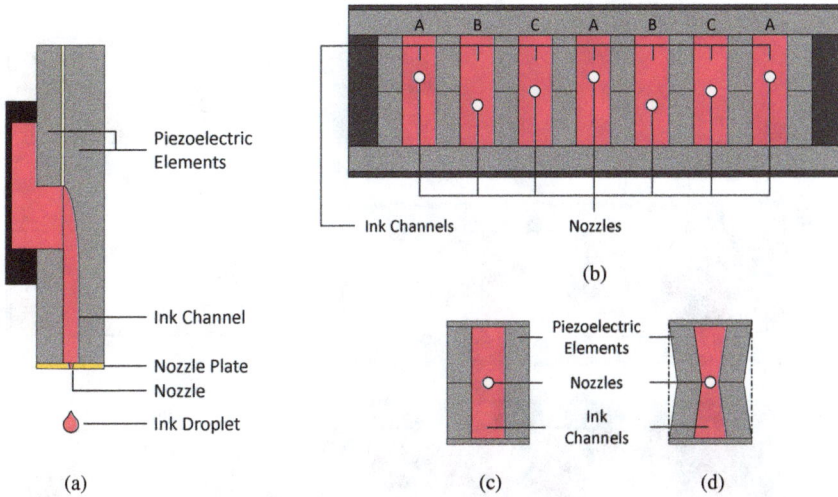

Figure 3.10. Schematic diagram of (a) cross-sectional area of the shear-mode piezoelectric drop-on-demand inkjet printhead parallel to the ink channel; (b) bottom view of the inkjet printhead at the nozzles surface; (c) cross-sectional view of an individual ink channel in the original state; and (d) cross-sectional view of an individual ink channel as the piezoelectric element contracts to eject an ink droplet [19–21].

The ink is fed into each ink channel from a common ink reservoir, and each channel has one nozzle for droplet ejection [21]. The ink channels are made with piezoelectric materials (see **Figure 3.10(c)**). When voltage is applied, the piezoelectric materials deform and thereby causing a contraction in each ink channel (see **Figure 3.10(d)**). The decrease in the ink channel's volume pressurises the ink within and causes the ejection of ink droplet through the nozzle [20]. As the ink channel returns to its original shape and volume, capillary action draws ink from the ink reservoir to refill the channel and prepares itself for the next droplet ejection [19].

The printhead utilises "the principle of shear mode piezo actuation with shared wall structure" [18], whereby multiple ink channels are arranged in parallel with each other (see **Figure 3.11(a)**).

The printhead also comes with the 3-cycle drive technology that allows timed ink-ejection control, by dividing the nozzles into three groups (e.g. group A, B and C) and eject the ink in three different phases [19]. During the first phase, all "A" ink channels within the printhead are activated first (see **Figure 3.11(b)**) to eject ink droplets onto the substrate to form part of the image. In the subsequent two phases, the "B" and "C" ink channels are activated sequentially in their respective phase

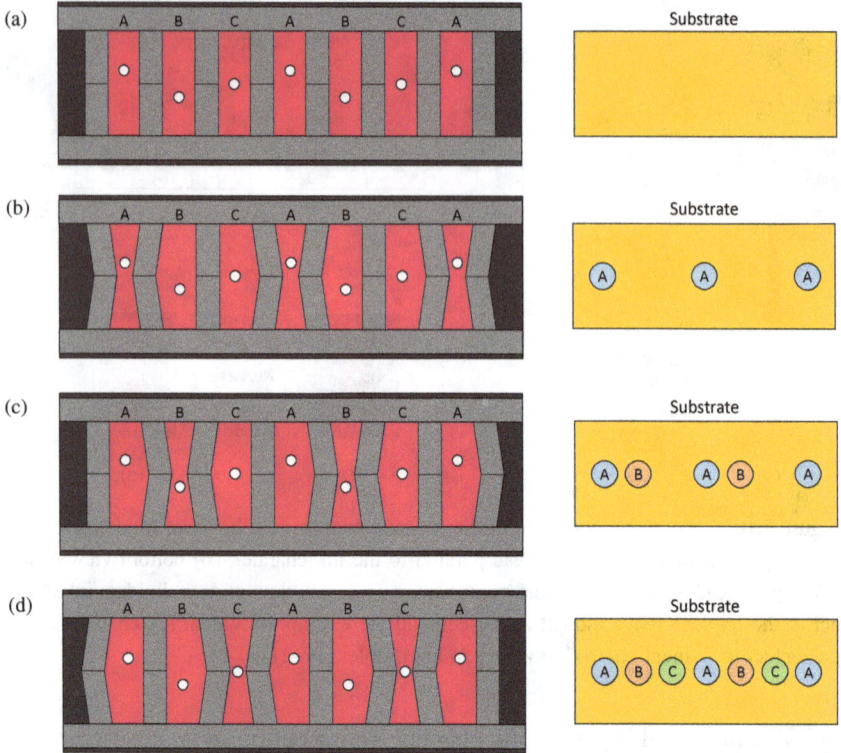

Figure 3.11.   Bottom view of the shear-mode piezoelectric drop-on-demand inkjet print-head with the 3-cycle drive technology and timed ink-ejection control: (a) all ink channels at original state, no ink deposition on substrate; (b) "A" channels activated and ink is ejected from the "A" channels onto the substrate; (c) "B" channels activated and ink is ejected from the "B" channels onto the substrate; and (d) "C" channels activated and ink is ejected from the "C" channels onto the substrate to form the completed image [19].

(see **Figure 3.11(c) and (d)**) to eject ink droplets to complete the image. This type of printing technology can achieve the higher-density nozzle array while reducing power consumption [19].

(b) *Printing process*

The printing process consists of sequential deposition of conductive silver nanoparticle ink and dielectric ink onto a removable sacrificial substrate placed on a heated tray [22], building up the multilayer AME boards or circuits layer-by-layer [23]. The heated tray can be heated up to 175°C and it helps to evaporate the solvents present in the silver nanoparticle ink [22].

The print system has two inkjet printheads, in which each printhead is dedicated for each type of ink. Each type of ink is precisely deposited in turn, according to the pattern of the specific layer [22]. The silver nanoparticle ink is used for printing various conductive features such as pads, traces and vias. The dielectric ink is used for printing electrical insulation and mechanical support structures for the circuits [22,23]. There is minimal post-processing required after the printing process, as the essential sintering and curing processes required by the inks are integrated into the printing process to obtain the desired electrical conductivity, adhesion and shape [23].

To begin, silver nanoparticle ink is deposited onto the sacrificial substrate to form a circuit pattern [24]. The entire circuit pattern immediately passes through an IR lamp for sintering when the printhead has finished depositing silver nanoparticle ink on the substrate for that particular layer [22]. This type of sintering technique is also commonly known as the infrared sintering technique. The silver nanoparticle ink is not electrically conductive initially and therefore requires a sintering process to turn the silver nanoparticle ink electrically conductive. The IR irradiations first evaporate the solvents from the ink and then sinter the silver nanoparticles together to form electrical contacts. A 1 kW IR lamp (quartz tube with coiled tungsten and halogen gas) with wavelength ranging from 0.75–1.4 μm is used for sintering the silver nanoparticle ink.

The dielectric ink is then deposited onto the areas that are not occupied by the previous circuit pattern [24]. The entire printed pattern immediately passes through the UV lamp when the printhead has finished depositing dielectric ink on the substrate for that particular layer [22]. During the UV curing process, the visible and UV light initiates a photochemical reaction in the dielectric ink that causes cross-linking and hardening of polymers. An ultraviolet (UV) curing lamp with a wavelength of 395 nm and curing intensity of 1 W/cm^2 is used for curing the dielectric ink.

The printing process repeats itself until the entire component or PCB is completed. Annotations, solder mask, drills and vias can also be included in the printed components. No etching, pressing and electroplating processes are needed in the printing process.

3.1.2.4 *Materials*

The AgCite™ Conductive Ink and Dielectric Ink are developed in-house in Nano Dimension's nano-ink facility. The AgCite™ Conductive Ink is formulated with a patent protected nanoparticle synthesis process.

(a) *AgCite™ Conductive Ink: silver nanoparticle ink*: The AgCite™ Conductive Ink is silver nanoparticle ink specifically optimised for inkjet deposition of conductive traces and patterns in DragonFly™ platforms. This ink has excellent compatibility with the matched dielectric ink and can also be used for printing of surface finishes and annotations on the printed components. The AgCite™ Conductive Ink also has 50% (w/w) silver content and the average particle size of its silver nanoparticles is around 70 nm. Its solvent has a low evaporation point and boiling point of 250°C. The AgCite™ Conductive Ink can achieve 5–35% trace conductivity relative to the bulk conductivity of copper, depending on the process parameters.

(b) *Dielectric Ink*: Nano Dimension's Dielectric Ink is specifically optimised for inkjet deposition of insulating layers, circuit structures and solder masking with the DragonFly™ platforms. The Dielectric Ink is also compatible with the AgCite™ Conductive Ink to provide essential electrical insulation, thermal resistance and mechanical support. It is stable across a wide range of frequencies. This ink contains a mixture of photo-initiators, monomers and oligomers, and is also ultraviolet (UV) curable in the low visible spectrum range. The dielectric properties of the ink are tabulated in Table 3.4.

Table 3.4. Dielectric properties for Nano Dimension Dielectric Ink (DI). (Courtesy of Nano Dimension — Electrifying Additive Manufacturing).

Frequency (MHz)	Dielectric Constant (e')	Dielectric Loss ($\tan(\delta)$)
200	2.80	—
500	2.81	0.004
1000	2.81	0.006
2000	2.80	0.011
5000	2.78	0.012
10,000	2.76	0.013
15,000	2.75	0.013
20,000	2.78	0.012

3.1.2.5 *Strengths and weaknesses*

The key strengths of Nano Dimension DragonFly LDM™ are:

(a) *In-house PCB prototyping*: Nano Dimension DragonFly LDM™ can allow complex PCBs to be fabricated in-house on demand, and hence protecting the PCB design data and its relevant intellectual property (IP) from external theft during the prototyping and manufacturing processes.

(b) *Time saving*: The Nano Dimension DragonFly LDM™ can print functional multilayer AMEs in a matter of hours or a couple of days depending on the circuit complexity, and hence reducing the time-bottlenecks faced in the prototyping and developing equivalent PCBs.

(c) *Inserted electronics*: The Nano Dimension DragonFly LDM™ can incorporate an inserted functional circuitry designs within complex part geometries without traditional manufacturing process constraints to help save cost, weight and space. Intricate interconnections, such as blind vias, buried vias and plated-through-holes, can also be fabricated within the printed boards at no additional expense.

(d) *Simultaneous printing with multiple materials*: The Nano Dimension DragonFly LDM™ has two printheads to allow simultaneous printing of conductive silver nanoparticle ink and dielectric polymer ink in a single print job.

(e) *Integrated sintering and curing processes*: The sintering and curing processes, for the conductive ink and dielectric ink respectively, are integrated into the printing process. This is a convenient and time-saving feature as it eliminates the need to transfer the printed patterns to other equipment for sintering or curing.

(f) *Automatic printhead self-cleaning feature*: The Nano Dimension DragonFly LDM™ has the automatic printhead self-cleaning feature to maintain the printing quality. The printheads are cleaned automatically every few hours, thereby reducing the need for tedious human interventions for printhead cleaning during long printing jobs, increasing uptime and improving yield.

The key weaknesses of Nano Dimension DragonFly LDM™ are:

(a) *Limited material choice*: Nano Dimension DragonFly LDM™ can only deposit Nano Dimension's conductive and dielectric inks, as these inks are optimised for the system to achieve best results.

Hence, this system may not be ideal for users who wish to experiment with their own customised materials. Besides, there are also limited material choices, in which only silver nanoparticle ink and dielectric nanoparticle polymer ink are used as conductive and dielectric inks respectively.

(b) *Flat substrate*: Nano Dimension DragonFly LDM™ cannot print directly onto conformal substrates as the inks can only be deposited onto flat 2D planar substrates. However, products with non-planar and conformal geometries can be printed despite the requirement for a flat substrate. The layer-by-layer deposition printing process can allow each layer to offset slightly from the previous layer to form complex, non-planar geometries. **Figure 3.12** shows a 3D printed molded inter-connect devices (MIDs) with non-planar geometry.

3.1.2.6 *Applications*

The Nano Dimension DragonFly LDM™ can be used in the following applications for printed electronics:

(a) *Multilayer rigid printed circuit boards*: The Nano Dimension DragonFly LDM™ can fabricate functional multilayer rigid AMEs in-house on demand. This allows faster prototyping of equivalent PCBs whereby the fabrication time can be significantly reduced from weeks to hours, and at the same time safeguarding the confidential IP in-house. For instance, the Nano Dimension DragonFly LDM™ took

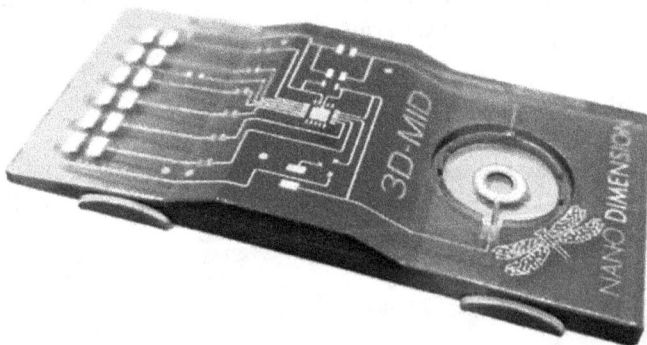

Figure 3.12.   3D printed molded interconnect devices (MIDs). (Courtesy of Nano Dimension — Electrifying Additive Manufacturing).

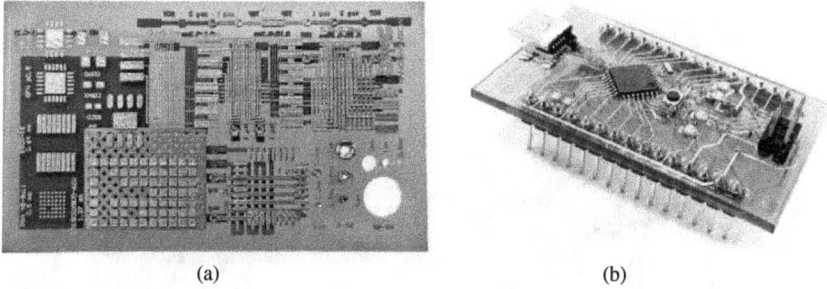

(a)   (b)

Figure 3.13. (a) 4-layer AME coupon printed by the Nano Dimension DragonFly LDM™; and (b) functional Arduino 2-layer AME prototype with soldered electrical components. (Courtesy of Nano Dimension — Electrifying Additive Manufacturing).

20 hours to print three pieces of the 4-layer (55 mm × 31 mm) AME coupon (see **Figure 3.13(a)**) from Gerber file. **Figure 3.13(b)** shows a functional Arduino 2-layer AME prototype with electrical components soldered onto the board.

(b) *Vertically stacked integrated circuits (ICs)*: The Nano Dimension DragonFly LDM™ can fabricate complex AMEs structures for vertically stacked ICs (see **Figure 3.14**). High density interconnects (HDI) and vias within the AMEs' structures allow multiple ICs with varying sizes, from small to large, to be stacked vertically, mounted, interconnected and assembled on top of each other within a single structure. Vertically stacked ICs can achieve higher circuitry density and smaller device footprint than traditional PCBs. The AME sample shown in **Figure 3.14** can mount up to four ICs: one small ball grid array (BGA) IC at the bottom, followed by two SMT ICs in the middle, and lastly by a big BGA IC with four rows of BGA contacts at the top. Only eight layers of conductive traces and vias were needed to interconnect these four ICs together. The Nano Dimension DragonFly LDM™ only took 25 hours to print four pieces of these AMEs samples.

(c) *3D printed inserted electronics*: Conductive traces, passive components and active components can be printed and embedded within complex non-planar part geometries with the Nano Dimension DragonFly LDM™ to save space and reduce weight. These conductive traces and electrical components can also be protected from humidity, vibrations and damages by embedding them within the

Figure 3.14.   Vertically stacked integrated circuits (ICs) in an AMEs structure. (Courtesy of Nano Dimension — Electrifying Additive Manufacturing).

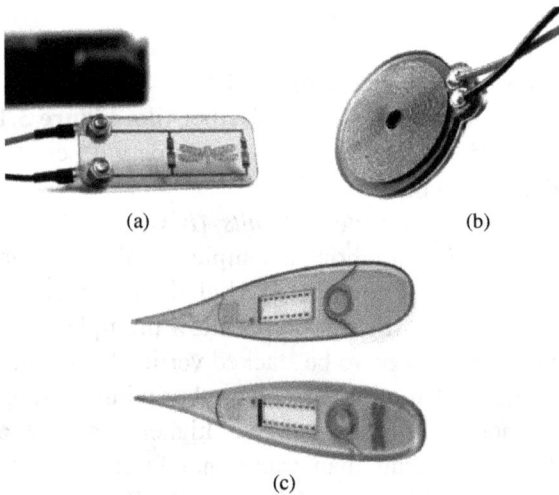

(a)   (b)

(c)

Figure 3.15.   (a) 3D printed embedded electronics; (b) 3D printed electromagnetic coil; and (c) 3D printed digital thermometer. (Courtesy of Nano Dimension — Electrifying Additive Manufacturing).

dielectric polymer matrix. **Figure 3.15(a)–(c)** show a 3D printed inserted electronics, 3D printed electromagnetic coil and 3D printed thermometer, respectively.

Other printed electronics applications by Nano Dimension DragonFly LDM™ reported also include printed antennas, RFIDs, sensors, AME circuits with integrated functional capacitors and transmission lines.

3.1.3 *BotFactory, Inc.*

3.1.3.1 *Company*

Founded in 2013, BotFactory Inc. is a New York City-based hardware company that focuses on developing all-in-one desktop electronic circuit printers capable of printing multilayer printed circuit boards (PCBs) and assembling electrical components. BotFactory has also won numerous prestigious awards and competitions, including the Technical Development Manufacturing award from IDTechEx in 2016. BotFactory, Inc is located at 4334 32nd Place, FL 3, RM 3Rb, Long Island City, NY 11101, USA.

3.1.3.2 *Products*

BotFactory Inc. provides an all-in-one desktop solution for printing and assembling electronic circuits by integrating additive manufacturing technologies, nanotechnology and image-recognition algorithms. BotFactory has two main product series: Squink Multilayer PCB Printer and BotFactory SV2 (Starter, Enhanced and Professional editions). Both products have the same overall user interface (UI) and the software interface. Both printers utilise thermal inkjet printing technology for depositing conductive silver nanoparticle ink and insulating polymeric ink directly onto a variety of flexible and rigid substrates, straight from the PCB Gerber files. The printing head is equipped with an ultraviolet (UV) light-emitting diode (LED) to cure the insulating polymeric ink instantly as it is deposited on the substrate. Single-layer or multilayer circuits can be printed directly onto Kapton and FR-4 substrates, or any flat and non-porous substrates that have been coated with the insulating ink and do not warp or melt under 100°C. Same inks and substrates are used for both printers. For each process (printing, dispensing and assembly), individual tool head needs to be swapped out manually for each desired function.

(a) *Squink Multilayer PCB Printer*
The Squink Multilayer PCB Printer is BotFactory's first-generation desktop PCB printer (see **Figure 3.16**), which can print multilayer circuits, dispense conductive pastes, and pick-and-place components directly onto the printed circuit boards. Its thermal inkjet printhead has 12 nozzles and can produce a minimum trace width of 250 μm and minimum trace pitch

Figure 3.16.   Squink Multilayer PCB printer. (Courtesy of BotFactory, Inc.).

of 760 μm for conductive lines. It can have a maximum working area of 85 × 127 mm for printing a two-layers PCB. The technical specifications of Squink Multilayer PCB Printer are summarised in Table 3.5.

(b) *BotFactory SV2*
BotFactory's second-generation desktop PCB printer, BotFactory SV2 (see **Figure 3.17**), is an enhanced version of the Squink Multilayer PCB Printer with additional features and improved functionalities. Similarly, the BotFactory SV2 can also print multilayer circuits, dispense conductive pastes, and pick-and-place components directly onto the printed circuit boards.

The BotFactory SV2 is equipped with better hardware and faster onboard computing power for better efficiency in printing, dispensing and assembly. The better inkjet printing technology in BotFactory SV2 allows

Table 3.5.   Specifications of the Squink Multilayer PCB Printer and BotFactory SV2. (Courtesy of BotFactory, Inc.)

Printer Model	Squink Multilayer PCB Printer	BotFactory SV2
Mechanical		
Frame: Size	44 × 44 × 40 cm	42 × 35.5 × 44.5 cm
Additional clearance req.	6 cm in front	7.5 cm in front, 10 cm in the back
Frame: Weight	9 kg	12 kg
Shipping box: Size	44 × 44 × 40 cm	55 × 43.2 × 68.6 cm
Shipping box: Weight	16 kg	22.5 kg
XYZ positioning resolution	10 microns	10 microns
Max. substrate size [X/Y]	152 × 152 mm	152 × 152 mm
Max. circuit size [X/Y]	127 × 127 mm	117mm × 152 mm
Heads: Weight	Less than 500 g	Less than 500 g
Electrical		
Power	DC 24V, 9A	DC 24V, 22A
Connection	Ethernet, Wi-Fi	
Printing		
Print technology	Thermal inkjet printing	Thermal inkjet printing
Number of nozzles	12	300
Ink types (curing method)	Conductive silver ink (Heat) Insulating polymeric ink (UV)	
Min. trace width	250 microns	200 microns
Min. trace pitch	760 microns	450 microns
Conductive ink sheet resistance	40 mOhms/square (*When printed with BotFactory's standard settings. Sheet resistance can go down to 10mOhms/square using custom settings)	
Number of layers	2	4
Part attachment	Conductive glue/epoxy	Conductive glue/epoxy and solder paste
Working print area single layer printing 2-layers printing	127 × 127 mm 85 × 127 mm	117 mm × 152 mm
Supported formats	GERBER RS-274X,. .jpg, .png, .tiff, .bmp	
Dispensing		
Paste technology	Dotted extrusion	Dotted and continuous extrusion
Paste types	Conductive glue/epoxy and solder paste (*Dispensing can be used independently on pre-made boards.)	
Speed	3000 dots per hour	1 cm/sec

(*Continued*)

Table 3.5.   (*Continued*).

Area covered	5 sq. inches of pad area per hour	Depends on the micro-dispensing tip used and the percentage of pad coverage
Extruded dot size	500 microns	200 microns
Curing method	Heat	
Max dispensing area	127 × 127 mm	117 mm × 152 mm
Supported formats	GERBER RS-274X, .jpg, .png, .tiff, .bmp	
Assembly (Pick and Place)		
PNP technology	Vacuum pickup, computer vision, automatic assembly	Vacuum pickup, computer vision, automatic assembly (*Manual mode also available)
Camera	Onboard (work area)	
Speed	4 parts per minute	
Tray: Types	Individual components	Tape cut and individual components
Number of tray slots/tape tray: number of slots × tape width	• 4 oversized slots, • 12 slots for parts 2512 (6.3 × 3.2 mm) or smaller • No support for rails	Tape Cut: • 6 × 8mm, 2 × 12mm, 1 × 16 mm Individual components: • 2 slots: 12 × 15 mm • 10 slots: 15 × 15 mm • 1 slot: 32 × 30 mm
Tray: Parts loaded	12 per batch	Individual components: • 13 per batch Tapes: • 24 cpnt (6 × 8 mm) • 6 cpnt (2 × 12 mm) • 2 cpnt (1 × 16 mm)
Min. part size	0603 [1608 Metric]	
Max. part size	SOIC-24 (10 × 16 mm)	20 × 20 mm
Supported formats	Centroid file (text file describing reference designator, footprint, X/Y location, and rotation. typ. CAD-generated)	
Software		
Installation	Software runs on-board, local installation not required	
System requirements	Web browser (platform-independent)	
Recommended browser	firefox, chrome	
Temperature		
Ambient: Operating	10–40°C	10–40°C
Heat bed: Operating	25–120°C	25–160°C
Heat bed: Max	130°C	170°C

Figure 3.17.   BotFactory SV2. (Courtesy of BotFactory, Inc.).

improvement in the printing precision and ability to print higher layer count. Its wider thermal inkjet printhead has 300 nozzles and can dispense smaller inkjet droplets. Hence, thinner conductive lines, as fine as 200 μm, can be printed at faster printing speed with an increased sweep. The ink cartridges also have large ink capacity and thus can allow extended printing time. It also has a maximum working area of 125 × 140 mm for printing multi-layers PCB. The work bed also has a more powerful heating source and fan for faster heating and cooling. When dispensing conductive pastes, the dispensing resolution is greatly improved in BotFactory SV2 with its new micro-tip and micro-extrusion mechanism where an extruded dot size of 200 μm can be achieved. There are also on-board cleaning stations for cleaning and maintaining ink cartridges and syringe tip automatically.

The BotFactory SV2 can pick-and-place components with its vacuum tips and assemble the components precisely on the printed PCBs with the help of computer vision and image processing. These vacuum tips can be interchanged automatically to adapt to different components part size. Assembly pick-and-place tray for surface-mount technology (SMT) strips is also available for easier assembly. The technical specifications of BotFactory SV2 are summarised in Table 3.5.

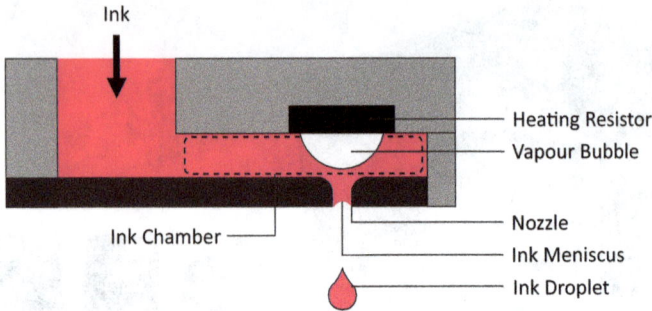

Figure 3.18.   Schematic diagram of a roof-shooter thermal DoD inkjet printhead [25–27].

3.1.3.3 *Working principles and process*

(a) *Working principles*

BotFactory's systems utilise the roof-shooter thermal inkjet technology for material deposition. **Figure 3.18** shows a typical roof-shooter thermal DoD inkjet printhead, in which the heater is located directly on top of the printhead's nozzle. The thermal inkjet printing process can be described in four major process steps: bubble nucleation, bubble growth, bubble collapse, and ink refill [25].

  (i) *Bubble nucleation*: When the thermal inkjet printhead receives a command for ink ejection, a short electrical current pulse is passed through the heating resistor instantaneously [25]. The heating resistor heats up rapidly and the generated heat vaporises the ink to the critical temperature of bubble nucleation [25,27]. This bubble nucleation process occurs rapidly in the first few microseconds [27].

  (ii) *Bubble growth*: The nucleated bubble quickly expands in volume and fills the entire ink chamber. The rapid expansion of the bubble results in a pressure pulse in the ink and thereby ejecting an ink droplet out from the nozzle onto the substrate [26,27].

(iii) *Bubble collapse*: The bubble immediately collapses once the ink droplet is ejected out from the nozzle [25,27]. The droplet breakoff from the ink meniscus at the nozzle is driven by the ink's upwards displacement and the meniscus's retraction during the bubble collapse [25].

(iv) *Ink refill*: The ink chamber is then refilled due to the surface tension of the ink meniscus in the nozzle [25], as the ink flows from the ink reservoir to the nozzle. The thermal inkjet printhead is now ready for another droplet ejection.

(b) *Printing processes*
This section will describe the fabrication process of a double layer PCB board with the Squink Multilayer PCB Printer.

(i) *Upload design files*: Various design files are required to be uploaded onto the Squink Multilayer PCB Printer for printing and assembly. These files allow the printer to print conductive and insulating layers, apply conductive pastes, and pick-and-place electrical components for assembly.

(ii) *Preparation and calibration*: To start the printing process, the inkjet tool head is first attached to the printer with the conductive ink cartridge inserted. A substrate is taped down onto the heated platen and the conductive traces are to be deposited directly onto the substrate. The inkjet tool head is then aligned with the substrate and a "print nozzle" test is run for checking any clogged nozzle before the actual circuit printing process.

(iii) *Circuit printing*: The printer begins with the circuit printing process after the calibration and checking processes. The bottom layer of the PCB board, which only comprises conductive traces, is directly deposited onto the substrate. The as-deposited conductive ink on the substrate is wet and needs to be sintered. The heated platen is then heated up for drying and sintering the wet conductive ink after the bottom layer is completed. The conductive ink cartridge is swapped out and replaced with the insulating ink cartridge. The insulating ink is used for printing the insulating layer on the bottom conductive layer. The printhead is equipped with an UV LED to cure the insulating ink instantly as it gets deposited on the substrate. The printing process is repeated for the insulating and conductive top layer until the entire circuit is completed. **Figure 3.19** shows a double-layer PCB printed by the Squink Multilayer PCB Printer.

(iv) *Dispensing conductive pastes*: The inkjet tool head is detached from the printer and is replaced with the glue head attachment. The conductive paste is first filled in a syringe and the syringe is then inserted into the glue head attachment. The glue head is calibrated and aligned to the target market first so that the conductive paste can be dispensed precisely at the exact locations. Parameters such as load amount and

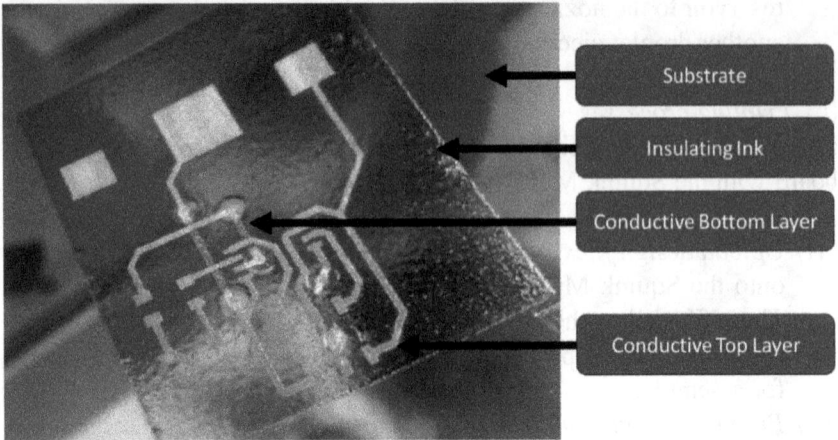

Figure 3.19.   A double-layer PCB printed by the Squink Multilayer PCB printer, showing the different layers within the printed PCB. (Courtesy of BotFactory, Inc.).

drop size can also be adjusted accordingly to the amount of conductive paste required.

(v) *Pick-and-place electrical components for assembly*: The glue head is detached from the printer and is replaced with the pick-and-place head attachment. The component tray alignment is first calibrated with the "Check Tray Alignment" test. The pick-and-place head is then calibrated and aligned to the target marker so that the electrical components can be placed accurately at the required locations. The electrical components are slotted in the component tray in the correct part orientations. These electrical components are picked up and shown to the upward-facing camera. Using image recognition, the orientation of the component is corrected according to the uploaded Gerber files and then placed directly on top of the previously dispensed conductive paste for assembly. The heated platen is heated up again for curing the wet conductive pastes to ensure the electrical components adhere to the PCB properly.

3.1.3.4 *Materials*

BotFactory provides their own conductive silver nanoparticle ink and insulating polymeric ink for the printing of conductive traces and insulating layers, respectively.

(a) *Conductive silver nanoparticle ink*: BotFactory's conductive silver nanoparticle ink is a low viscosity conductive silver nanoparticle ink specifically optimised for inkjet printing. This ink approximately has 30–50% of silver and 20–40% of 2,2'-oxydiethanol in its mixture. It has a relative density ranging from 1.4–1.7, pH ranging from 4.5–7, and viscosity ranging from 4–20 cP. The silver nanoparticle ink has to be sintered by heat after deposition to achieve high electrical conductivity in the printed patterns.

(b) *Insulating polymeric ink*: BotFactory's insulating polymeric ink is a type of non-viscous ultraviolet (UV) curable acrylic photopolymer specifically optimised for inkjet printing and needs to be cured by UV light after printing. This ink is a mixture of acrylate and urethane monomers, ethanol and photo-initiator. It is compatible with BotFactory's conductive silver nanoparticle ink for multilayer PCBs fabrication.

3.1.3.5 *Strengths and weaknesses*

The key strengths of BotFactory systems are:

(a) *Wide range of substrates*: BotFactory systems can deposit the conductive silver nanoparticle ink and insulating polymeric ink directly onto any planar substrates such as FR4, polyimide and photopaper.

(b) *Easy to use*: BotFactory systems are easy to use and no prior training is required. BotFactory's software interface will guide the users step-by-step through the entire fabrication process.

(c) *Low cost*: BotFactory systems are low-cost and affordable, which is ideal for rapid prototyping multilayer PCBs in labs or classrooms.

(d) *Multilayer PCBs prototyping*: Circuit printing, solder paste dispensing, and components assembly can be completed within a BotFactory system. BotFactory systems also have a built-in heated platform for sintering conductive ink and UV LED for curing insulating polymeric ink. This is a convenient feature as it eliminates the need to transfer the as-deposited inks to another equipment for sintering or curing.

The key weaknesses of BotFactory systems are:

(a) *Printing on flat planar substrates only*: BotFactory systems are unable to print on conformal substrates and hence, may limit its applications for 3D printed electronics.

(b) *Limited materials only*: There are limited material choices, in which only silver nanoparticle ink and UV curable acrylic photopolymer are used as conductive and dielectric inks respectively.

3.1.3.6 *Applications and case studies*

BotFactory systems can be used in the following applications for printed electronics:

(a) *Rapid prototyping PCBs*: BotFactory systems are capable of printing multilayer PCBs and assembling electrical components for rapid prototyping PCBs. **Figure 3.20** shows that a prototype PCB can be printed by BotFactory SV2 on FR-4 substrate directly in 20 minutes, whereas a similar conventional PCB prototype took 2 weeks to fabricate. BotFactory SV2 also demonstrates the successful printing of a 4-layer PCBs (see **Figure 3.21(a)**). **Figure 3.21(b)** shows a fully assembled 3D printed PCB with electronic components and copper rivets through-holes.

(b) *Flexible electronics*: BotFactory SV2 and Squink Multilayer PCB Printer can also deposit silver nanoparticle ink and assemble electronic components directly onto flexible substrates for flexible electronics applications. **Figure 3.22(a)** shows assembled electronic components on flexible printed circuits. **Figure 3.22(b)** shows a fully

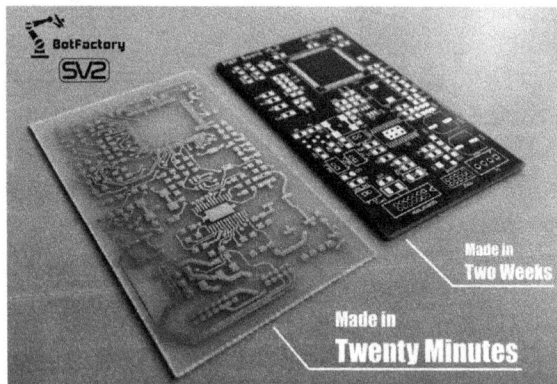

Figure 3.20.   Prototype PCB printed by BotFactory SV2 in 20 minutes vs. a conventional PCB prototype fabricated in 2 weeks. (Courtesy of BotFactory, Inc.).

(a)   (b)

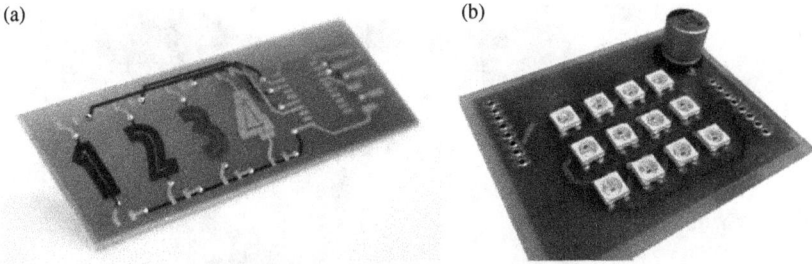

Figure 3.21. (a) 4-layer PCBs printed by BotFactory SV2; (b) Printed PCB with assembled electronic components and copper rivets through-holes. (Courtesy of BotFactory, Inc.).

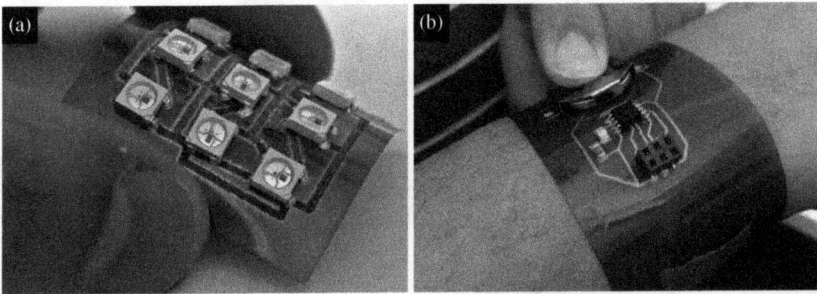

Figure 3.22. (a) Assembled electronic components on flexible printed circuits; (b) Flexible printed circuits for wearable electronics application. (Courtesy of BotFactory, Inc.).

assembled printed circuit with electronic components on a flexible polyimide substrate for wearable electronics application.

(c) *3D printed edge connectors*: The BotFactory SV2 can be used for 3D printing edge connectors. The 3D printed edge connector still can function well after 1000 mating cycles, hence demonstrating durability for prototyping purposes. **Figure 3.23(a)** shows the 3D printed edge connectors, whereas **Figure 3.23(b)** shows a printed edge connector plugged into a power adapter to demonstrate functionality.

(d) *Disposable 64-well papertronic sensing array*: Tahernia *et al.* [28] fabricated a disposable paper-based 64-well microbial fuel cells (MFCs) papertronic sensing array (see **Figure 3.24(a)**) for rapid screening of electroactive micro-organisms. The electrically conductive anodic and cathodic wiring layers (see **Figure 3.24(b)**) of the

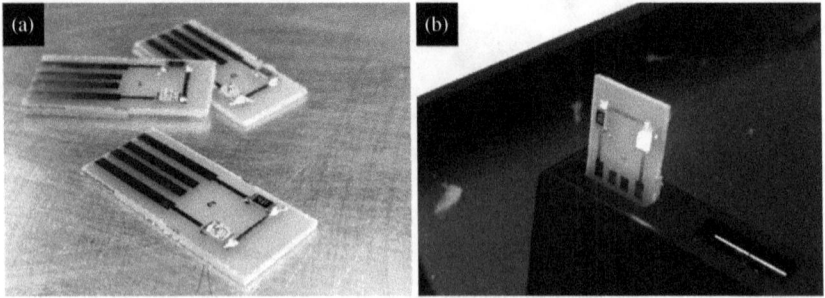

Figure 3.23.   (a) 3D printed edge connectors; (b) A 3D printed edge connector plugged into a power adapter to demonstrate functionality. (Courtesy of BotFactory, Inc.).

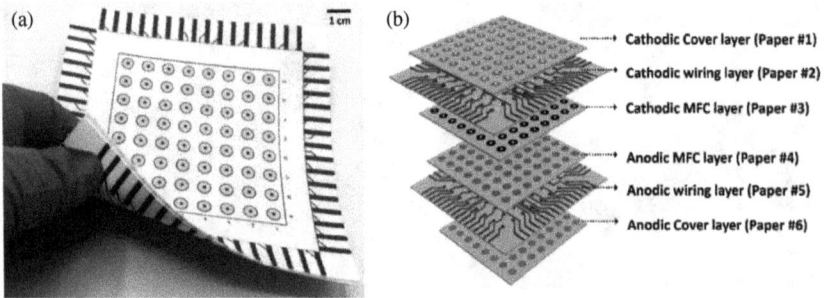

Figure 3.24.   (a) A disposable paper-based 64-well MFCs papertronic sensing array; and (b) individual layers of the papertronic 64-well MFC array device. Adapted from Ref. [28], Copyright (2019), with permission from Elsevier.

sensing array were printed on paper by using Botfactoy's Squink Multilayer PCB Printer to deposit silver nanoparticle ink.

This application demonstrated paper-based printed circuit boards can be effectively integrated into a biosensing array without messy electrical connections, while effectively characterising microbial electrochemical activities. In addition, this paper device can be safely disposed away after usage through incineration to prevent the risk of bacterial infection.

3.1.3.7 *Research and development*

BotFactory Inc. specifically addresses the latest developments in fabricating flexible and non-traditional PCBs and aims to improve their printers' capabilities to make full PCB fabrication easy with a reliable and fully automated process. BotFactory Inc. is still working to provide a tool to verify the files before execution and evaluate results after the execution and use unsupervised Artificial Intelligence (AI) models to detect and fix problems automatically.

3.2 Aerosol-Based Printing

Aerosol-based printing, also known as aerosol-based direct-write, is a type of printing technique which can generate and deposit a focused collimated aerosol beam [29] for direct material deposition on substrates. The aerosol-based printing technique is one of the printing techniques that can produce fine printing features in sub-10 μm resolution [30]. Optomec, Inc. and Integrated Deposition Solutions, Inc. (IDS, Inc) are some of the companies that utilise the aerosol-based deposition technology for material deposition and they will be further discussed in this section.

3.2.1 *Optomec, Inc.*

Optomec Inc. is one of the very few providers of additive manufacturing systems for high-performance applications. Optomec was founded in 1982 and it has been devoted to providing additive manufacturing solutions in two main areas: Aerosol Jet® process for 3D printed electronics and LENS® (Laser Engineered Net Shaping) directed energy deposition process for metal additive manufacturing. Apart from printing 3D parts, Optomec's technology can also deposit materials onto existing 3D parts that were created by traditional manufacturing techniques. The first commercial Aerosol Jet® system for printed electronics was released in 2004. The corporate headquarters of Optomec, Inc. is located at 3911 Singer Blvd N.E., Albuquerque, NM 87109, USA.

3.2.1.1 *Products*

Aerosol Jet® systems are designed specifically for the fabrication, development, enhancement and repair of high-performance electronic and biological devices for end-use products in consumer electronics, aerospace, automotive, defence, displays, semiconductor packaging and life sciences. A wide range of functional materials can be deposited by the

Aerosol Jet® process for fabricating high-resolution electronic circuits and components.

To meet specific market needs, Optomec offers four different Aerosol Jet® system models for printed electronics applications: Aerosol Jet® FLEX System, Aerosol Jet® 5X System, Aerosol Jet® Print Engine and Aerosol Jet® HD System. The technical specifications and key features of the various Aerosol Jet® systems are summarised in Tables 3.6 and 3.7 respectively.

(a) *Aerosol Jet® FLEX System*
The Aerosol Jet® FLEX system is a digitally driven, modular print solution for 2, 2.5, and 3D printed electronics applications. This system comes with a 350mm × 250mm heated vacuum chuck and has 3-axis of coordinated print motion with +/− 2 microns motion repeatability accuracy. It can be upgraded with a full 5-axis tilt-and-rotate trunnion to give 5-axis of coordinated motion. The Aerosol Jet® FLEX system can print features sizes ranging from 10 to 250 microns with various fine feature deposition print heads. Wider features can be achieved with the optional wide feature deposition print head add-on. The Aerosol Jet® FLEX system is also equipped with interchangeable closely-coupled print modules, consisting of material cassettes and associated print heads. Different materials can be filled in material cassettes to facilitate fast material change. These print modules can provide up to four hours of uninterrupted run time and can be changed easily to reduce machine downtime during long print runs. Optional infrared (IR) laser and ultraviolet (UV) cure modules for *in-situ* curing can also be added on to the Aerosol Jet® FLEX system.

(b) *Aerosol Jet® 5X System*
The Aerosol Jet® 5X System (see **Figure 3.25**) is a digitally driven, modular print solution which can support 5-axis of coordination motion for 3D printed electronics applications. This system has a 200 mm × 300 mm × 200mm print envelope. Similarly, this system can also print features sizes ranging from 10 to 250 microns with various fine feature deposition print heads. Optional features, such as the IR laser module, UV cure module, interchangeable flat heated vacuum platen, interchangeable wide feature deposition printhead and atomisation cassettes, can be added on to the Aerosol Jet® 5X System.

Table 3.6.   Specifications of various Aerosol Jet® system models. (Courtesy of Optomec, Inc.).

Specifications	Aerosol Jet®			
	FLEX System	5X System	Print Engine	HD System
Minimum line width	10 μm at 20 μm pitch (Materials and Surface Dependent)			20 μm
Layer thickness	100 nm–up to 5 μm (single print pass)			
Ultrasonic atomizer	1–15 cP			1–15 cP (Standard)
Pneumatic atomizer	1–500 cP			1–600 cP (optional feature)
Material droplet size diameter	1–5 μm			
Nozzle stand-off height	Up to 5 mm (nozzle tip to substrate surface)			
Printing area (mm)	350 × 250 × 300 mm (x,y,z)	200 × 200 × 300 mm (x,y,z)	Not available	315 × 425 × 30 mm (x,y,z)
Positional accuracy (μm)	±10 μm (100 mm range)		Not available	X/Y: ±30 μm @ 3σ Z: ±10 μm @ 3σ
Positional repeatability (μm)	±2 μm (x,y,z axis)		Not available	X/Y: ±10 μm @ 3σ Z: ±5 μm @ 3σ

(Continued)

Table 3.6. (*Continued*)

Specifications	Aerosol Jet®				
	FLEX System	5X System	Print Engine	HD System	
		Rotational pivot axis			
Rotational Positional Accuracy	(Option) Add AJ5x A/B Axes	80 arc sec	Not available	Not available	
Rotational Repeatability		03 arc sec			
Pivot axis Positional Accuracy		80 arc sec			
Pivot axis Repeatability		03 arc sec			
System approx. weight (kg)	1088		150	900	
System dimensions (mm)	1020 × 1375 × 2240 (D × W × H)		19" rack mount units	1260 × 770 × 1450 (D × W × H)	
Electrical requirements	110/220V, 50 or 60Hz, 40 Amps (10 amps continuous operation, typical)		110 or 220V, Single phase, 50/60 Hz, 10 Amps	200–250V single phase 50/60Hz	
Gas input to system	345–425 kPa (50–60 psi), >99.9% nitrogen gas, at 20 slpm				

Table 3.7. Key features of various Aerosol Jet® system models. (Courtesy of Optomec, Inc.).

Key Features

Aerosol Jet® **FLEX System**	• Fine feature deposition printhead with integrated divert shutter • Closely coupled interchangeable atomizers • Wide feature print head supporting millimetre feature sizes • Repeatable recipe driven dispense • Optional 4th and 5th-axis automation capabilities • Optional *in-situ* laser and UV curing
Aerosol Jet® **5X System**	• 3D capabilities • Deposition onto 3D substrates • 5-axis of coordinated motion • Wide feature print head supporting millimetre feature sizes • Repeatable recipe driven dispense • Optional *in-situ* laser and UV curing
Aerosol Jet® **HD System**	• Available in 20 μm, 50 μm and 100 μm configurations • Configurable conveyor stations • Body recognition function • Direct printing of conformal 3D interconnects • Optional *in-situ* laser and UV curing • CE standard compliance
Aerosol Jet® **Print Engine**	• Scalable networked print engine or module expansion ○ Capable of supporting four or more simultaneous printheads by networking multiple Print Engines to meet high volume production requirements • Independent process control for each print module • Rapid change print module • Quick release ink cassettes for rapid changeover • High material output for volume production • CE standard compliance • Easy integration into automation platforms • Industry standard communication interface protocols • Easy to disassemble print components for periodic cleaning • Integrable production automation platform • Improved process stability and process run-time • High reliability and uptime • Low operating cost • Process diagnostic capabilities • Process monitoring and data collection interface

Figure 3.25.   Aerosol Jet® 5X system. (Courtesy of Optomec, Inc.).

(c) *Aerosol Jet® HD System*

The Aerosol Jet® HD System (see **Figure 3.26**) is a compact, configurable production platform suitable for high volume manufacturing of printed electronics. It can dispense a wide range of functional materials on various substrates and produce features as fine as 20 μm. The Aerosol Jet® HD System also allows direct printing of conformal 3D interconnects and conformal coatings on non-planar surfaces. Its system features also include high accuracy linear motors, bi-directional single lane board conveyor, removable 300 mm heated vacuum platen, digital computer-aided design (CAD) to path programming, and body recognition vision

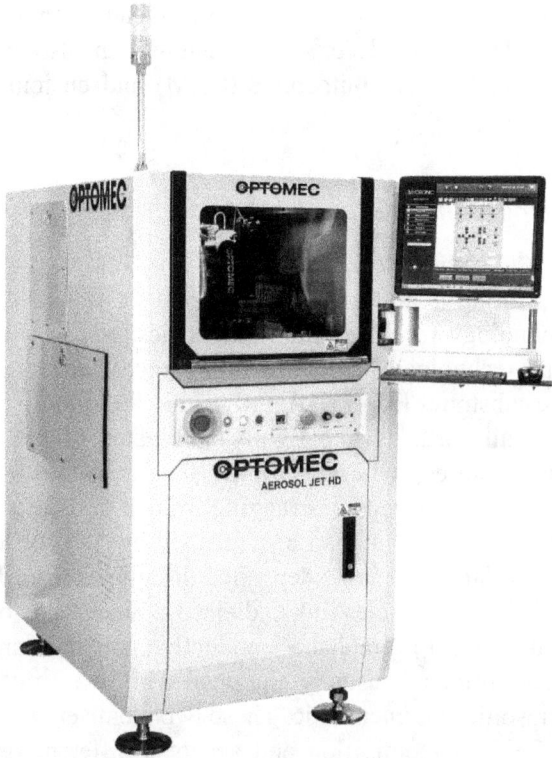

Figure 3.26.   Aerosol Jet® HD system. (Courtesy of Optomec, Inc.).

alignment and automatic visual inspection. Optional IR laser and UV cure modules for *in-situ* curing can also be added.

(d) *Aerosol Jet® Print Engine*

The Aerosol Jet® Print Engine is a scalable and robust Aerosol Jet® system, which is ideal for low to high volume printed electronics manufacturing. It features closely coupled print cassettes and sophisticated process controls. Ultrasonic and pneumatic atomisers are integrated into a single ink cassette to increase material run-time and improve process stability. The ink cassettes also have a quick release feature for rapid material changing and easy to disassemble print components for facilitating cleaning.

Each Aerosol Jet® Print Engine can support two simultaneous printheads, and multiple Print Engines can be linked together to simultaneously support four or more printheads for high volume manufacturing. The Aerosol Jet® Print Engine can also be integrated with original equipment manufacturer (OEM) and custom automation platforms.

3.2.1.2 *Working principles and process*

(a) *Aerosol Jet® Printing Process*
Aerosol Jet® printing is an emerging, non-contact, fully digital material deposition technology for 3D printed electronics, which utilises aerodynamic focusing technology to deposit inks directly, accurately and precisely onto the substrates [31–33]. Aerodynamic focusing is a mechanism for focusing small particles into tightly collimated beams [34], which allows printing features to be as fine as 10 microns. Diverse types of inks and materials with viscosities ranging from 1–500 cps [31] and particle size smaller than 500 nm can be deposited by Aerosol Jet® printing. Some of the materials that can be deposited through Aerosol Jet® printing include metallic nanoparticles inks, dielectric materials, biomaterials, carbon nanotubes (CNT), graphene, conductive polymers and ceramics. The Aerosol Jet® printer is usually equipped with two different types of atomisers: ultrasonic and pneumatic. These two atomisers utilise different technologies for inks atomisation and accept a different range of inks viscosities. The ultrasonic atomiser utilises high-frequency sound waves, whereas the pneumatic atomiser utilises pressurised gas, to atomise inks into fine aerosols. The ultrasonic atomiser is recommended to only atomise inks with viscosities ranging from 1–15 cps, whereas the pneumatic atomiser can atomise inks with viscosities ranging from 1–500 cps. Some Aerosol Jet® printing systems also come with 5-axis printing (for instance, Optomec Aerosol Jet® 5X), which allows the printing of conductive patterns directly onto 3-dimensional (3D) surfaces. Hence, promoting interesting applications such as printed electronics on conformal surfaces [2,31–33,35–44].

(i) *Aerosol Jet® printing process with ultrasonic atomiser*
The Aerosol Jet® printing process with the ultrasonic atomiser can be illustrated in four main steps (see **Figure 3.27**): (1) ink atomisation, (2) aerosol delivery, (3) aerosol collimation and aerosol focusing, and (4) ink deposition on the substrate.

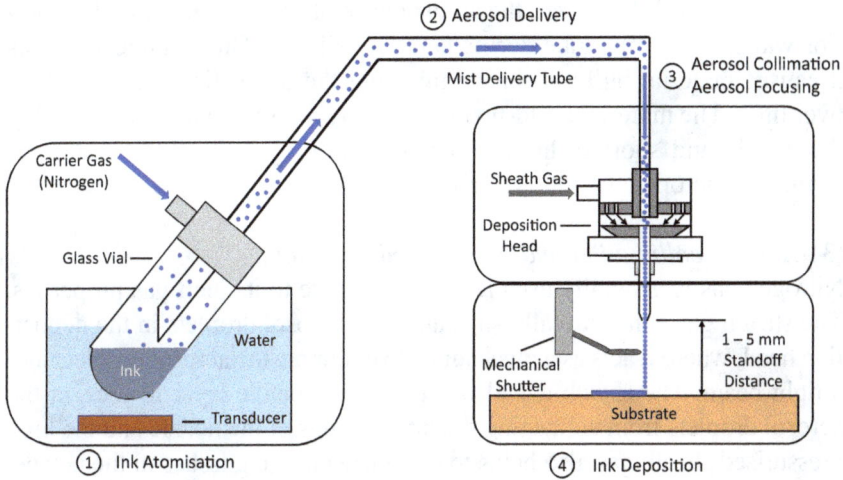

Figure 3.27. Schematic diagram of the Aerosol Jet® printing process with the ultrasonic atomiser [48].

(1) *Ink atomisation*

High-frequency ultrasonic energy (1.6–2.4 MHz [45]) generated by the ultrasonic transducer drives the ink atomisation in an ultrasonic atomiser, in which the sound energy propagates through the ink and creates capillary waves on the ink surface [46]. These capillary waves cause small ink droplets to break off from the ink surface, and the ink is atomised into fine aerosol droplets in due process. The typical size of the atomised aerosol droplets ranges between 1–5 microns [47] as any droplets larger than 5 microns fall back into the ink reservoir during the atomisation process due to the effects of gravity. As the typical size of the atomised aerosol droplets ranges from 1–5 microns, the maximum allowable particle size within the droplets is 500 nm. Note that the droplet size is dependent on the ultrasound frequency, ink surface tension and density of the ink, which directly affects the capillary wavelength [46]. The atomisation power required for atomisation of the ink is also dependent on the ink's amount and viscosity [46].

(2) *Aerosol delivery*

A positive pressure carrier gas, typically nitrogen gas, delivers the atomised aerosol droplets from the glass vial through the mist tube to the nozzle head [49]. However, the aerosol droplets can condense on the interior walls of the mist tube during aerosol delivery. This is undesirable as it causes material build-up in the tube and the deposition rate to reduce over time. The material build-up in the mist tube can contaminate and clog the nozzle and shorten the printing lifetime [46]. Note that there is no virtual impactor in the ultrasonic atomiser setup.

(3) *Aerosol collimation and aerodynamic focusing*
Nitrogen gas is generally used as sheath gas due to its inert gas properties. The sheath gas concentrically surrounds the aerosol droplets in the deposition head, where it acts as a mechanism to collimate the aerosol droplets into a tight beam. The sheath gas also acts as a protective layer to prevent the aerosol droplets from contacting the inner linings of the nozzle and the high pressurised sheath gas also helps to eliminate clogging issues in the nozzle. The aerosol droplets, alongside with the sheath gas, are discharged through the converging nozzle which gives the fine features of high-resolution printing of Aerosol Jet® printer through aerodynamic focusing [46]. Note that the focusing ratio, F_R (defined as the sheath gas flow rate (ShGFR) to the carrier gas flow rate (CGFR)) has a direct influence on the line morphology of the as-printed line [45]. The focusing ratio, F_R can be written as

$$F_R = \frac{\text{Sheath Gas Flow Rate (ShGFR)}}{\text{Carrier Gas Flow Rate (CGFR)}} \qquad (3.5)$$

(4) *Ink Deposition on Substrate*
The nozzle is usually placed at a standoff distance ranging from 1–5 mm from the substrates. The focused beam of aerosol droplets, together with the sheath gas, exit the nozzle and impact directly onto the substrates. Hence, the Aerosol Jet® printing technique is classified as a non-contacting printing technique as the nozzle does not contact the substrates during the printing process. As the jet of aerosol droplets is continuous, a mechanical shutter is required to restrict the flow of the aerosol droplets from impacting the substrates on the areas where printing is not required. The printing of the patterns is made possible by controlling the nozzle using a precise motion controller system [2,31–33,35–43,46].

(ii) *Aerosol Jet® Printing process with pneumatic atomiser*
The Aerosol Jet® printing process with the pneumatic atomiser can be illustrated in four main steps (see **Figure 3.28**): (1) ink atomisation, (2) aerosol

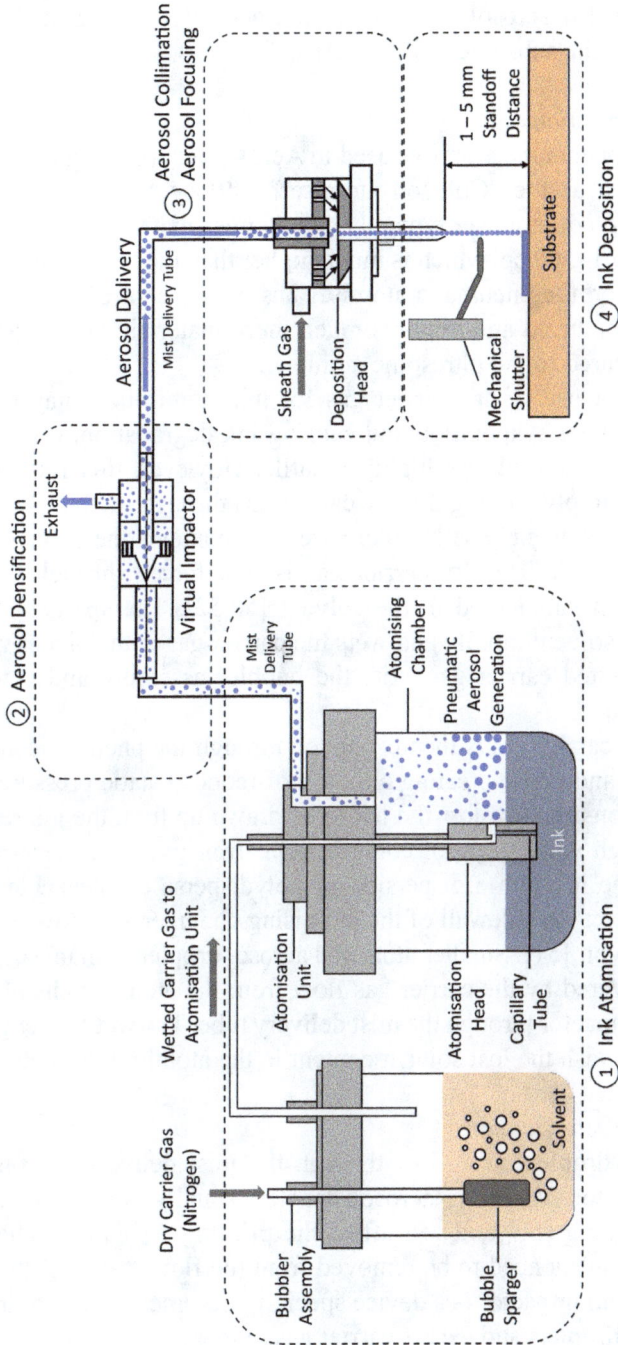

Figure 3.28. Schematic diagram of the aerosol jet printing process with the pneumatic atomiser [50].

densification, (3) aerosol collimation and aerosol focusing, and (4) ink deposition on the substrate.

(1) *Ink Atomisation*

The type of pneumatic atomiser used in Aerosol Jet® printing is also commonly known as the "Collison atomiser" [49] or "Collison nebuliser" [51]. This type of atomiser typically can atomise materials with viscosity ranging from 1–500 cp, which is much higher than the ultrasonic atomiser [49]. However, the pneumatic atomiser has poorer control of the aerosol uniformity produced and requires much more materials to start atomisation as compared to the ultrasonic atomiser.

The direct use of dry carrier gas for ink atomisation may result in excessive solvent evaporation and causing ink degradation, especially if the solvents in the ink are highly volatile. However, this issue can be resolved by simply passing the dry carrier gas through the bubble sparger to wet it with solvent. The bubbler assembly contains the solvent that is present in the ink. The dry carrier gas is first forced through a bubble sparger that is submerged in the solvent [50,52]. The sparger helps to vaporise the solvent rapidly and wets the carrier gas with solvent vapours [50]. The wetted carrier gas exits the bubbler assembly and enters the atomisation unit.

The wet carrier gas is then expanded through the pneumatic atomiser nozzle [50], and hence creating a region of reduced static pressure above the ink suction tube to allow the ink to be drawn up from the ink reservoir [49]. The high velocity jet of compressed carrier gas then shears the top surface of the ink into a dispersion of polydisperse droplets. Larger ink droplets impact the sidewall of the atomising chamber and flow back into the ink reservoir [51]. Smaller atomised aerosol droplets remain suspended and get delivered by the carrier gas flow from the atomiser chamber into the virtual impactor through the mist delivery tube. The wet carrier gas also helps to replenish the lost solvent content in the atomised droplets.

(2) *Aerosol Densification*

The aerosol droplets now flow through the mist delivery tube into the virtual impactor. Very fine aerosol droplets usually contribute to overspray in printing [50] and can affect the printing resolution. Therefore, these fine droplets need to be removed from the flow.

The virtual impactor is a device specially designed for removing very fine aerosol droplets and excess carrier gas to increase the aerosol density.

The exhaust is attached to an outlet of the virtual impactor to apply a suction perpendicular to the flow direction [50]. Hence, the aerosol droplets are separated based on their inertia. Fine aerosol droplets have low inertia and are carried away by the suction flow [49,50]. The larger aerosol droplets have higher inertia and are not affected by the suction flow. The larger aerosol droplets pass through the virtual impactor and get delivered to the deposition head.

(3) *Aerosol collimation and aerodynamic focusing*
Nitrogen gas is generally used as sheath gas due to its inert gas properties. The sheath gas concentrically surrounds the aerosol droplets in the deposition head, where it acts as a mechanism to collimate the aerosol droplets into a tight beam. The sheath gas also acts as a protective layer to prevent the aerosol droplets from contacting the inner linings of the nozzle and the high pressurised sheath gas also helps to eliminate clogging issues in the nozzle. The aerosol droplets, alongside with the sheath gas, are discharged through the converging nozzle which gives the fine features of high-resolution printing of Aerosol Jet® printer through aerodynamic focusing [46].

(4) *Ink deposition on substrate*
The nozzle is usually placed at a standoff distance ranging from 1–5 mm from the substrates. The focused beam of aerosol droplets, together with the sheath gas, exit the nozzle and impact directly onto the substrates. As the jet of aerosol droplets is continuous, a mechanical shutter is required to restrict the flow of the aerosol droplets from impacting the substrates on the areas where printing is not required. The printing of the patterns is made possible by controlling the nozzle using a precise motion controller system [2,31–33,35–43,46].

(b) *Aerosol Jet® working principles*
Mahajan *et al.* [53] empirically observed that the thickness and line width of the as-printed line are significantly influenced by these process variables: carrier gas flow rate, nozzle diameter, focusing ratio and stage speed. Table 3.8 shows how the line thickness and line width of the as-printed line changes with each increasing independent process variable. Therefore, it is critical to understand the physical principles of Aerosol Jet® printing to find out how each process variables affect the line geometry of the as-printed line.

Table 3.8.   Line thickness and line width of the as-printed line with each increasing independent process variable [53].

Process Variable	Line Thickness	Line Width
Carrier gas flow rate	Increases	No change
Nozzle diameter	Decreases	Increases
Focusing ratio	Increases	Decreases
Stage speed	Decreases	Decreases

(i) *Aerosol beam collimation*

The sheath gas act as a collimation mechanism, in which it concentrically surrounds and collimates the aerosol droplets laden carrier gas into a tight aerosol beam upon entering the nozzle head [46]. Therefore, the sheath gas flow rate, carrier gas flow rate and the nozzle diameter have a direct impact on the diameter of the Aerosol Jet® beam, D_a.

Assuming a parabolic velocity profile for fully developed laminar pipe flow for a simple non-mixing volume displacement, the ratio of the diameter of the Aerosol Jet® beam, D_a to the diameter of the nozzle, D_n can be expressed in terms of the focusing ratio, F_R [46]:

$$\frac{D_a}{D_n} = \sqrt{1 - \sqrt{\frac{F_R}{1 + F_R}}}. \tag{3.6}$$

Hence, the diameter of the exiting Aerosol Jet® beam can be expressed as a function of D_n and F_R:

$$D_a = D_n \sqrt{1 - \sqrt{\frac{F_R}{1 + F_R}}}. \tag{3.7}$$

(ii) *Aerodynamic Focusing*

Aerodynamic focusing is a mechanism that further collimates the aerosol beam, whereby the aerosol beam is discharged through a converging nozzle [46]. Thus, giving the fine features and high-resolution printing in Aerosol Jet® printing.

The dimensionless Stokes number, St can be used to describe the effectiveness of the aerodynamic focusing effect [46]:

$$St = \frac{\rho_p d_p C_c}{18\mu} \frac{U}{D_n},$$ (3.8)

where ρ_p is droplet density, d_p is droplet diameter, C_c is the slip correction factor, μ is the fluid viscosity, U is the average flow velocity, and D_n is the nozzle diameter. Typically, the Stokes number should range between 0.5–1.5 for optimal aerodynamic focusing. A low Stokes number may result in ineffective focusing, whereas a high Stokes number (> 2) can usually lead to over-focusing [46].

In actual printing, the Stokes number depends mostly on the droplet size, nozzle diameter and flow rate [46]. Larger deposition nozzle may cause the Stokes number to fall below the optimal aerodynamic focusing range and leads to ineffective focusing. Hence, smaller droplets and smaller deposition nozzle diameter are usually preferred for obtaining higher printing resolution [46].

(iii) *Line geometry of the as-printed line*
Assuming no mass loss, the principle of mass conservation can be applied to ink deposition of the Aerosol Jet® printing process (see **Figure 3.29**) in which the mass flow rate at the nozzle exit, $\dot{m}_e$ is equal to the mass flow rate, $\dot{m}_s$ where the Aerosol Jet® beam impinges on the substrate [53]. This continuity equation can be written as:

$$\dot{m}_e = \dot{m}_s,$$ (3.9)

$$\rho_e A_e V_e = \rho_s A_s V_s,$$ (3.10)

where ρ_e, A_e, V_e are the density, cross-sectional area and velocity of the exit aerosol beam, and ρ_s, A_s, V_s are the density, cross-sectional area and velocity of the as-printed line. Note that the velocity of the as-printed line is also defined as the stage velocity, V_s. Rearranging the terms, the thickness, t and line width, w of the as-printed line can be expressed as:

$$wt \approx A_s = \left(\frac{\rho_e}{\rho_s}\right)\left(\frac{V_e}{V_s}\right)A_e.$$ (3.11)

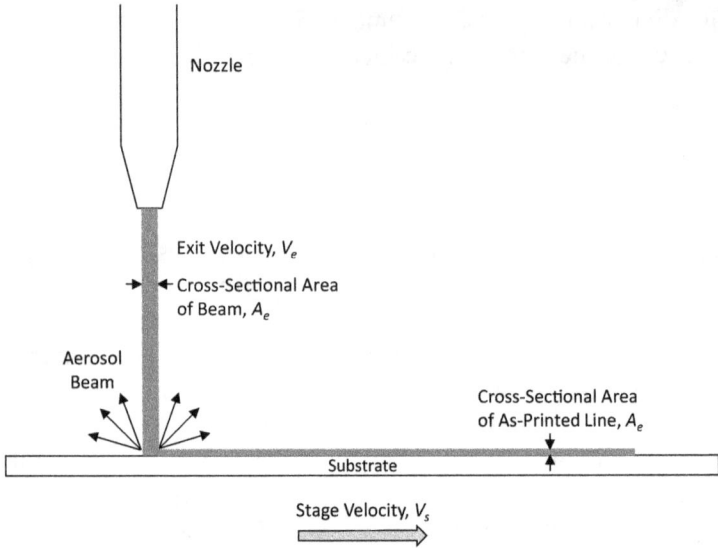

Figure 3.29.   Schematic of ink deposition of the Aerosol Jet® printing process [53].

Note that the cross-sectional area of the as-printed line, A_s is different from the cross-sectional area of the line after sintering. This is because the cross-sectional area of the printed line can change after sintering due to evaporation, vaporisation of the organic additives and densification of the metallic nanoparticles [53].

3.2.1.3 *Materials*

Aerosol Jet® systems can dispense a wide range of functional materials, such as pure solvents or liquids, dispersions and solutions. The material properties of the inks can directly affect the achievable minimum feature size. However, the feature size can be increased by using a wide nozzle printhead or repeatedly print with multiple passes. The detailed list of supported materials is tabulated in Table 3.9.

Particulate-based inks, such as metallic nanoparticle inks, typically should have the following properties for successful dispensing in *Aerosol Jet®* systems:

- The ink's viscosity should range between 1–500 cP at ambient temperature. The ink also can be heated to ensure its viscosity fall within this range.

Table 3.9. Supported materials by Optomec *Aerosol Jet*® systems. (Courtesy of Optomec, Inc.).

Metal Inks	Resistor Inks	Non-Metallic Conductors
• Gold • Platinum • Silver • Nickel • Copper • Aluminium	• Carbon • Ruthenate	• Single-walled carbon nanotubes • Multi-walled carbon nanotubes • PEDOT:PSS
Dielectrics and Adhesives	**Semiconductors**	**Others**
• Polyimide • Polyvinylpyrrolidone (PVP) • Teflon AF • SU-8 • Adhesives • Opaque coatings • UV adhesives • UV acrylics	• Organic semiconductors • Single wall carbon nanotubes • Organic semiconductors	• General solvents, acids and bases • Photo and etch resists • DNA, proteins, enzymes

- Newtonian or shear-thinning fluids are preferred.
- The solvents should have high boiling points and low vapour pressure.
- The particle size is preferred to be less than 200 nm and should not exceed more than 500 nm. The solids content can range from 5–70%.
- Particulate-based inks with multiple solid components should be homogenously dispensed throughout the ink.

3.2.1.4 *Strengths and weaknesses*

The key strengths of *Aerosol Jet*® systems are:

(a) *Ability to print on conformal surfaces*: Some of the *Aerosol Jet*® printing systems are equipped with 5-axis motion, such as the *Aerosol Jet*® 5X system, allowing deposition of conductive inks directly onto orthogonal and non-planar surfaces [38,39]. Hence, favouring full optimisations of available spaces in electronic devices to facilitate additional cost, space and material savings.

(b) *Wide variety of substrates can be used*: Aerosol Jet® systems can deposit materials on any flat planar or 3D substrates.

(c) *Deposits a wide range of materials*: A wide range of materials can be deposited with the Aerosol Jet® printing process including metals, ceramics, polymers, biomaterials, dielectric materials, nanotubes, and graphene [31]. The pneumatic atomiser of the Aerosol Jet® printer can atomise inks with viscosities ranging from 1–500 cps [33,54].

(d) *High-resolution printing*: The Aerosol Jet® printing process can print high-resolution patterns with high accuracy of profiles and layer thickness [31], in which it can achieve a minimum line width as fine as 10 microns [2]. Furthermore, it has the considerable potential to improve further its current printing resolution, which is already higher than that by inkjet printing technology. Thus, Aerosol Jet® printing is now the most suitable technology for additive manufacturing of high quality and high-performance electrical components in the near future [2].

The key weaknesses of *Aerosol Jet®* systems are:

(a) *Ink compositions instability*: Dry carrier gas used in the Aerosol Jet® printing process can destabilise the ink composition by removing solvents through evaporation [45,50,52], and hence changing print characteristics and reducing run times. This issue is particularly unfavourable as the ink composition is unable to remain consistent throughout the Aerosol Jet® printing process to give reproducible, consistent prints. Furthermore, extensive solvent evaporation can cause pre-drying of the ink and undesirable material accumulations in the atomiser assembly [45,50], which makes printing problematic.

An integrated bubbler, filled with the same type of solvent, can be employed to stabilise the ink composition by compensating solvent evaporation losses [50,52]. The dry carrier gas is first forced through a sparger submerged in the solvent instead of carrying the atomised droplets to the mist tube directly [52]. The sparger helps to vaporise the solvent rapidly and wets the carrier gas with solvent vapours. The wet carrier gas then carries atomised droplets to the mist tube while replenishing the lost solvent content in the atomised droplets. In addition, the ink formulation can also be formulated with a larger percentage of low-volatility co-solvents to slow down solvent evaporation [46].

(b) *Effects of process parameters on line morphology*: The Aerosol Jet® printing process parameters directly influence the line morphology of the as-printed line. Moreover, different ink formulations have different physical properties, and the compositions of the inks can drift during a long printing process [45]. Therefore, it is not an easy and direct task to optimise the Aerosol Jet® printing process parameters for ink, and it needs an accumulation of knowledge and experiences.

(c) *System drift*: System drift is one of the commonly faced issues faced in Aerosol Jet® printing, in which the printed line morphology changes over time even though the printing parameters are held constant during the printing process [55]. The system drift issue is especially prominent during a long printing session, where the line width of the printed lines is either observed to reduce or increase significantly over time. This issue is highly dependent on the ink used and the printing system. Michael *et al.* [55] suggested that the cause of the line width of the printed lines increasing during extended printing duration is due to the accumulation of ink within the atomiser assembly components. On the other hand, the cause of the line width of the printed lines decreasing during extended printing duration is probably due to contaminants in the atomiser assembly clogging the nozzle head. However, the system drift issue can be solved by cleaning the entire atomiser assembly thoroughly periodically.

3.2.1.5 *Applications*

Aerosol Jet® systems can be used in the following applications for printed electronics:

(a) *Printing conductive traces, passive and active components*: Conductive traces and patterns can be printed directly on a wide variety of substrates, including planar and 3D substrates. Passive and active components such as resistors, capacitors, inductors and thin film transistors can also be printed by *Aerosol Jet®* systems.

(b) *Flexible electronics*: *Aerosol Jet®* systems can deposit functional inks directly onto flexible substrates for flexible electronics applications.

(c) *Printed antennas*: The *Aerosol Jet®* printing process can print functional antennas directly onto 3D structures for significant space utilisation and weight reduction. **Figure 3.30** shows an antenna directly

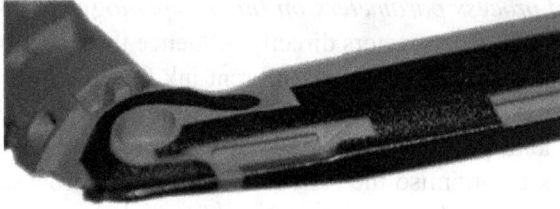

Figure 3.30.   3D printed antenna on mobile phone case. (Courtesy of Optomec, Inc.).

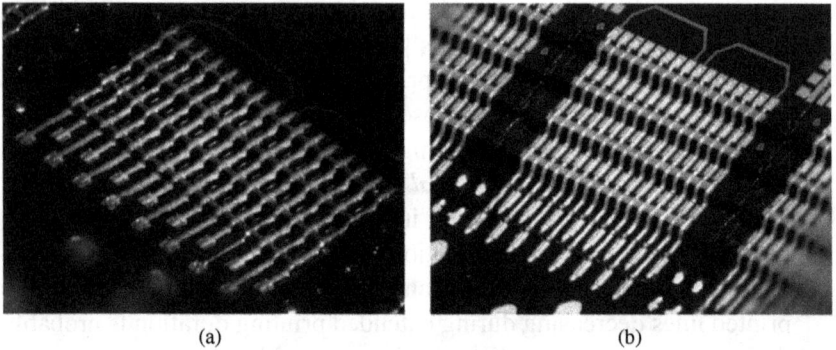

| (a) | (b) |

Figure 3.31.   Printed interconnects for multi-die stacks: (a) rounded profile; and (b) flat profile. (Courtesy of Optomec, Inc.).

printed onto an injection moulded plastic mobile phone case by the *Aerosol Jet®* process.

(d) *Printed interconnects*: The *Aerosol Jet®* printing process can print interconnects directly onto 3D structures for various multi-chip packaging applications. This non-contact printing process can significantly reduce chip damage, decrease overall package height and pad to pad spacing, and lower the number of individual interconnects needed as compared to traditional wire bonding. **Figure 3.31** shows printed interconnects for multi-die stacks with rounded and flat profiles.

3.2.2 *Integrated Deposition Solutions, Inc.*

3.2.2.1 *Company*

Founded in 2013, Integrated Deposition Solutions, Inc. (IDS, Inc) is focused on developing its next-generation aerosol-based printing technology, *NanoJet™* (formally known as MycroJet). IDS has licensed and adapted the aerosol-based additive manufacturing technology from Sandia National Laboratories (SNL License #CO2686; U.S. Patent 6,348,687) for direct-write electronic (DWE) printing applications. Currently, IDS has filed two U.S. patents: *Methods and Apparatuses for Direct Deposition of Features on a surface using a Two-Component Microfluidic Jet*; and *Apparatuses and Methods for Stable Aerosol Deposition Using an Aerodynamic Lens System*. IDS, Inc is located at 5901 Indian School Rd NE. Suite #125, Albuquerque, NM 87110, USA.

3.2.2.2 *Products*

The *NanoJet™* printing technology whose jetting mechanism is based on aerosol collimation and aerodynamic focusing, to allow stable and reliable printing process while producing superior printed line-edge quality in a compact, cartridge-based print head. The *NanoJet™* technology is robust, cost-effective, simple to use, allow quick material change and has low maintenance requirements. The *NanoJet™* printing technology can produce fine features with line width ranging from 10 μm–200 μm and can also dispense a wide range of functional materials including conductive inks, resistive inks, dielectric ink and other functional materials. The printing technology of *NanoJet™* can also be conveniently retrofitted onto current print platforms and incorporated into new OEM products.

(a) *NanoJet™ Print Heads*

There are two types of *NanoJet™* printhead options available: *NJ Aerosol Modular Print Head* and *Multi-Material NJ Aerosol Print Head* (see **Figure 3.32**). Both printheads are point of use aerosol generation. These printheads generate aerosol continuously at point of use. These printheads can also allow easy system integration into other customised print platforms for deposition and printing applications.

The *NJ Aerosol Modular Print Head* (see **Figure 3.32(a)**) is a cartridge-based printhead which only supports one material printing. Its ink cartridge is removable for ease of use and reusable for better cost-effectiveness. This printhead consists of various components which include the cooled mounting plate, removable ink cartridge, atomiser base, aerosol focusing flow cell and dispense tip (see **Figure 3.33**). The aerosol focusing flow cell and dispense tip are the critical components for aerosol focusing.

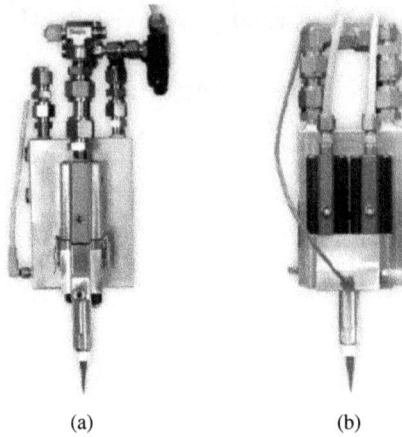

(a)   (b)

Figure 3.32.   (a) NJ Aerosol Modular Print Head; and (b) Multi-Material NJ Aerosol Print Head. (Courtesy of Integrated Deposition Solutions).

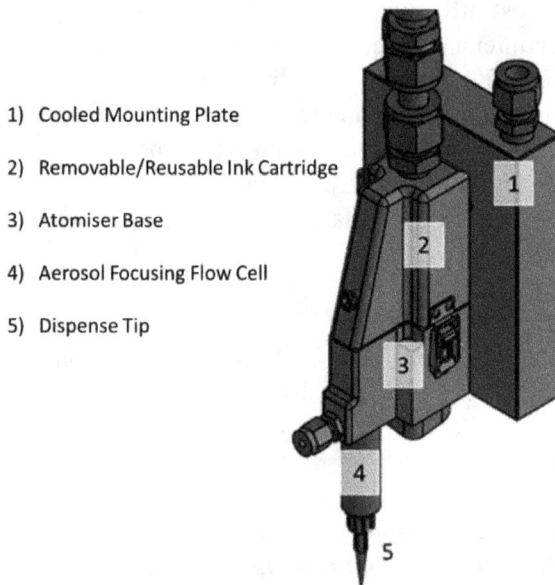

1) Cooled Mounting Plate

2) Removable/Reusable Ink Cartridge

3) Atomiser Base

4) Aerosol Focusing Flow Cell

5) Dispense Tip

Figure 3.33.   Essential components of the NJ Aerosol Modular Print Head. (Courtesy of Integrated Deposition Solutions).

The *NJ Aerosol Modular Print Head* can print fine lines less than 20 μm wide with 40 μm pitch.

The *Multi-Material NJ Aerosol Print Head* (see **Figure 3.32(b)**) comes with two interchangeable and removable ink cartridges. This print-head can support printing of up to two different materials, in which printing can be done either independently or simultaneously with each material. Hence, it is highly advantageous for printing applications that require sequential material printing or material mixing in real-time. The printed material compositions can also be constantly varied as required during multi-material aerosol printing.

(b) *NanoJet™ Desktop Printer*
The *NanoJet™* desktop printer (see **Figure 3.34**) is a fully functional printing platform equipped with the *NanoJet™* aerosol-based printing technology. The *NanoJet™* desktop printer is a relatively low-cost and affordable system that is specially targeted for research and low-volume production in printed electronics applications. The printer features industrial computer numerical control (CNC) motion control, print process controls, process

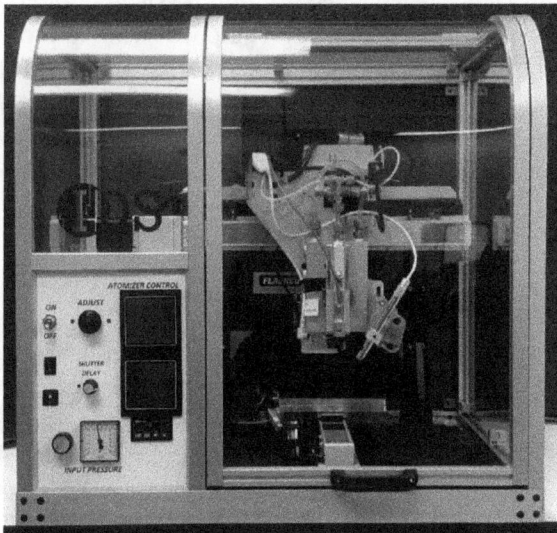

Figure 3.34.   *NanoJet™* desktop printer. (Courtesy of Integrated Deposition Solutions).

vision system and a 150 × 150 mm heated vacuum platen. Generic computer-aided manufacturing (CAM) toolpath generators are also used in the *NanoJet*™ desktop printer to offer end-users flexibility.

There are two different types of printhead options available for the printer: *NJ Aerosol Modular Print Head* and *Multi-Material NJ Aerosol Print Head*. The printhead has a travelling range of 300 × 150 × 100 mm and has a working distance of 2–10 mm. The printed line widths can range between 10–1000 μm and achieve less than 5% variation. The single line pass thickness is material dependent and can range from 100 nm–4 μm. The printing process is also tested for stability, and data has shown that the printer can maintain a stable printing process for more than 4 hours without the operator's assistance. The technical specifications of the *NanoJet*™ desktop printer are summarised in Table 3.10.

(c) *NanoJet*™ *Aerosol Print System*

The *NanoJet*™ aerosol print system (see **Figure 3.35**) is a custom print platform specialised for aerosol printing of flexible and rigid electronics. It comes with 3 axes of coordinate motions and has a travelling range of 500 × 500 × 200 mm with 10 μm positional accuracy. A *NJ Aerosol Modular Print Head* is attached to the z-axis motion carriage. The *NanoJet*™ aerosol print system is equipped with a process monitor camera and high magnification alignment camera for real-time process monitoring and print tool alignment respectively. It also comes with a heated vacuum platen, which can heat up until 100°C, for securing and heating the samples during the printing process. The technical specifications of the *Nanojet*™ desktop printer are summarised in Table 3.11.

3.2.2.3 *Working principles and process*

The *NJ Aerosol Modular Print Head* (single material *NanoJet*™ printhead) comprises of an atomiser, flow cell and a process control system (see **Figure 3.36**). The ink is first filled in a removable ink cartridge and is then atomised into small poly-dispersed aerosol droplets by an ultrasonic atomiser. Carrier gas is pumped into the atomiser through the carrier gas input port. The angled aerosol exit channel and baffle inhibit entrainment of large aerosol droplets and fluid into the aerosol exit channel and aerosol delivery line [56]. Small aerosol droplets are delivered by the carrier gas into the flow cell through the aerosol delivery line.

Table 3.10. Specifications of *NanoJet*™ Desktop Printer. (Courtesy of Integrated Deposition Solutions).

Deposition technology	*NanoJet*™ Aerosol-Based Printing Technology
Build volume	$300 \times 150 \times 100$ mm
Heated vacuum platen	150×150 mm
Printheads options	*NanoJet*™ Aerosol Modular Print Head
	Multi-Material *NanoJet*™ Aerosol Print Head
Inks	Polymers
	Metal Nanoparticle inks
	Resistive Inks
	Magnetic Inks
Line widths	10–1000 μm
Single pass line thickness	100 nm–4 μm
Working distance	2–10 mm
Printing speed	5–50 mm/s
Mean-time between assist	> 4 hours
Line pitch	Standard deviation of +/−5 microns
Line width	< 5% variation
Power supply	120/240VAC, 10/5A, 50/60 Hz
Gas	Dry air or nitrogen gas at 10 psig/ 0.7 Bar, 200 sccm
Cooling	2 lpm flow, 60 W min
Maintenance	Material change over < 1 min.
	Print nozzle change over < 1 min.
Special features	Industrial CNC control
	Generic CAM toolpath generation
	Plug and play operation
	Point of use aerosol generation
	Removeable Ink cartridges
	Compact printhead

The flow cell which comprises of an aerodynamic lens system and a flow chamber housing (see **Figure 3.37**), can significantly reduce the output aerosol beam's diameter with the combined influence of the aerodynamic lens system and sheath gas flow [56,57]. The sheath gas flow is

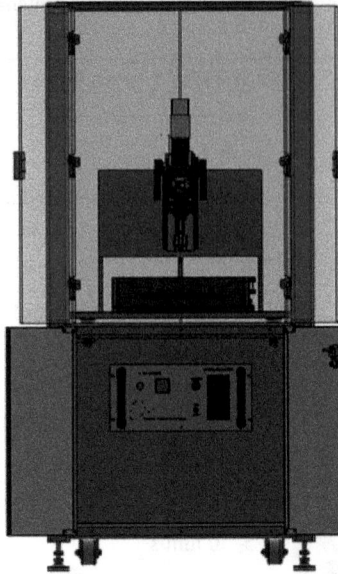

Figure 3.35. *NanoJet*™ Aerosol Print System. (Courtesy of Integrated Deposition Solutions).

Table 3.11. Specifications of *NanoJet*™ Aerosol Print System. (Courtesy of Integrated Deposition Solutions).

Deposition technology	*NanoJet*™ Aerosol-Based Printing Technology
X/Y travel range	500 × 500 mm
Z travel range	200 mm
Positional accuracy	10 μm
Printheads options	*NanoJet*™ Aerosol Modular Print Head with integrated process control module
Operating system	Windows
Heated platen	Up to 100°C
Cameras	Process Monitor Camera
	Alignment Camera

mainly used for collimating the aerosol flow into a tight beam and keep the aerosol droplets from contacting with the inner linings of the flow cell.

A typical aerodynamic lens system usually has at least two aerodynamic lenses for enhancing focusing of polydispersed aerosol droplets

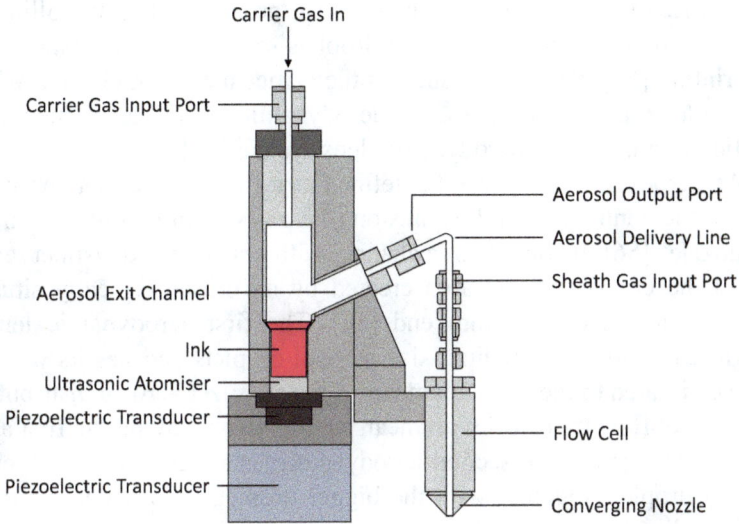

Figure 3.36. Schematic diagram of the NJ aerosol modular print head [56].

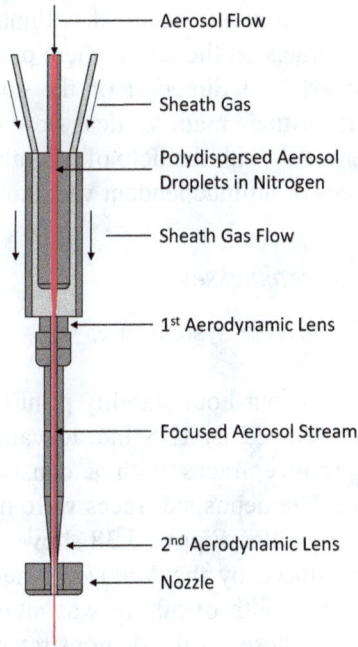

Figure 3.37. Schematic diagram of the flow cell [57].

(droplet size ranging from 200 nm–5 μm), producing highly collimated print stream, and minimising small droplets satellites and overspray during printing [56]. Parameters such as the orifice diameter, channel width, channel length, and the number of aerodynamic lenses can influence the functional range of the aerodynamic lens system [56].

An aerodynamic lens can be defined as a flow device that "produces at least one contraction and expansion of a gas stream before entering an exit nozzle" [56]. It consists of a channel with a sudden and distinct reduction in the cross-sectional area created by an orifice typically situated at the channel's downstream end [56]. The first aerodynamic lens is designed to collimate medium-size aerosol droplets and has its working range calibrated to the mean diameter of the atomizer aerosol distribution. The aerosol flow trajectories are near-parallel upon leaving the first aerodynamic lens [57]. The second aerodynamic lens then coaxially focuses the tiny droplets together with the bigger droplets to produce a narrow, highly collimated aerosol beam after exit [56]. The collimated aerosol beam exits the nozzle's orifice and deposits the ink directly onto the substrate to produce high resolution traces of line widths varying from 10–1000 μm, with minimal overspray and satellites [56]. A diverter valve is used to interrupt the continuous beam of collimated aerosol droplets from impacting the substrates on the areas where printing is not required. The diverter valve diverts the direction of the combined aerosol and sheath gas flows off from their main vertical flow channel into another exhaust channel for removal, with the help of a combination of both pressure and vacuum sources or an independent vacuum source.

3.2.2.4 *Strengths and weaknesses*

The key strengths of *NanoJet*™ systems are:

(a) *High print stability*: A four-hour stability print test with silver nanoparticle ink was conducted by IDS Inc. to validate *NanoJet*™ systems' ability to produce traces with a consistent line width and well-defined edges. The deposited traces were measured optically to determine their line widths. **Figure 3.38** shows the line width of the deposited traces produced by the *NanoJet*™ desktop printer against time. An average line width of 50 μm was attained with a standard deviation of ± 4 μm. These results demonstrated that the *NanoJet*™ desktop printer can achieve high print stability for more than four

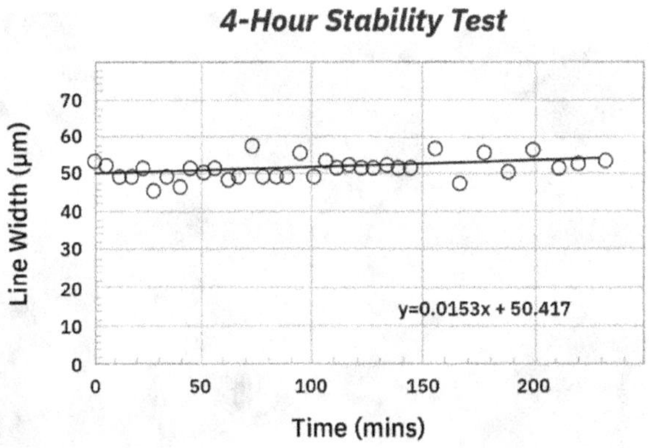

4-Hour Stability Test

Figure 3.38.   Line width of the deposited traces produced by the *NanoJet*™ desktop printer against time. (Courtesy of Integrated Deposition Solutions).

hours of unattended, continuous printing in producing high-resolution traces.

(b) *Lower cost of ownership*: *NanoJet*™ systems are comparably cheaper than other aerosol-based printing systems that are currently available for commercial use.

(c) *Wide variety of substrates can be used*: *NanoJet*™ systems can deposit materials on any flat planar or 3D surfaces (see **Figure 3.39**).

(d) *High-resolution printing*: The *NanoJet*™ printing process can print high-resolution patterns with high accuracy of profiles and layer thickness, in which it can produce traces with minimum line width as fine as 10 microns.

(e) *Dual material printing*: The *Multi-Material NJ Aerosol Print Head* can allow printing with up to two different materials, in which printing can be performed either independently or simultaneously with each material. Hence, it is highly advantageous for printing applications that require sequential material printing or material mixing in real-time. The printed material compositions can also be constantly varied as required during multi-material aerosol printing.

The key weaknesses of *NanoJet*™ systems are:

(a) *Limited printable materials*: *NanoJet*™ ultrasonic atomiser can only atomise fluidic materials within a specific viscosity range, ranging

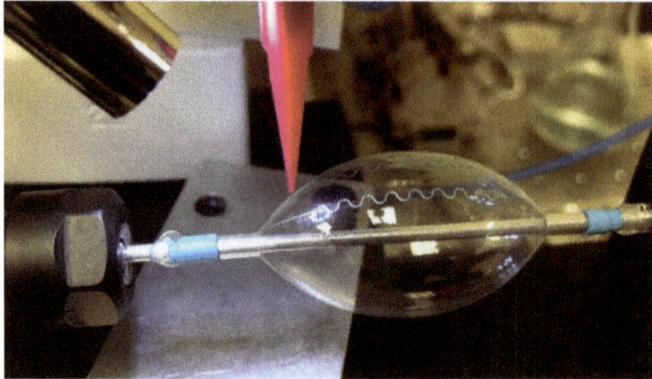

Figure 3.39. Silver nanoparticle ink deposited directly onto a 3D substrate. (Courtesy of Integrated Deposition Solutions).

from 1–20 cPs [58]. Nevertheless, a wide range of materials such as metallic nanoparticle inks, polymeric inks, resistive inks and dielectric inks can be deposited.

3.2.2.5 *Applications*

NanoJet™ systems can be used in the following applications for printed electronics:

(a) *Printing conductive traces and passive components*: Conductive traces and patterns can be printed directly on a wide variety of substrates, including planar and 3D substrates. Passive components such as resistors, capacitors [58] and inductors can also be printed by *NanoJet*™ systems. **Figure 3.40** shows a printed circuit pattern printed directly onto a polycarbonate substrate.

(b) *Multi-layers printed circuits*: Sequential material printing can be performed with the *Multi-Material NJ Aerosol Print Head* for fabricating multi-layers printed circuits. For instance, dielectric ink can be deposited over existing conductive traces for insulation so that another layer of conductive ink can be deposited directly on top to fabricate multi-layers printed circuits.

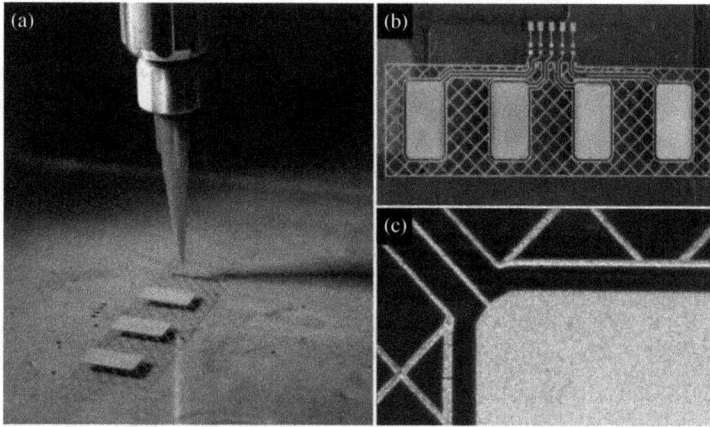

Figure 3.40.   (a) A *NanoJet*™ system depositing silver nanoparticle ink onto a polycarbonate substrate; (b) a printed circuit pattern on polycarbonate substrate; and (c) an enlarged image of the printed circuit pattern. (Courtesy of Integrated Deposition Solutions).

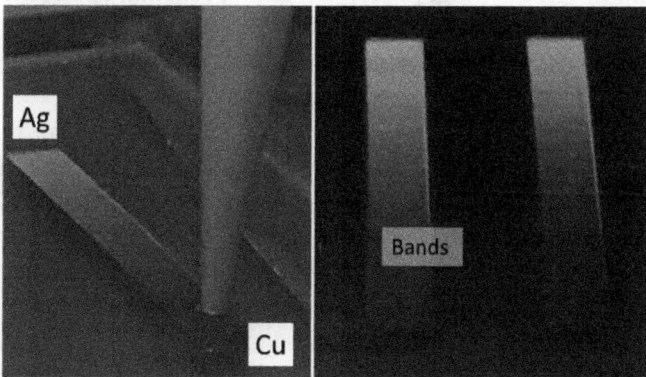

Figure 3.41.   An aerosol printed gradient consists of copper at the bottom which slowly transitioning to silver at the top. (Courtesy of Integrated Deposition Solutions).

(c) *Aerosol printed gradients*: The *Multi-Material NJ Aerosol Print Head* can allow printed material compositions to be constantly varied as required during multi-material aerosol printing. **Figure 3.41** shows an aerosol printed gradient consisted of copper at the bottom which slowly transitioning to silver at the top.

3.3 Extrusion-based Printing

Extrusion-based printing, as the name implies, is a printing technique which forces out a continuous flow of materials through a nozzle [59,60]. The extrusion-based printing technique is simple to use, inexpensive, has a lesser risk of clogging [61], and can dispense materials from a wide viscosity range [62]. However, this printing technique is considerably slower and has poorer resolution, as compared to inkjet printing [62]. Extrusion-based printing can be further categorised into these various types, based on the type of extruder used: filament-based extrusion printing, plunger-based extrusion printing, screw-based extrusion printing and pneumatic-based extrusion printing [59,60,62].

(a) *Filament-based extrusion printing*
Filament-based extrusion printing, also commonly known as fused filament fabrication (FFF) or fused deposition modelling (FDM), is one of the most popular additive manufacturing technologies used in low-cost 3D printers for prototyping, modelling, and production applications [63]. It is also simple to use, safe, low cost and has a wide variety of materials available for printing [60]. **Figure 3.42** shows the schematic diagram of the filament-based extrusion printing process.

The materials used for filament-based extrusion take the form of a filament, which are normally housed in a spool. A pair of feed rollers first feed the filament is into the liquefier (extrusion head) and the heating elements heat the filament to semi-liquid state [47]. The filament is then forced out at a smaller diameter through the heated nozzle and gets deposited in thin layers on the building platform, one layer at a time [47,60]. The deposited material rapidly solidifies since the ambient temperature is much lower than its melting temperature. When a layer is completed, the extrusion head moves one layer up and begins printing the next layer [47]. Some of commercially available conventional, non-filled thermoplastics filaments that are used in FFF 3D printers include acrylonitrile butadiene styrene (ABS), acrylonitrile styrene acrylate (ASA), polyamide (PA), polycarbonate (PC) and polylactic acid (PLA) [60].

Conductive nanofillers [64], such as carbon black (CB) [65], CNTs [66], graphene [67,68] and metallic particles [69], can be added into polymeric matrixes [70] to produce FFF-printable functional, electrically conductive composite filaments. The type, concentration and morphology of conductive nanofillers present in these electrically conductive composite filaments can have a direct influence on their electrical

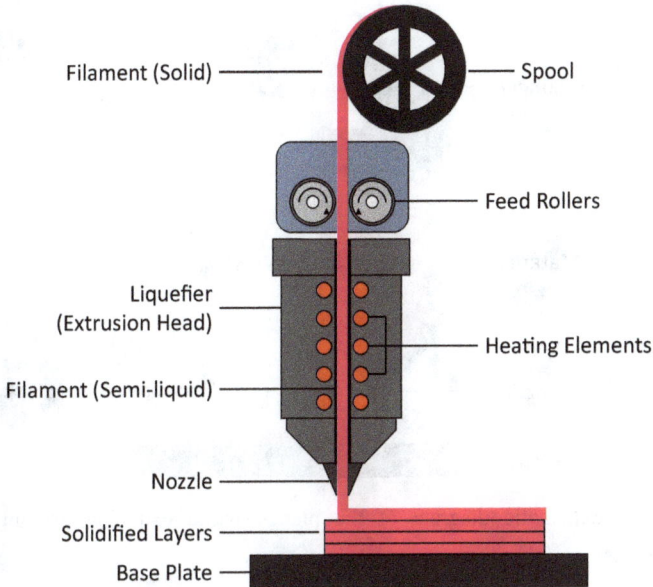

Figure 3.42. Schematic diagram of the filament-based extrusion printing process [47,60].

properties. Their mechanical properties are also dependent on the structure and type of polymer matrix used [69].

Electrically conductive composite filaments can be used in 3D printed electronics applications, for fabricating functional electrical components, interconnects, sensors and devices [65,69,71]. In addition, electrically conductive composite filaments and conventional thermoplastic filaments may also be printed together in a single print job to fabricate structural electronics, whereby the functional electrical circuits and components are embedded inside the polymeric structures [69]. However, the commercially available functional filaments are currently limited in choices for 3D printed electronics applications [69]. Electrically conductive composite filaments with more functionalities, such as conductive, capacitive, resistive and semi-conductive properties, may be developed in near future to cater for wider applications.

(b) *Plunger-based extrusion printing*
Plunger-based extrusion printing, also known as piston-driven extrusion-based printing [62] or ram extrusion [72], utilises the use of a

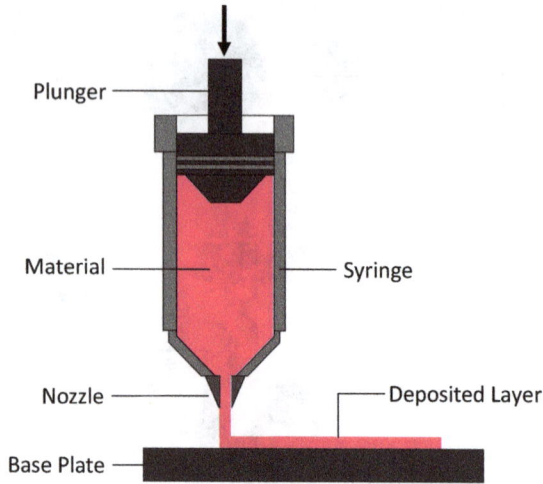

Figure 3.43.   Schematic diagram of the plunger-based extrusion printing process [60,65,72].

ram-driven plunger or piston to push the material out from the nozzle (see **Figure 3.43**) [60]. This extrusion technique is very straightforward. The material flowrate can be simply controlled by regulating the plunger's downwards velocity [72]. The extrusion process starts when a force is exerted on the plunger and stops when this exerting force is released [72].

(c) *Screw-based extrusion printing*
The rotating action of the rotary screw in screw-based extrusion printing creates a pumping action which drives the materials downwards to the extrusion nozzle (see **Figure 3.44**) [60]. The material flowrate can be controlled by regulating the angular velocity of the rotary screw. Hence, it is relatively more challenging to achieve precise control of material flow in screw-based extrusion printing as compared to plunger-based extrusion. The extrusion process stops when the rotary screw stops its rotation [60,72].

(d) *Pneumatic-based extrusion printing*
The pneumatic-based extrusion printing technique applies air pressure directly onto the material as a driving force, to push the material out

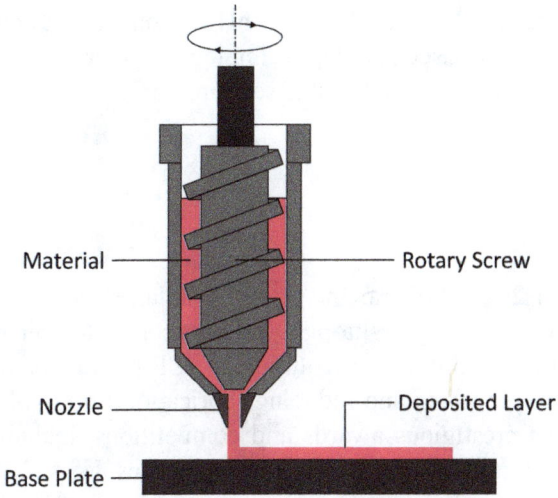

Figure 3.44. Schematic diagram of the screw-based extrusion printing process [60].

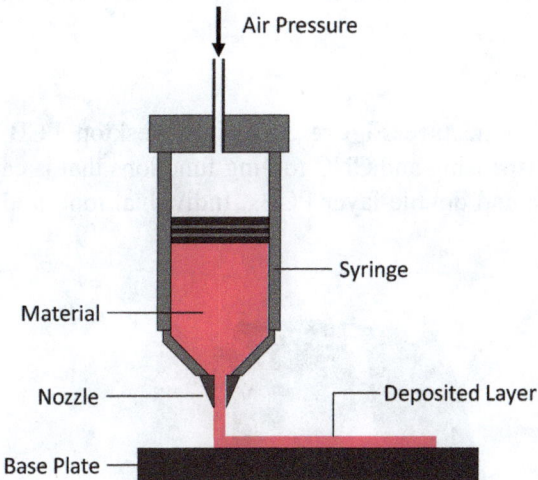

Figure 3.45. Schematic diagram of the pneumatic-based extrusion printing process [62,73].

from the extrusion nozzle (see **Figure 3.45**) [62]. The material flow can be controlled by regulating the air pressure [73]. This printing technique may face significant delays at the start and the end of the extrusion process due to the time required for compression of the gas volume [73].

Voltera, Inc. and Neotech AMT GmbH are some of the companies that utilise the extrusion-based printing technology for material deposition and they will be further discussed in this section.

3.3.1 *Voltera, Inc.*

3.3.1.1 *Company*

Established in 2013, Voltera, Inc. is a Canadian start-up company that specialises in developing desktop 3D printers for prototyping printed circuit boards (PCBs), aiming to revolutionise PCB production by streamlining fabrication processes and reducing fabrication time. Voltera has also won numerous prestigious awards and competitions, including the Best Product Award in IDTechEx Printed Electronics USA 2017, Editor's Choice for Popular Science Greatest Inventions of 2015, international winner for the James Dyson Award 2015, international winner for TechCrunch Hardware Battlefield 2015. Voltera, Inc. is located at 113 Breithaupt St., Suite 100, Kitchener, ON N2H 5G9, Canada.

3.3.1.2 *Products*

The Voltera V-One (see **Figure 3.46**) is a desktop PCB printer with direct-write dispensing and CNC tooling functions that is capable of fabricating single and double-layer PCBs. Individual tool modules, such as

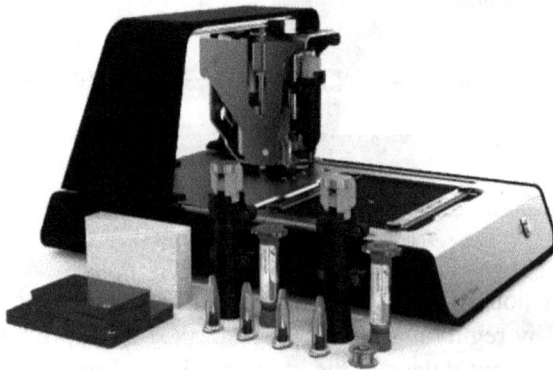

Figure 3.46.   Voltera V-One PCB printer. (Courtesy of Voltera, Inc.).

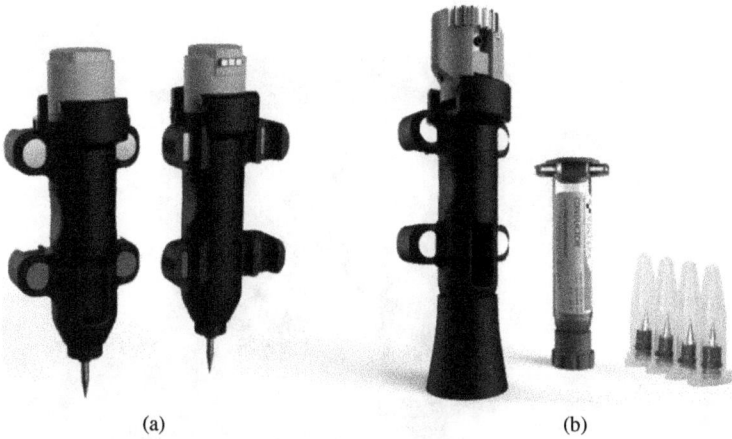

(a)　　　　　　　　　　　　　　　(b)

Figure 3.47.　(a) Precision probes; and (b) Interchangeable dispenser and nozzles for conductive inks and solder pastes.

the precision probe head (**Figure 3.47(a)**) and dispensing heads for conductive inks and solder pastes (**Figure 3.47(b)**), need to be swapped out manually from the tool holder for each desired function. These tool modules are magnetically interchangeable and can facilitate fast and easy changing in between different processes (calibration, circuit printing and paste dispensing).

The Voltera V-One has a print area of 13.5 × 11.4 cm, and resolution of 10 μm, 10 μm and 0.625 μm in the x, y and z axes respectively. Minimum trace size as fine as 0.2 mm can be achieved. Active height mapping of the substrate is done with the precision probe head prior to the printing process, to ensure the dispensing heads maintain a constant printing height of 100 μm during the printing and dispensing processes.

Voltera also offers four different types of nozzles: 225-micron nozzles, 150-micron nozzles, 100-micron nozzles and disposable nozzles. The 225-micron nozzles are specifically designed for accurate deposition of Voltera's conductive inks and solder pastes, while the 150-micron nozzles can be used for pushing down the printing resolution further. For users looking to dispense their own low-viscosity materials, the 100-micron nozzles can be used.

The optional V-One Drill attachment unit (see **Figure 3.48**) can be added on to the Voltera V-One for making through-holes and vias in double sided PCBs on FR1 substrates. It has 25 W drilling power, maximum

Figure 3.48.   V-One Drill attachment unit. (Courtesy of Voltera, Inc.).

spindle speed of 13,000 rpm with a runout of 75 μm and uses Excellon (TXT) drilling data. Various precision drill bit sizes can be used depending on the desired through-hole's size. The technical specifications of the Voltera V-One PCB printer and V-One drill attachment unit are summarised in Tables 3.12 and 3.13 respectively.

3.3.1.3   *Working principles and process*

(a) *Mechanical displacement syringe extrusion*
The working principle of the Voltera V-One ink dispenser involves applying pressure on the material through the nozzle for extrusion [60]. Its design is based on a typical mechanical displacement syringe extrusion system [74] which uses rotary gears and a motor-controlled drive mechanism to depress the plunger at a controlled rate [75] (see **Figure 3.49**).

The dispensing syringe is first filled with ink and then inserted into the syringe sheath. The syringe sheath helps to support the syringe structurally, by physically restricting movements, mitigating offsets and inhibiting warping of the syringe during the printing process [75]. The rotary gears are rotated by the motor-controlled drive mechanism, and this rotary movement of the rotor gears is converted to downward linear motion of threaded rod through the motion conversion system. The drive piston attached to the threaded rod then exerts a force that pushes the mechanical

Table 3.12.   Specifications of the Voltera V-One PCB printer. (Courtesy of Voltera, Inc.).

Printer model	Voltera V-One PCB Printer
Printer Properties	
Technology	Direct-Write Dispensing
	Magnetically Interchangeable Print Head
Height control	Active Height Mapping with Detachable Probe
	20 um Repeatability
Print area	13.5 × 11.4 cm
XYZ resolution	10 μm; 10 μm; 0.625 μm respectively
Feature size	
Minimum line dimensions*	0.2 mm Space/Trace
Printing	
Minimum trace width	0.2 mm
Minimum passive size	0402
Minimum pin-to-pin pitch	0.65 mm
Resistivity	12 mΩ/sq @ 70 um Height
Supplied substrate material	FR4
Maximum board thickness	3 mm
Soldering	
Minimum pin-to-pin pitch	0.5 mm
Solder paste alloy	Sn42/Bi57.6/Ag0.4
Solder wire alloy	Sn42Bi57Ag1
Soldering iron temperature	180–210°C
Print Bed	
Print area (L × W)	128 mm × 116 mm
Max. heated bed temperature	240°C
Footprint	
Dimensions (L × W × H)	90mm × 257mm × 207mm
Weight	~7kg
Software Requirements	
Operating systems	Windows 7, 8, 10 (64bit), OSX 10.11+
Compatible file format	Gerber
Connection type	Wired USB

Table 3.12.   (*Continued*)

Hardware	
Maximum dimensions	$40 \times 26 \times 21$ cm
Weight	7 kg
Environmental conditions	
Operating temperature	15–30°C
Maximum altitude	2000 m
Relative humidity	80%
Pollution degree	2
Maximum temperature	240°C
Ramp rate	2°C/s
Power Requirements	
Voltage	100–120 V AC
Current	4.7 A 50/60 Hz
Power	575 W

* With standard 225-micron nozzle (inside diameter)

Table 3.13.   Specifications of the Voltera V-One drill attachment unit. (Courtesy of Voltera, Inc.).

Device model	Voltera V-One Drill
Drilling	
Supported material	FR1
Spindle speed (maximum)	13,000 RPM
Power	12V, 25W
Drilling data	Excellon (TXT)
Drill Bits	
Runout (TIR)	0.075 mm
Shank diameter	3.175 mm
Bit diameter (maximum)	2.0 mm
Bit length (maximum)	38.1 mm

plunger downwards and creates a linear downward displacement [75,76]. The rate and amount that the plunger is displaced, refer to the plunger feed-rate and displacement respectively [76]. The plunger successively pushes the ink downwards, resulting in ink compression within the dispensing syringe and creates yield stress for ink extrusion. The ink exits the

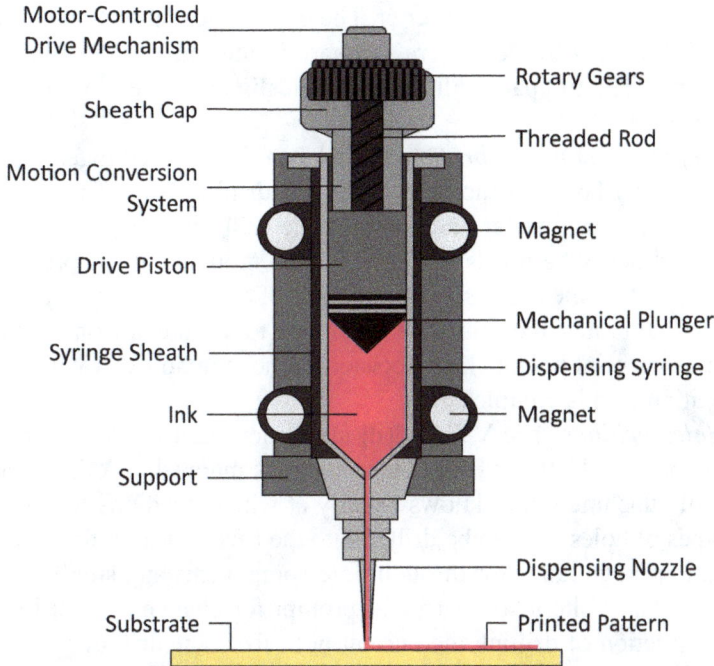

Figure 3.49.   Schematic diagram of a mechanical displacement syringe extruder (with reference to Voltera V-One ink dispenser's design) [60,74,75].

dispensing nozzle and deposits directly onto the substrate. To discontinue ink extrusion, the motor-controlled drive mechanism rotates the rotary gears in the reverse direction to displace the threaded rod upwards and alleviating the pressure applied on the mechanical plunger by the drive piston [75,76].

The volumetric output is directly proportional to the plunger displacement, while the flow rate is directly proportional to the feed-rate at which the plunger is displaced. The optimal feed-rate is highly dependent on the initial thixotropy and viscosity of the ink, and hence the operating parameters can vary for various inks [76].

(b) *Fabrication process of a double-sided PCB board*
This section will describe the fabrication process of a double-sided PCB board with the Voltera V-One PCB printer.

(i) *Upload design file*: Circuits can be designed with various compatible CAD tools such as Altium, Cadence, DipTrace, EAGLE, KiCad,

Mentor Graphics and Upverter. The design file is then to be exported and uploaded to the Voltera software in the form of the Gerber file format. The toolpath will be automatically calculated by the Voltera software.

(ii) *Preparation and calibration*: A FR1 board is secured together with a sacrificial layer on the heated bed for drilling later. The sacrificial layer acts as a buffer and prevents the drill bits from damaging the heated bed. The precision probe is then mounted to the printer. The precision probe defines the boundaries of the printable area and maps and calibrates the substrate's surface to ensure a consistent 100 microns print height. The precision probe is removed when this calibration step is completed.

(iii) *Holes drilling*: The V-One drill attachment unit is now mounted to the printer. As the substrate previously is mapped in *XYZ* space, the drill attachment unit knows exactly at which positions to drill. Two types of holes need to be drilled into the FR1 substrate for this application: large holes for through-hole components and small holes for vias. The Voltera software will prompt for change of drill bit upon completion of drilling the current hole size. The drill debris is carefully cleaned out from the substrate while avoiding any adjustments to the substrate's position.

(iv) *Circuit printing*: The drill head is swapped out and replaced with the ink dispenser. The ink dispenser is required to be calibrated prior to printing, to determine the resolution of the print. After calibration, the ink dispenser deposits the conductive ink directly onto the top layer of the FR1 substrate. The Voltera V-One has the selective printing feature which can allow rework and design modifications. The as-deposited conductive ink on the substrate is wet and needs to be sintered. The sacrificial layer is first removed from the heated bed. The substrate is then carefully flipped over and placed on the baking ledges of the clamp. The heated bed is then heated up for drying and sintering the wet conductive ink.

The substrate is clamped onto the heated bed with the bottom layer facing up, and the software will automatically select two drilled holes for alignment of the substrate. The probing, printing and sintering processes are repeated for the bottom layer, and after which the double-sided printed circuit board is now completed. The double-layer PCB can now proceed for riveting and assembly of electrical components.

(v) *Riveting*: Rivets are fitted manually with specialised rivet tools into the various drill holes of the completed printed board. A multimeter is then used to check and ensure good connections between the rivets and the conductive traces.

(vi) *Paste dispensing, mounting electrical components and reflow*: Similar to the previous printing process, the printed board needs to be clamped onto the heated bed and calibrated with the precision probe. The printer first aligns the orientation of the board and locates the position of the printing pads. The printer validates the pad alignment again before the deposition of solder paste. It is critical to do the alignment of the board properly to ensure precise deposition of solder paste at the desired locations. The solder paste dispenser is then mounted to the printer and the printer dispenses the solder paste directly onto the conductive pads. Electrical components are then mounted manually on the conductive pads. The heated bed heats up for reflowing when all the electrical components are fully mounted on the printed board.

3.3.1.4 *Materials*

Voltera sells several types of silver-based inks and solder pastes for various applications. However, users can also fill their materials in various empty standard 5cc syringe cartridges.

(a) *Silver-based ink*
Voltera currently provides two types of silver-based inks for the printing of conductive traces: conductor2 ink and flexible2 ink. The conductor2 ink is used for normal printing of conductive traces and pads, whereas the flexible2 ink is specifically developed for flexible circuits applications. Both inks have approximately 60–75% concentration of silver and viscosities ranging from 5,000–10,000 cP at 25°C.

The flexible2 ink is compatible with polyimide, polyethylene terephthalate (PET), and most flexible polymer substrates. It is recommended to sinter the flexible2 ink immediately after printing, at 160°C for 30 minutes, for best flexibility performances.

(b) *Solder pastes*
Voltera currently provides three types of solder pastes: T4-Sn42/Bi57.6/Ag0.4 solder, T5-Sn42/Bi57.6/Ag0.4 solder and T5-Sn63/Pb37 solder.

Table 3.14.   Material compatibility of the various solder pastes. (Courtesy of Voltera, Inc.).

Material Compatibility	Sn42/Bi57.6/Ag0.4 Solder	T5-Sn63/Pb37 Solder
Standard ink	✔	✘
Flexible ink	✔	✘
Copper PCBs	✔	✔
Hot air solder levelling (HASL) PCBs	✔ (brittle)	✔

Both T5-Sn42/Bi57.6/Ag0.4 solder and T5-Sn63/Pb37 solder have smaller particle sizes than T4-Sn42/Bi57.6/Ag0.4 solder, and they can allow more consistent dispensing with less clogging. Either the 225-micron nozzle or the150-micron nozzle can be used to dispense T5-Sn42/Bi57.6/Ag0.4 solder and T5-Sn63/Pb37 solder. The material compatibility of these solders with various inks and surfaces are tabulated in Table 3.14.

3.3.1.5 *Strengths and weaknesses*

The key strengths of Voltera V-One PCB printer are:

(a) *Low cost*: The Voltera V-One PCB printer is low-cost and affordable, which is ideal for rapid prototyping single and double-layer PCBs in labs or classrooms.

(b) *Wide range of materials can be used*: The Voltera V-One PCB printer can dispense a wide range of ink, paste or viscous fluids. For users who wish to experiment their own customised materials, they also can fill their materials in empty standard 5cc syringe cartridges. Empty ultraviolet (UV) blocking cartridges are also available for UV sensitive fluids.

(c) *Different substrates can be used*: The Voltera V-One PCB printer can deposit materials on any planar substrates or gently curved surfaces. However, Voltera's conductive inks typically need to sinter at 200°C for 30 minutes. Hence, the substrates used must withstand the sintering temperature. FR4, glass, ceramics and polyimide are some of the substrates that can be used with the Voltera V-One and its conductive inks.

(d) *Built-in heated platform*: Voltera V-One has a built-in heated platform for sintering conductive inks and reflow solder pastes. This is a convenient feature as it eliminates the need to transfer the as-deposited printed conductive traces to another oven or equipment for sintering and reflowing. In addition, the temperature can be controlled directly and create a customised heating profile by inputting the desired sintering temperature and time. The maximum allowable sintering temperature is 240°C and maximum sintering time is 60 minutes.

(e) *Dispensing solder paste on factory-fabbed boards*: The Voltera V-One can be a useful tool for in-house assembly of electrical components on factory-fabbed PCBs to reduce cost and waiting time. It can be used for dispensing solder paste and reflow soldering directly on factory-fabbed PCBs.

The key weaknesses of Voltera V-One PCB printer are:

(a) *Only capable of fabricating single and double-layer PCBs*: The Voltera V-One is not suitable for rapid prototyping multilayer PCBs which have more than two layers, as it can only fabricate single and double-layer PCBs.

(b) *No pick-and-place function for assembly*: The Voltera V-One does not have the pick-and-place capability for assembling electrical components onto PCBs. Electrical components must therefore be placed on the PCBs manually. Manual of electrical components on PCBs is tedious, time-consuming, highly skill-dependent, and risks making a mistake in positioning polarised components with the correct orientation.

3.3.1.6 *Applications*

The Voltera V-One PCB printer can be used in the following applications for printed electronics:

(a) *Printed electronics on various substrates*: Apart from printing single layer and double layer PCBs (see **Figure 3.50(a)**), the Voltera V-One can also deposit conductive inks onto different substrates for various printed electronics applications. **Figure 3.50(b)** shows conductive traces printed on a glass substrate.

(b) *Flexible electronics*: Although not specifically designed for this purpose, Voltera V-One users often use the machine to can deposit

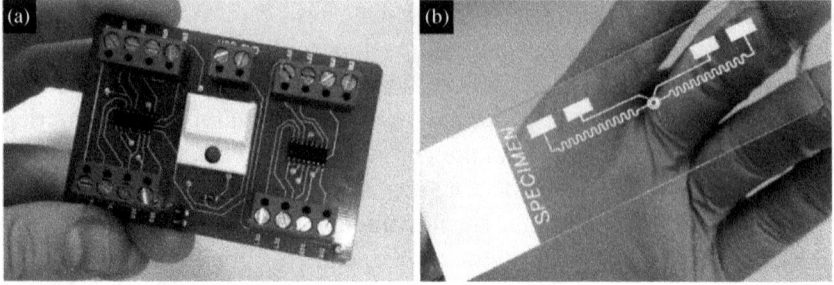

Figure 3.50.   (a) Printed circuits board; (b) Conductive traces printed on a glass substrate. (Courtesy of Voltera, Inc.).

Figure 3.51.   (a) Printed conductive traces on polyimide substrate; (b) Flexible heater. (Courtesy of Voltera, Inc.).

conductive inks directly onto flexible substrates for flexible electronics applications. **Figure 3.51(a)** shows printed conductive traces on a polyimide substrate, whereas **Figure 3.51(b)** shows a flexible heater. Abera *et al.* [77] demonstrated the fabrication of a flexible electrochemical immunosensor for aflatoxin M1 (AFM1) detection in milk. The Voltera V-One was used to deposit silver and silver chloride pastes on the PET substrate to fabricate the various electrodes.

3.3.2   Neotech AMT GmbH

3.3.2.1   Company

Neotech AMT GmbH is a leading developer of production level systems for 3D printed electronics (3DPE). The company began pioneering 3DPE developments in 2009 and installed the world's first 5 axis 3DPE system in 2010. This system expertly combines hardware and software elements into a unique solution capable of producing complex mechatronic systems, from prototyping through to high volume manufacture. The first mass-production capable system of type 45X was built in 2012, with EU/US/CN patents granted in 2015. Commercial mass production started on Neotech systems in late 2015 in mobile devices. In July 2019, Neotech was the winner of the 2019 TüV Süd — Innovation prize (Germanywide competition). Neotech continues to develop opportunities in this newly emerging technological field. Neotech AMT GmbH is located at Petzoltstr. 3, 90443, Nuremberg, Germany.

3.3.2.2 *Products*

Neotech AMT GmbH 3D print systems are multi-functional additive manufacturing tools for use in research and development (R&D), product development and high-volume manufacture. They consist of 5-axis motion platforms that house diverse functionalisation and pre-/post-base processing tools.

(a) *Print Systems*
The print systems allow the user to deposit every class of ink or paste used in commercial printed electronics (PE). Furthermore, the user is not tied to any one proprietary print technology — an important consideration when implementing research results for customers in a production environment. A broad range of pre- and post-processing technologies are offered to extend the materials handling capability. The available printing, functionalisation, pre-processing and post-processing tools that are customisable with Neotech's print platforms are tabulated in Table 3.15.

The systems have 5-axis range of print motion with a tailored CAD/CAM module, Motion 3D, that can print both 2D and also complex 3D structures. The pre-/post-process and functionalisation tools are interchangeable to allow the user to set up varied process routes without unduly compromising the motion range. Each print, pre-process and post-process module can be added either in the factory or at a later date in the field as when the customer needs to develop. The system is designed to be future secure allowing the customer to upgrade with the latest Neotech developments.

Table 3.15. List of available printing, functionalisation, pre-processing and post-processing tools that are customisable with Neotech's print platforms. (Courtesy of Neotech AMT GmbH).

Print Platforms	Print/Functionalisation Tools	Pre/Post-Processing Tools
• PJ15X*	• Piezo jetting	• CNC machining
• 45X G3*	• Aerosol-based jetting	• Plasma cleaning
• Motion 3D CAD/ CAM SW)	• Inkjet (single nozzle)	• Near-infrared (NIR) light sintering
	• Inkjet (multi-nozzle)	• Pulsed and continuous wave (CW) laser sintering
	• Dispensing	• Laser ablation
	• FDM	• Ultraviolet (UV) curing
	• Surface-mount device (SMD) pick-and-place	• Adaptive tool path vision system

Note: *Custom platforms available

The PJ15X base configuration model (see **Figure 3.52**) is a rapid prototyping system for 3D printed electronics, which allows complex 3D printing by combining piezo-actuated jetting technology with 5-axis motion control. The piezo jet print module in this system can dispense materials with ink viscosities ranging from 50–200,000 cp and suspensions with particle size up to 50 μm. Conductive nanoparticle inks, micro-particle inks, resistive inks, dielectric inks and biological reagents can be deposited on complex, non-planar substrates accurately. The base system can also be upgraded with other printing and post-processing tool modules, depending on the customers' needs and applications. For instance, FDM and dispensing printing modules can be integrated into the PJ15X for fabricating "fully additive" 3D electronics.

The 45X G3 (see **Figure 3.53**) is a patented high throughput, mass production system that integrates advanced motion and software controls with high performance print technology, which allows complex 3D printing of electronic devices. The 45X G3 is equipped with 5 axes of simultaneous motion and four print modules, which supports parallel concurrent printing of multiple parts. The choice of print modules can be customised for each system to meet each customer's requirements, which includes inkjet, high-speed piezo-driven drop-on-demand and

Figure 3.52.   Neotech AMT PJ15X. (Courtesy of Neotech AMT GmbH).

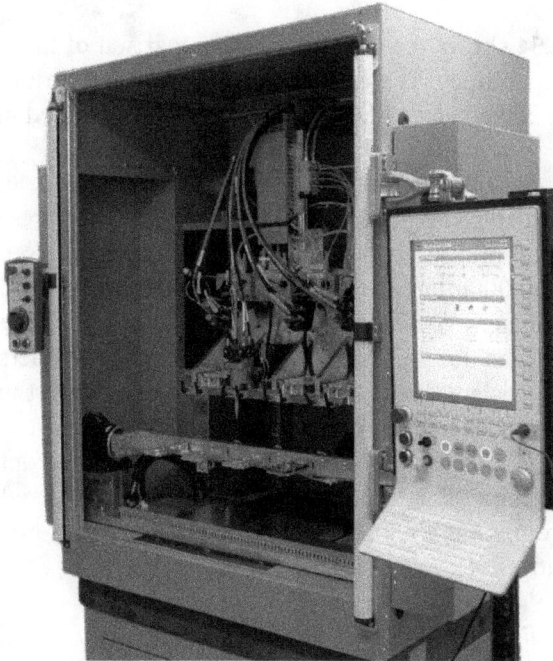

Figure 3.53.   Neotech AMT 45X G3. (Courtesy of Neotech AMT GmbH).

aerosol-based processes. Hence, wide materials range with viscosity up to 200,000 cp can be supported such as conductive inks, adhesives, dielectrics, etchants and dopants. The technical specifications of Neotech AMT PJ15X and 45X G3 are summarised in Table 3.16.

(b) *Printing tools for ink deposition*
Neotech works with the widest range of electronic printing tools and has unique expertise in the selection and implementation for critical applications. The interchangeable printing technologies that cover the main classes of printing electronics that process the maximum range of functional materials. Neotech has various type of print modules for selection:

  (i) Piezo-jet dispensing module
 (ii) Piezo-jet dispensing module with dispense adapter
(iii) Aerosol-based jetting print module
 (iv) Single nozzle inkjet print module
  (v) Multiple nozzles inkjet print module
 (vi) Extrusion print module

Figure 3.54 shows the performance comparison of the various print technology capabilities. The critical parameters of the various print modules available in Neotech AMT print platforms are tabulated in Table 3.17.

(i) *Piezo-jet dispensing module*
The piezo-jet dispensing module (see **Figure 3.55**) features non-contact drop-on-demand printing and high-speed processing of viscous inks and pastes up to 200.000mPas. It can process both nanoparticle inks as well as micron scale materials. The piezo-jet dispensing module has a compact modular print head design, which is simple to clean and operate. It is also suitable for consistent day-to-day printing and can support up to 4 hours of continuous printing. Each piezo-jet dispensing module come with two different nozzle sizes (50 μm and 100 μm). The piezo-jet controller stores up to 10 print recipes that can be linked together and triggered by the print platform's programmable logic controller (PLC) during printing, which gives a high level of control of the printing process.

(ii) *Piezo-jet dispensing module with dispense adapters*
The piezo-jet dispensing print heads are also supplied with dispense adapters to allow attachment of Luer-lock nozzles, which enable precision

Table 3.16.   Specifications of the PJ15X and 45X G3. (Courtesy of Neotech AMT GmbH).

Model Name	PJ15X	45X G3
Motion module		
Print speed	100 mm/s max	1000 mm/s max.
Motion range (X-Y-Z)	400 × 300 × 140 mm	600 × 500 × 250 mm
X, Y and Z – Axes motion accuracy	—	+/- 5 μm
X, Y and Z – Axis repeatability	+/- 10 μm	+/- 2 μm
A&B axis position accuracy	Angle deviation 0° 1' 20"	< 1.5 arcmin
A&B axis repeatability	Angle deviation 0° 0' 6"	< 6 arcsec
Standalone system dimensions (X-Y-Z)	769 mm × 834 mm × 1370 mm	1400 × 1055 × 2165mm
Standalone system weight	350 kg	1250 kg
Electrical requirements	240V/10A or 110V/10A	400V AC, 50–60Hz, 26A and Circuit Protection 32AH400V
Compressed air required	4 bar minimum	8 bar minimum
System features	Full 3D functional printing capability	Simultaneous 5 axis or indexed printing (X/Y/Z linear + A & B rotational) with free definition of the print sequence to optimize cycle times
	3DPE combining structural build, surface mount devices and printed circuits	"Motion 3D" tool path generation software for printing complex geometries
	Intuitive and easy to use software	Tool Centre Point mode for consistent print velocities on variable curve surfaces
	Cost effective prototyping to low volume manufacturing	Line, contour and area fill with arbitrary geometry with automatic island recognition
		Virtual simulation of the printing path
		Editing possibility for the NCP/G-Code data
		Robust, low maintenance platform with industrially proven CNC technology
		Simple operator interface with remote service via Team Viewer

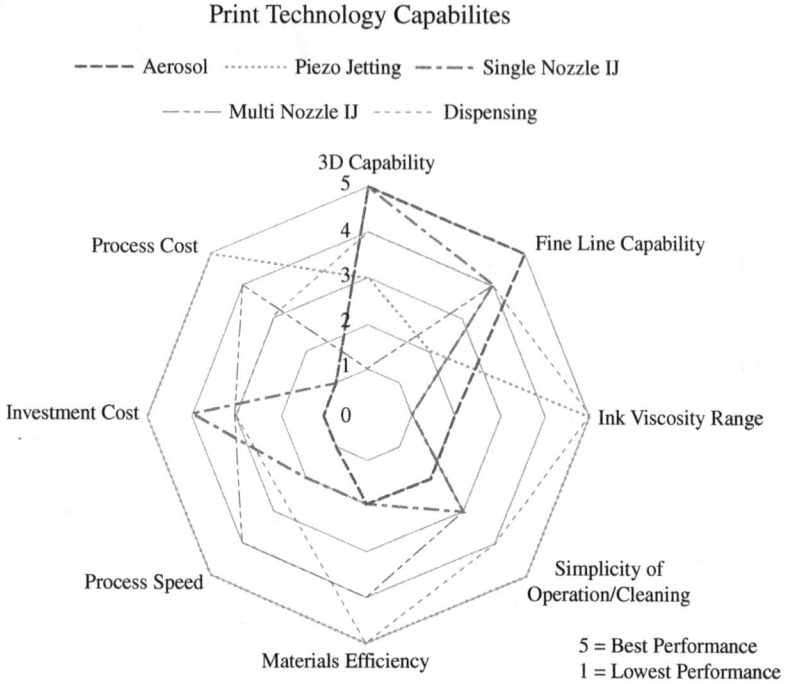

Figure 3.54. Performance comparison of the various print technology capabilities. (Courtesy of Neotech AMT GmbH).

dispensing of function inks and pastes. This also helps to extend the piezo-jet functionality to allow media viscosity up to 1,000,000 mPas and printing line widths below 20 μm. The programmable piezo-jet parameters (tappet stroke) and small diameter fluid assembly diameter combined to enable extremely fine dosing of materials.

For fine features dispensing (line width less than 200 μm), it is critical to control the standoff distance between the dispensing nozzle and part surface to achieve consistent printing. The positioning accuracy of the nozzle tip above the part surface needs to be within +/−10 % of the nozzle diameter. Since the nozzle diameter defines the line width, it is necessary to scan the part surface accurately and regulate the nozzle standoff. For piezo-jet dispensing, a surface scan/height control module, consisting of confocal scanning system, is supplied (Confocal Sensor IFS 2405–Micro Epsilon GmbH). The confocal sensor holds the nozzle standoff height to be within +/−10% of the nozzle diameter that is needed for dispensing operations. This sensor works with transparent, opaque and glossy substrates.

Table 3.17.   Comparison of various print modules available in Neotech AMT print platforms. (Courtesy of Neotech AMT GmbH).

Parameters	Piezo-Jet Dispensing	Piezo-Jet Dispensing with Dispense Adapter	Aerosol-Based Jet	Single Nozzle Inkjet	Multiple Nozzles Inkjet	Extrusion[a]
			Type of Print Modules			
Heatable head	Up to 90°C	—	—	—	—	25–75°C
Operation |frequency	Max. 4000 Hz	—	—	1–30 kHz (Material dependent) Typically, up to 5000 Hz	—	—
Viscosity range	50–200,000 mPas	Up to 1,000,000 mPas	Up to 20 mPas	Up to 20 mPas		50–200,000 mPas
Surface tension			—	20–70 dynes/cm		—
Droplet volume	0.5 nL			5 pL–0.5 nL		—
Particle size	Nano–micron scale	Nano–micron scale	Nano scale	Nano scale	Nano scale	Nano–micron scale
Printed line width[b]	300–1000 μm (150 μm)	20–1000 μm	20–500 μm	100–1000 μm	50–1000 μm	50–1000 μm
Typical thickness	> 20 μm		0.5–10 μm	1–10 μm	1–10 μm	2–10+ μm
Typical print speed	15–100 mm/s		1–10 mm/s	1–10 mm/s	5–50 mm/s	10–20 mm/s
Materials printed	• Conductive inks • Dielectric inks • Adhesives • Positive temperature coefficient (PTC) resistor pastes • Ceramics		• Conductive inks • Resistive inks • Dielectric inks	• Conductive inks • Dielectric inks • Conductive polymers • Adhesives		• Conductive inks • Dielectric inks • Adhesives • PTC resistor pastes • Ceramics • Epoxies

Notes: (a) Under development; (b) Material dependent

Figure 3.55.   Neotech AMT piezo-jet dispensing module. (Courtesy of Neotech AMT GmbH).

(iii) *Aerosol-Based Jetting Print Module*
The aerosol-based jetting print module utilises the *NanoJet*™ (NJ) technology, which is capable of printing fine line features with line width ranging from 20μm to 500μm (single pass). The aerosol-based jetting print module is compact with internal shuttering allowing full 3D print flexibility with long term, stable performance for up to 4 hours continuously. The ink cartridges are removable and replaceable and hence, allowing simple operation and easy cleaning. A wide variety of material with viscosity up to 20 cp can be printed. Commercially available print nozzles can be used too.

(iv) *Single Nozzle Inkjet Print Module*
Neotech systems are offered with an array of up to four single nozzle inkjet heads that can process multiple materials. 3D inkjet is a drop-on-demand, 3D capable inkjet print head that can operate with standoff distances of 1–2 mm from the part surface. A broad range of materials can be processed, including bioactive and optically active materials. It also

has a large 25 ml capacity removable glass reservoir to allow easy off-line fluid storage and refrigeration.

(v) *Multiple Nozzles Inkjet Print Module*
This module can allow classical multi-nozzle inkjet printing on 3D surfaces. Currently, FUJIFILM Dimatix print heads are used due to their wide use on printed electronics, but other manufacturers' print heads (such as Xaar and Ricoh) can also be substituted.

(vi) *Extrusion (based on Endless Piston Principle) Print Module*
The extrusion print module utilises the Vipro-Head5 print head from Viscotec GmbH to dispense viscous one component fluids and pastes of electrical or even structural functionality (e.g. conductive inks, pastes, epoxy, acrylate, light-curing adhesives, silicone, resin, waxes and abrasive pastes) with a viscosity ranging from 50–200,00 mPas. The materials are conveyed purely volumetrically via a dispensing pump based on the endless piston principle. This print head has a dosing rate ranging from 0.5–6 ml/min, can dispense 0.5 ml per revolution minute and has a minimum rotation step size of 0.15°. Fluctuating process parameters, due to temperature fluctuations, are levelled out to give consistent and precise print results even at high printing speed.

3.3.2.3 *Working principles and process*

This section will discuss the working principles of piezo-jet dispensing and extrusion-based on the endless piston principle. The working principles of the aerosol-based jet printing and inkjet printing are discussed previously in the previous sections.

(a) *Piezo-Jet Dispensing*
The working principle of piezo-jet dispensing is based on the inverse piezoelectric effect to drive the piezo-jet dispenser. A small amount of ink is held between the tappet and nozzle of the piezo-jet dispenser. The piezo actuator oscillates up and down at high frequency, generating kinetic energy that is used to eject droplets at the target position under a controlled pressure [78].
Figure 3.56 shows the schematic diagram of the piezo-jet dispensing module, in which it consists of a piezoelectric stack actuator, a syringe barrel and an injector subsystem [78,79]. The tappet, nozzle, lever and

Constant Pneumatic Pressure

Valve Communication Cable

Valve Power Cable

Syringe Barrel

Ink

Piezo Actuator Module

Electronic Module

Piezoelectric Stack Actuator

Fluid Inlet

Fluid Chamber

Fluid Body Assembly

Nozzle Head

Lever

Return Spring

Tappet

Nozzle

Injector Subsystem

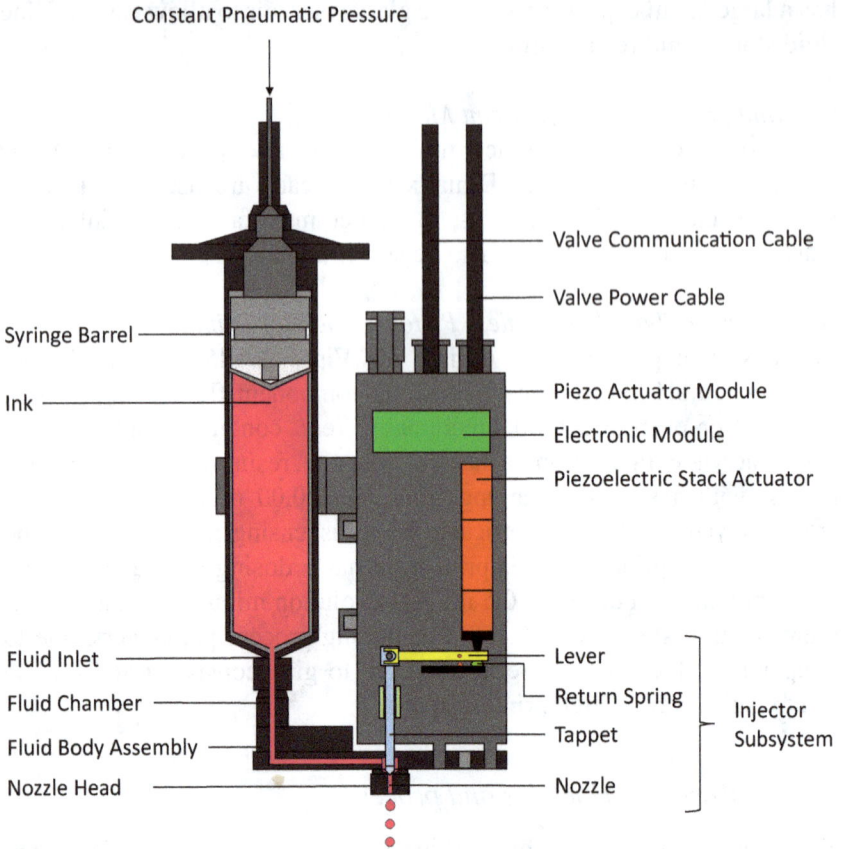

Figure 3.56.   Schematic diagram of the piezo-jet dispensing module [78–80].

return spring are some of the essential parts of the injector subsystem. The piezo-jet dispenser is initially closed as the tappet seats tightly in the nozzle to form a closed chamber (see **Figure 3.57(a)**). The syringe is subjected to a constant pneumatic pressure and the ink fills up the entire fluid chamber via the fluid inlet [79].

Piezoelectric stack actuators, which comprises of stacks of thin piezoceramic layers, are commonly used for linear actuations as a stack can obtain a free stroke displacement substantially larger than a single piezoelectric chip while requiring low drive voltage and maintaining short response time. A voltage is applied to excite the piezoelectric stack actuator [80,82], and the piezoelectric stack actuator expands in length due to the inverse piezoelectric effect. This movement is imparted to the tappet

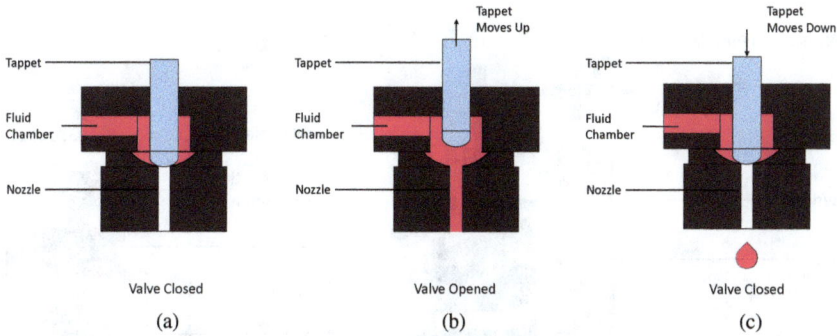

Figure 3.57.   Schematic diagrams of the piezo-jet dispenser's nozzle where (a) valve is closed before dispensing process, (b) valve is opened during the dispensing process; and (c) valve is closed and a droplet is ejected from the nozzle [81].

via the lever. The piezoelectric stack actuator pushes the lever down as it extends, and hence lifting the tappet up away from the nozzle. The return spring also gets compressed as the tappet moves up. The ink quickly fills up the gap between the tappet and nozzle under feed pressure [80,83] (see **Figure 3.57(b)**).

The actuator returns to the neutral state as it de-energises and the compression force in the return spring pushes the lever back up. Elastic potential energy stored in the return spring is immediately converted into kinetic energy of the tappet [83]. A large shear force is also generated to overcome the viscous forces of the ink. The tappet falls back to the nozzle and ejects out a droplet [82] (see **Figure 3.57(c)**). The fluid chamber is now closed as the tappet seats tightly in the nozzle back again [80]. The entire process is repeated for subsequent jetting.

(b) *Extrusion (based on Endless Piston Principle)*
The working principle of the Vipro-Head5 print head is based on the endless piston principle for material extrusion [84]. This type of material extruder belongs to the category of rotating displacement pumps, in which a pressure difference is built up when material is conveyed from the suction end to the discharge end [84].

Figure 3.58 shows a schematic diagram of a material extruder based on the endless piston principle. This type of material extruder, also commonly known as the helical rotor screw pump, is completely pressure-tight and self-sealing [84]. In this device, a steady pneumatic pressure is exerted on the syringe to feed the ink directly into the print head.

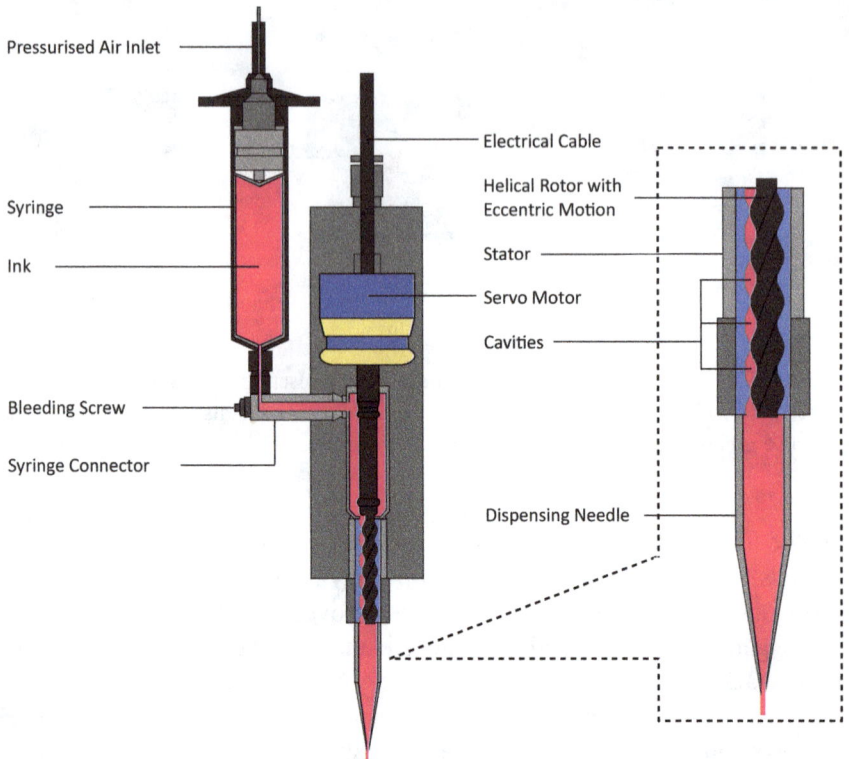

Figure 3.58.   Schematic diagram of a material extruder based on the endless piston principle [85,87].

The servo motor, soft polymeric stator, helical rotor with eccentric motion and dispensing needle are some of the essential components of the print head [85]. The servo motor drives the rotation of the helical rotor.

The complex geometries of the helical rotor and stator allow them to taper and overlap each other and hence making it possible for the rotor to seal against the stator (see **Figure 3.58**) to create a series of alternate cavities [85,86]. The input pneumatic pressure exerted on the syringe does not impact the extrusion process because of the seal between the rotor and stator and can achieve true positive displacement [86]. The controlled rotary motion of the rotor pushes the ink into these cavities and conveys the ink out to the dispensing needle to deliver a non-pulsing and continuous extrusion [84]. These cavities retain their form and volume during the extrusion process, and therefore a constant volume can be extruded proportionally to the angle of rotation per revolution [86]. The volumetric output is directly proportional to the number of revolutions, while the flow rate is directly proportional to the rotational speed [86]. As the rotating direction of the rotor can be reversed, the ink can be cleanly broken off in a controlled manner to prevent dripping [84,86].

3.3.2.4 *Materials*

A wide range of functional materials can be covered by the different interchangeable printing modules in Neotech systems. The viscosity range, surface tension and allowable particle size for each type of printing module are tabulated in Table 3.17. The dispensing print module (piezo-jet dispensing print head) can dispense the widest range of materials as compared to other print modules, with viscosity up to 1,000,000 mPas. Some of the common functional materials that can be dispensed by the various print modules are conductive inks, dielectric inks, adhesives, PTC resistor pastes and ceramics.

3.3.2.5 *Strengths and weaknesses*

The key strengths of Neotech AMT GmbH systems are:

(a) *Printing on complex parts*: Fully integrated with advanced 5-axis motion control for optimised motion, Neotech systems also have dedicated CAD/CAM system to allow simple creation of machine code for printing on complex parts. Full machine simulation with collision detection and CAM check functions in the Motion 3D software can also be applied prior to printing.
(b) *Various functionalisation tools available*: Various functionalisation tools can also be integrated into print platforms, such as the SMD pick-and-place (P&P) and FDM modules. These tools can allow the fabrication of embedded electronics, with both structural bodies and conductive traces fully printed in a single print.

The P&P module consists of industrial-grade P&P head with component alignment camera, tape feeders and parts tray holder. In conjunction with the Motion 3D software, it enables P&P of standard SMD components in 3D space. By using the Adaptive Tool Path

Vision system, direct-write of SMD interconnects can be achieved. The FDM print module prints the complete range of thermoplastic filaments commonly used for FDM with full 5 axis capability.

(c) *Wide range of materials*: A wide range of functional materials can be covered by the various interchangeable printing modules in Neotech systems. Users can customise different print modules in the print platform to fit their preferred choices of materials and applications.

The key weakness of Neotech AMT GmbH systems are:

(a) *Limited printing speed*: The limitation on the systems generally relates to the speed of the printing processes that vary from 5mm/s to 100 mm/s. Even at the higher end of the spectrum, certain applications are not economical compared to conventional manufacturing techniques. However, this limitation can be overcome by leveraging the use of a multi-station printing platform to combine various processes to improve the overall fabrication speed. Besides, there is also a need to develop the processes to enable First Time Right (FTR) manufacture due to process complexity.

3.3.2.6 *Applications*

There are two basic process routes exist for the Neotech systems in printed electronics applications:

• *Classical moulded interconnect devices (MID)*: In this process route, printed electronics functionality and Surface Mount Devices (SMDs) are added to the external 3D surfaces of parts manufactured by traditional methods such as injection moulding, composite, casting, machining etc. This methodology is used where parts can be scaled to high production.
• *"Fully additive" electronics*: In this process route, electronics printing is combined with 3D printing or additive manufacturing of the structural body with the electronics added internally and/or externally. This completely digitally-driven manufacture enables an additional level of uniqueness to product designs but is currently used only in low volume and prototyping.

Figure 3.59.   3D printed antenna for mobile devices. (Courtesy of Neotech AMT GmbH and LITE-ON Mobile Mechanical SBG).

The market for 3D PE is emerging across many market segments with applications in automotive, communications, medical, aerospace and industrial electronics. In the MID field, products use 3D PE to combine circuitry, antennas, sensors and heating capability. Some of the examples include:

(a) *3D printed antenna*: **Figure 3.59** shows a 3D printed antenna for mobile devices, in which the cell phone antenna is directly printed onto the moulded phone casing with silver-based inks and then oven sintered. As compared to traditional methods, this digitally-driven process can allow reduce device product thickness and improve design flexibility without the need for any plating and hard tooling processes. There are also significant reductions in the lead time for new product development, product design changes and initial sample delivery.

(b) *3D printed circuits and sensor structures*: Circuits (see **Figure 3.60**) and sensor structures (see **Figure 3.61**) can be directly printed onto 3D mechanical structures with different functional inks to help simplify the process chain. **Figure 3.61** shows filling level sensors printed on a prototype PA6 form moulded tank. Two capacitive sensors that register the filling level are printed within the circuitry and SMDs (microcontroller, LEDs and resistors) are mounted to complete the device.

Sensing structures are also being utilised for healthcare applications. After suffering from a stroke, patients often suffer unilateral motor dysfunction resulting in weak finger strength, weak grip

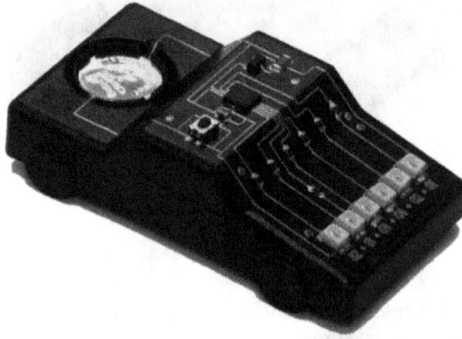

Figure 3.60.   3D printed circuits on MID. (Courtesy of Neotech AMT GmbH).

Figure 3.61.   3D printed MID sensor circuit on a form moulded tank. (Courtesy of Neotech AMT GmbH).

and poor circulation. The rehabilitation ball (see **Figure 3.62**) has printed circuits and embedded electronic components on its curved and flexible surfaces. It is to be held in the palm for close-and-open exercises, and effectively increases finger strength and stroke recovery. The device also provides real-time feedback of the patient's grip strength and monitors the training process for patients

(c) *"Fully additive" electronics*: "Fully additive" electronics are less well-developed than current MID designs, and usually used in new product development and prototyping. **Figure 3.63** shows a functioning egg timer completely fabricated in a single print using the "fully additive" methodology. Its structural body is printed with FDM, SMDs are added with the pick-and-place module, and circuits and touch sensors are printed directly onto the device. The various print heads operate sequentially, swapping over sequentially to complete the part build.

Figure 3.62.   Rehabilitation ball developed by EverYoung BioDimension Corporation. (Courtesy of Neotech AMT GmbH and EverYoung BioDimension Corporation).

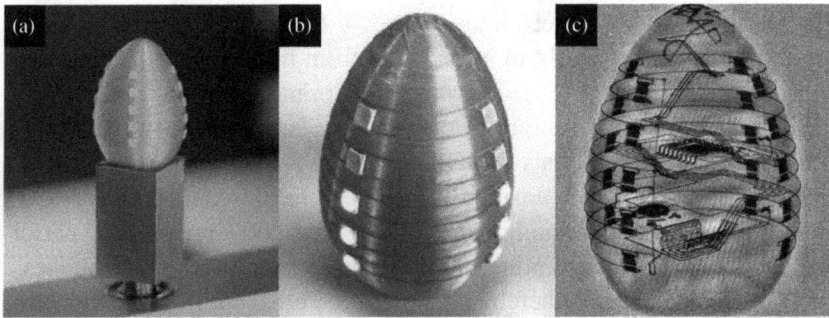

Figure 3.63.   (a) "Fully additive" printed egg timer device mounted in the print system; (b) Finished device showing the LEDs indicating the remaining time; (c) 3D X-ray image of the device revealing the embedded circuits and SMDs. (Courtesy of Neotech AMT GmbH and University of Erlangen-Nuremberg Institute for Factory Automation and Production Systems (FAPS)).

3.3.2.7 *Research and development*

Neotech AMT GmbH specifically relates its current R&D directions to:

(a) *Cost reduction through increasing processing speeds and efficiency.*
(b) *Enabling First Time Right production for the complete volume range from one-off custom parts to high volume manufacturing*: For "Fully additive" electronics there is also the need to develop the processes to enable First Time Right manufacture due to process complexity. The

interaction of multiple processing tools requires closed-loop controls to ensure part quality.

(c) *Expansion of the application range*: The application range is being expanded by developing additional combinations for processing steps reliably and cost-effectively.

3.4 Electrohydrodynamic Printing

Electrohydrodynamic (EHD) jet printing is an advanced, high-resolution printing technique that requires the use of the electric field to creates the electrohydrodynamic phenomena [88] that produce fluid flow for deposition onto substrate [89]. This unique printing technique is capable of producing tiny droplets and very fine features (sub-micrometre range) with the use of smaller nozzles [90]. In addition, EHD printing can deposit a wide variety of inorganic and organic materials and used for 3D printed electronics and bioprinting applications [91].

The inkjet printing technique utilises a "push" mechanism whereby small droplets of materials are pushed out from the nozzles [92]. The nozzle's diameter usually can determine the printing resolution [93] and thereby, smaller nozzles are preferred for better printing resolution. However, ejecting materials with very small nozzles are usually infeasible since high-pressure levels are required to overcome the capillary forces [94]. Hence, limiting the minimum nozzle size for good printability performances. Unlike inkjet printing, the EHD jet printing technique utilises a "pull" mechanism [92] whereby materials are pulled out from a conductive nozzle with the application of an electrical field [94]. It is relatively easier to "pull" than "push" materials out from a small nozzle [92,94]. Therefore, nozzles with inner diameters as small as 100 nm [94] can be used for reducing droplet sizes [90] and achieving high-resolution printing.

The schematic diagram of an EHD printer is shown in **Figure 3.64**. An EHD printing generally comprises of these five main sub-systems [91,92]: material supply system, motion system, power supply system, visualisation system and computer control system.

(1) *Material supply system*: The material supply system typically comprises a back-pressure supply device [94], such as a syringe pump, capable of precise material delivery to the nozzle [91,92]. The back-pressure supply can be directly controlled by the pneumatic regulator to regulate the output material flow rate through the nozzle [89].

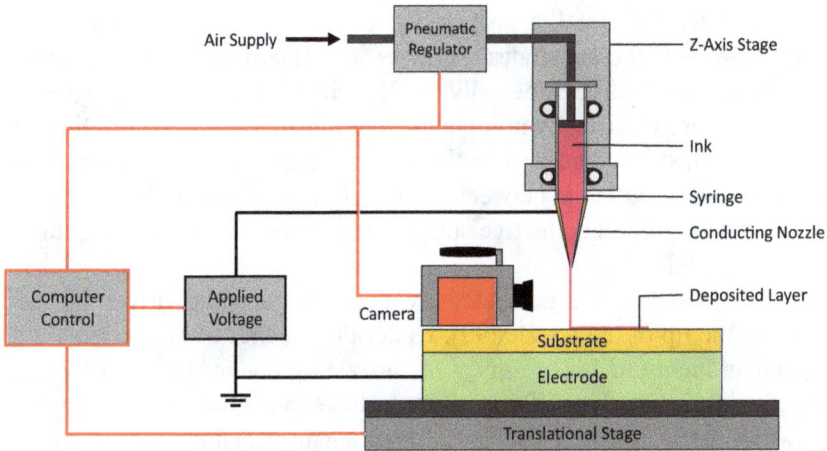

Figure 3.64. Schematic diagram of an EHD printer [95–98].

(2) *Motion system*: The motion system provides 3-axis motion for printing. The printhead is mounted on a Z-axis stage, which allows the printhead to adjust its standoff distance from the substrate. The substrate is placed on the planar translational stage that is movable in the X-Y horizontal plane for patterning [91,92].

(3) *Power supply system*: A voltage is applied between the conducting nozzle and collection substrate by the power supply system to initiate material flow from the nozzle using electrohydrodynamics [91,92].

(4) *Visualisation system*: The visualisation system typically comprises of a camera and a light source, is used for observing droplet formation and process monitoring [91,92]. Hence, the process parameters can be adjusted during the printing process to achieve the desired modes of EHD printing.

(5) *Computer control system*: The computer control system is used for controlling the material supply system, motion system, power supply system and visualisation system simultaneously during the printing process [92].

(a) Working Principle

For EHD printing, both nozzle and substrate must be electrically conductive for the power supply system to generate an applied voltage potential between them [91,92]. The outer surfaces of the nozzle and substrate can be coated with a metallic coating for enhanced electrical conductivity [98]. However, the substrates' surfaces are usually preferred not to be electrically conductive, especially for 3D printed electronics applications. In order to use electrically non-conductive substrates in EHD printing, an electrode must be placed directly under the substrate to generate the desired voltage potential between the nozzle and substrate [96]. Note that the electrically non-conductive substrates must be within a certain thickness range [98].

The back-pressure generated by the syringe pump drives the ink towards the tip of the nozzle [99]. Under the absence of an electric field, a pendent meniscus [88] forms at the nozzle tip due to surface tension at the liquid-air interface [91,100]. A voltage potential is then applied between the nozzle and substrate, resulting in accumulation of mobile ions at the surface of the liquid meniscus [91,94,100]. The Coulombic repulsion between ions induces a tangential stress on the meniscus' surface [94,98,100], and deforms the meniscus into a conical shape, known as the Taylor cone [91,94,98]. When the surface tension is overcome by the electrostatic stresses at sufficiently high electric fields, droplets are then ejected from the Taylor cone [88,91,94,98]. Simultaneous coordination of ink droplets ejection and substrate positioning allows patterns to be printed directly onto the substrates [94]. The applied voltage potential, offset height and back pressure can directly affect the printing conditions. The jetting frequency, f in EHD printing can be expressed as a function of voltage potential, V and offset height, h:

$$f = K\left(\frac{V}{h}\right)^{\frac{3}{2}}, \tag{3.12}$$

where K is a scaling constant dependent on nozzle diameter, applied back pressure, viscosity of the ink, and permittivity of free space [89].

(b) Various Jetting Modes

The EHD jetting modes are directly influenced by the flow rate and electric field strength. These various jetting modes are plotted in the graph as shown in **Figure 3.65**: "dripping" mode, "pulsating jet" mode, "cone jet" mode and "complex jet" modes [93,95]. The applied voltage between the conducting nozzle and the translational stage determines the electrical field strength [89].

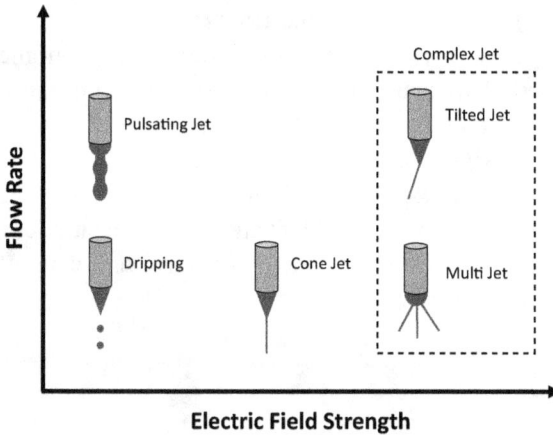

Figure 3.65.   Various EHD jetting modes under different electric field strength and flow rate [89,94].

(1) *"Dripping" mode*: The "dripping" mode is the most basic EHD jetting mode, in which the jetting phenomenon is same as droplet ejection from a nozzle without the influence of the electrical field. The "dripping" mode normally occurs when both electric field strength and flow rate are low [93,95,101]. As both electrical and gravitational forces overcome the surface tension forces, the droplets will fall from the nozzle and form a spherical shape [101]. The "dripping" mode is typically not used for industrial EHD printing as the patterning process is too slow and the droplet size is too large to give fine resolution printing.

(2) *"Cone jet" mode*: A continuous cone-jet is ejected from the liquid meniscus when electric field strength and flow rate for a particular ink are properly matched [93,95]. This is one of the jetting modes commonly used in EHD printing, known as the *"cone jet"* mode. The *"cone jet"* mode can produce very fine printing resolutions as the jet is very small in diameter and is very suitable for sub-micron or micropatterning. In order to have a steady cone-jet, a minimum flow rate (Q_m) is required to maintain a steady flow. The minimum voltage that yields a steady cone-jet at Q_m is known as the Taylor voltage (V_m) or the critical voltage, and it is dependent on the applied back pressure and ink properties [93,95,102,103].

(3) *"Pulsating jet" mode*: When the flow rate is greater than Q_m and the voltage is less than V_m, the steady cone-jet phenomenon cannot be sustained. This can induce repetitive formation and relaxation of the Taylor cone and then result in the ejection of streams of distinct droplets, also known as the "pulsating jet" mode [93,95].

(4) *"Complex jet" mode*: At very high electric field strength, complex jet" behaviours such as "tilted jet" and "multi-jet" modes occur. These "complex jet" modes are difficult to control and cannot be used for EHD printing [93,95].

3.4.1 *Enjet, Inc.*

3.4.1.1 *Company*

Since its establishment in 2009, Enjet Inc has been devoting to provide innovative manufacturing solutions to customers on printing and coating technologies. Over the years, Enjet Inc has developed and supplied many high-resolution printers, atmospheric plasma equipment, coaters and dispensers based on their patented *i*Electrohydrodynamic (*i*EHD) jet technology for companies in the information technology (IT) sector. The *i*EHD technology is a near-field jet printing technology that uses electrostatic forces for high resolution and high precision dispensing of functional inks with high stability and reliability. With Enjet's high precision *i*EHD nozzles, electrical pulse control technology and precise manufacturing platform, *i*EHD-based printers can produce ultra-fine droplets (nanoscale diameter), lines and complex patterns. Enjet Inc is located at 45, Saneop-ro 92 Beon-gil, Gwonseon-gu, Suwon-si, Gyeonggi-do, Republic of Korea.

3.4.1.2 *Products*

(a) *iEHD jet technology*
The *i*Electrohydrodynamic (*i*EHD) jet technology whose jetting mechanism is based on electrohydrodynamics to allow stable and reliable dispensing and liquid ejection. Based on the *i*EHD jet technology, Enjet has developed various types of systems for continuous jet printing, DoD jet printing and spray coating (see **Figure 3.66**).

Enjet *i*EHD jet printers are designed to produce high-resolution patterns that are less than 1 *μm,* which include dots, lines and complex

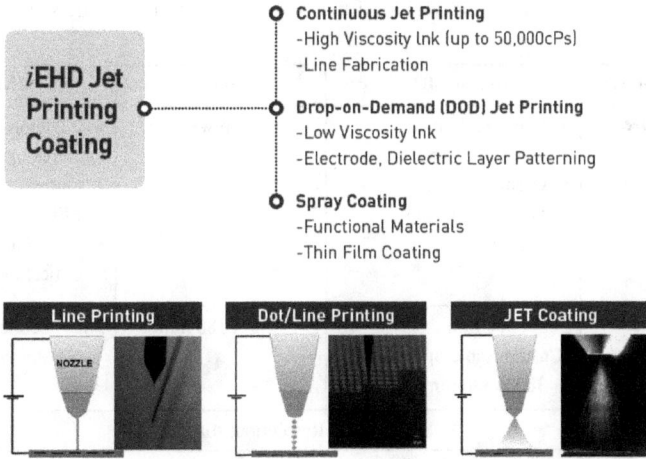

Figure 3.66.   Various types of Enjet systems for continuous jet printing, DoD jet printing and spray coating. (Courtesy of Enjet, Inc.).

shapes. With precise motion, electric signal, pump and valve controls, Enjet *i*EHD jet printers can eject super-fine droplets and deposit patterns on 2D and 3D substrates with low viscosity materials. Enjet *i*EHD jet printers can also dispense high viscosity materials continuously, producing fine lines with line width less than several micro-metres. The technical specifications and key features of the various Enjet *i*EHD jet printers are summarised in Table 3.18.

(b) *eNanoJet Printer*

The eNanoJet printer's jetting mechanism is based on the *i*EHD printing technology and novel nozzle designs to achieve ultra-fine patterns and nanoscale droplets. It can produce features sizes with line widths ranging from 1–100 μm and dispense materials with viscosities ranging from 1–1,000 cPs. The basic system has a printable size of 300 mm × 300 mm, and customisable to 380 mm × 380 mm. It has a printing resolution of 0.1 μm, printing accuracy of ± 1.5 μm and a maximum printing speed of 400 mm/s.

The eNanoJet printer (see **Figure 3.67**) is equipped with a high-resolution top-view vision camera to allow users to observe the printing process in real-time and perform printing adjustments to achieve high precision dispensing (see **Figure 3.68**). The nozzle auto-alignment feature in the printer ensures the printhead modules and nozzles are precisely

Table 3.18.   Specifications of various Enjet's *i*EHD jet printer models. (Courtesy of Enjet, Inc.).

Machine Model	eNanoJet	eNanoECO	MX2
Machine size/ type	• Full size with cover frame and tower lamp • Granite base and vibration isolator • Top FFU	• Tabletop with cover frame	• Tabletop size Optional upgrades • Optical table • FFU • Anti-vibration optical table
Printable size	300 × 300 mm (Customisable up to 380 × 380 mm)	130 × 180 mm	300 × 220 mm (Customisable)
XY motion	Resolution: 0.1 μm		
	Accuracy: ± 1.5 μm		Accuracy: ± 2 μm
Maximum speed	400mm/s		400mm/s (Customisable)
Triggered printing	Yes	No	No. Customisable.
Vision	Realtime top view printing observation		Offset-top view
	Dual screen: Two cameras can be shown simultaneously	—	
Nozzle cleaner	Included	Not Included	
Evaporator	Included	Optional	
Pressure to nozzle	Controllable, up to 100 kPa	No	Controllable, up to 200 kPa
Machine dimension (W×D×H)	1360 × 1980 × 1830 mm (Frame size)	640 × 530 × 660 mm	1575 × 1175 × 1820 mm (Table + Frame size)
Weight	2,800 kg	50 kg	500 kg

aligned for optimal printing performances after each change. Its unique SmartSignal (patent pending) feature can allow stable, reliable and repeatable printing by adjusting the signal to nozzle while inspecting the drop size and predicting ink change.

The eNanoJet printer can be used for many high-resolution printing or pattering applications, such as PCB, semiconductors, displays and

Figure 3.67. eNanoJet printer. (Courtesy of Enjet, Inc.).

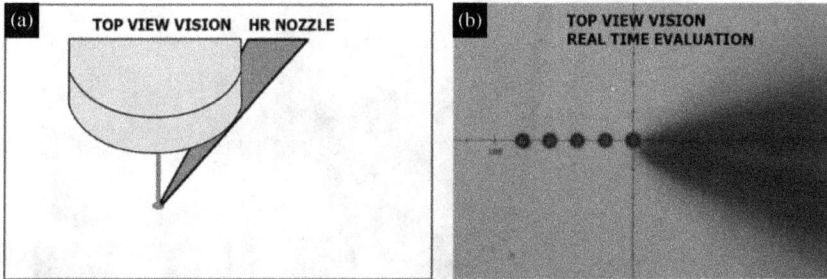

Figure 3.68. (a) Schematic diagram of the top view vision camera and the HR nozzle; and (b) Top view vision real-time evaluation. (Courtesy of Enjet, Inc.).

bio-medical devices fabrications. **Figure 3.69(a)–(c)** show the silver ink deposited by the eNanoJet printer on thin-film transistor (TFT) substrate, and these printed lines can achieve line width as fine as 0.5 μm. **Figure 3.69(d)** shows the organic ink deposited by the eNanoJet printer on the glass substrate with 5 μm line width.

(c) *MX2 Multi-function/purpose printer*
The MX2 multi-function/purpose printer (see **Figure 3.70**) is a modulated platform with interchangeable printhead module (see **Figure 3.71**) for precision dispensing applications, such as precise dot and line dispensing, underfill and electrospinning. The basic system has a printable size of 300 mm × 220 mm and a maximum printing speed of 400 mm/s. It has a

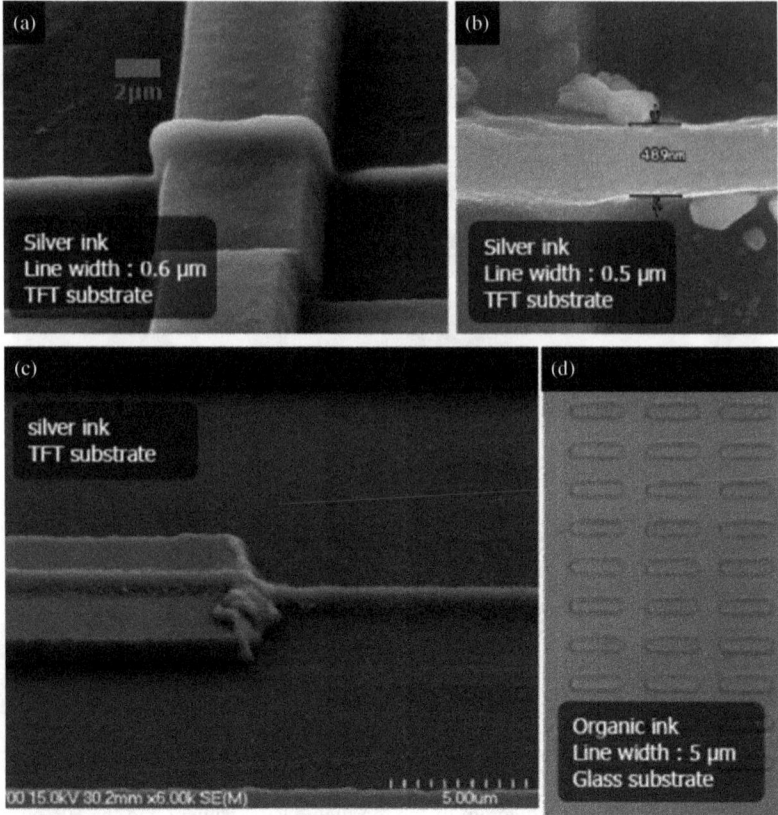

Figure 3.69.   (a) Silver ink deposited by the eNanoJet printer on thin-film transistor (TFT) substrate (0.6 μm line width); (b) Silver ink deposited on TFT substrate (0.5 μm line width); (c) Silver ink deposited on TFT substrate; and (d) Organic ink deposited by the eNanoJet printer on a glass substrate (5 μm line width). (Courtesy of Enjet, Inc.).

printing resolution of 0.1 μm and an accuracy of $\pm$ 2 μm. The MX2 is also equipped with a top-view vision camera for real-time inspection of print patterns and a flatness detection sensor to allow printing on non-flat surfaces (see **Figure 3.71**). Optional features, such as the optical table, cover frame and top fan filter unit (FFU) can be added on to the MX2 printer.

The modular nature of the printer can enable users to use various types of printhead modules and nozzles, depending on the requirements of the applications. Different combinations of the printhead module and the nozzle can directly affect the printer's dispensable ink viscosity range.

Figure 3.70.   MX2 multi-function/ purpose printer. (Courtesy of Enjet, Inc.).

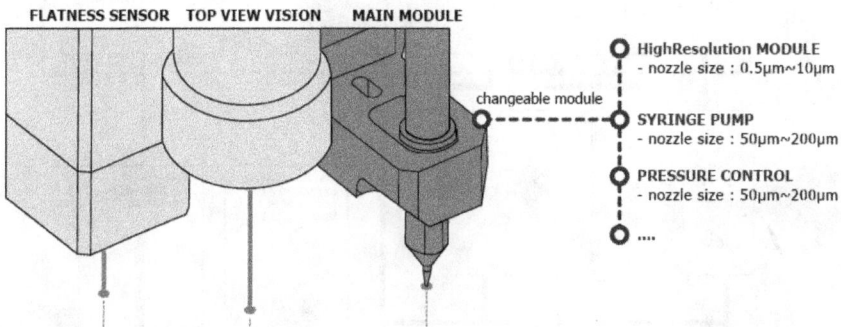

Figure 3.71.   Schematic diagram of various essential components of the MX2 printer. (Courtesy of Enjet, Inc.).

The selection guide for various MX2 printhead modules and nozzles are shown in **Figure 3.72**.

The syringe pump and pressure control printhead modules can dispense materials with viscosities ranging from 1–100,000 cPs, whereas the active valve and high-resolution (HR) printhead modules can dispense materials with viscosities ranging from 1–50,000 cPs. The nozzle sizes available for syringe pump and pressure control printhead modules range from 50–200 μm, while the nozzle sizes available for the HR module

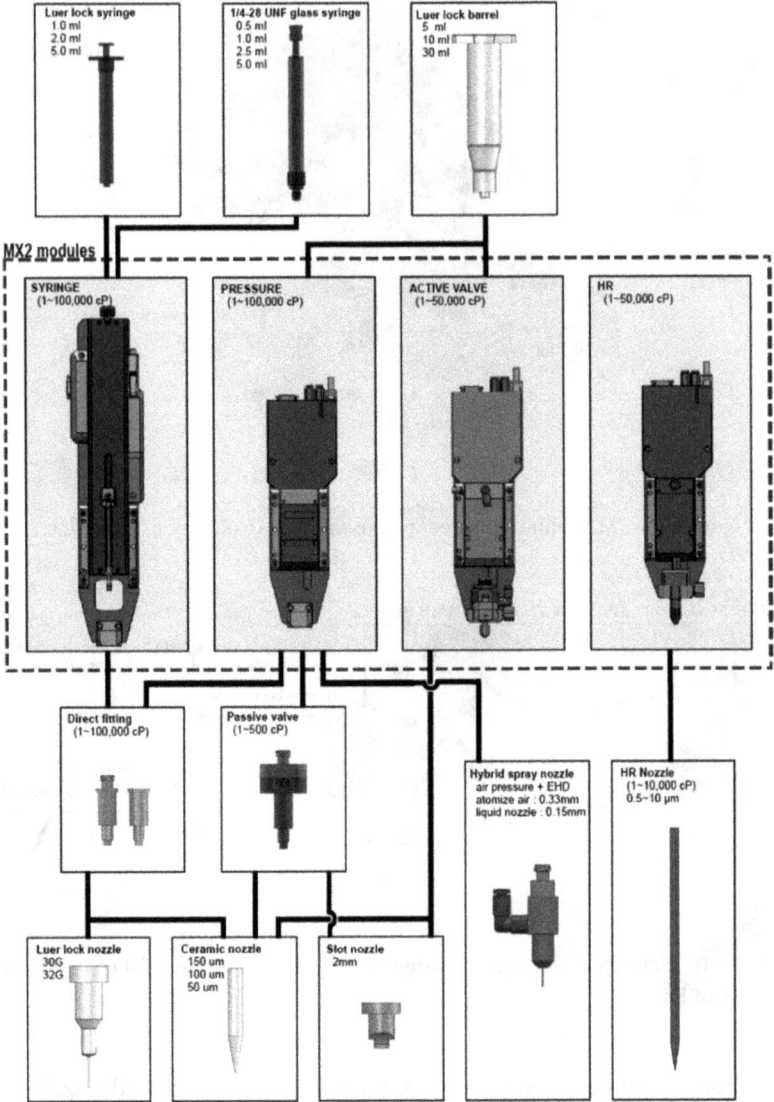

Figure 3.72.   MX2 printhead modules and nozzles selection guide. (Courtesy of Enjet, Inc.).

range from 0.5–10 μm. The nozzle auto-alignment feature in the printer ensures the printhead modules and nozzles are precisely aligned for optimal printing performances after each change. **Figure 3.73(a)** shows the Loctite-3318LV material deposited by the MX2 printer and achieved

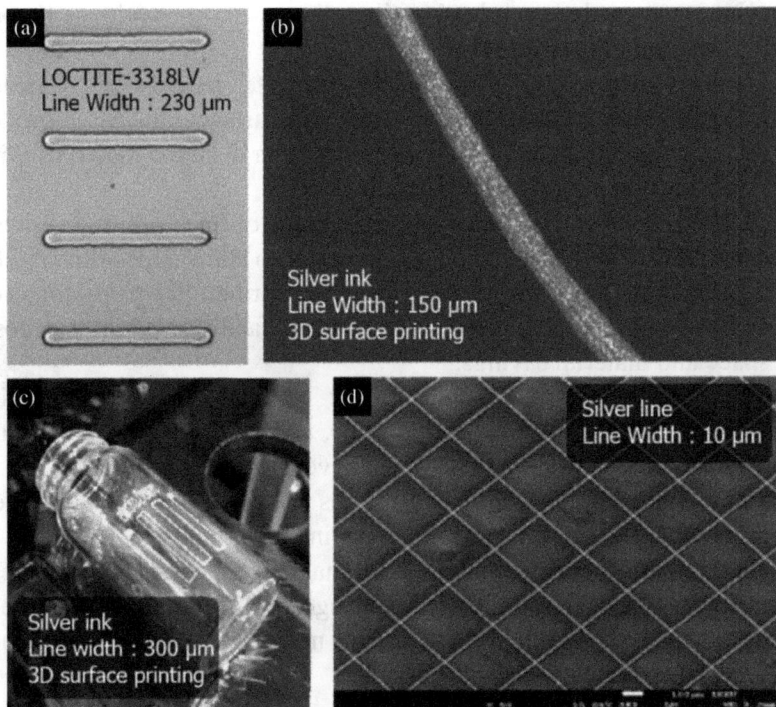

Figure 3.73. (a) Line dispensing of LOCTITE-3318LV with 230 μm line width; (b) 3D surface printing with silver ink (150 μm line width); (c) 3D surface printing of silver ink on a cylindrical glass bottle (300 μm line width); and (d) printed silver-mesh electrodes (10 μm line width). (Courtesy of Enjet, Inc.).

230 μm line width. The MX2 printer can print directly onto conformal surfaces with conductive silver ink (see **Figure 3.73(b) and (c)**). The MX2 printer can also fabricate silver-mesh electrodes directly on the substrate with 10 μm line width (see **Figure 3.73(d)**).

3.4.1.3 *Strengths and weaknesses*

The key advantages of Enjet systems are:

(a) *High resolution printing*: Enjet systems can achieve very high-resolution printing with high accuracy, in which structures with line width less than 1 μm can be achieved. For instance, the eNanoJet printer can produce conductive silver lines with line width as fine as 0.5 μm (see **Figure 3.69(b)**).

(b) *Wide viscosity range*: Enjet *i*EHD jet printer models can dispense materials with wide viscosities range. For instance, the MX2 printer with the syringe pump printhead module can dispense materials with viscosities ranging from 1–100,000 cPs.

(c) *Ability to print on conformal surfaces*: Some of Enjet printing systems are equipped with flatness detection sensor to allow deposition of inks directly onto non-flat surfaces. Hence, favouring full optimisations of available spaces in electronic devices to facilitate additional cost, space and material savings.

(d) *Real-time inspection of print patterns*: The shape or size of the print patterns can be evaluated in real-time, allowing adjustments on-the-fly automatically or manually to achieve specific requirements.

(e) *Quick changing of nozzles*: The nozzles in Enjet systems are designed to be replaced quickly to reduce machine downtime.

(f) *Smart signal (patent pending)*: The SmartSignal (patent pending) feature can predict the ink viscosity change with the analysis of inspection data, to decide the best signal for nozzle input to achieve highly repeatable dispensing.

The key weaknesses of Enjet's systems are:

(a) *Low throughput*: Most Enjet systems are single-nozzle type, and hence they may only deliver limited throughput. Nevertheless, Enjet is still amid testing several prototypes with multiple print nozzles.

(b) *Limited functional inks can be used*: Due to the small nozzle sizes, Enjet systems might not be able to dispense suspensions which contain large particles. Hence, only limited functional inks can be used in their systems. Nevertheless, Enjet is also actively developing their own particle-less conductive and functional inks that are compatible with their systems.

3.5 Other Micro-dispensing Technologies

This section discusses some of the other micro-dispensing technologies that are not categorised under the inkjet printing, aerosol-based printing, extrusion-based printing and electrohydrodynamic printing categories. Sonoplot, Inc. utilises the ultrasonic dispensing technology for material deposition, whereas nScrypt Inc. utilises a positive pressure pump with patented precision dispensing tip and high-precision computer-actuated valve for material deposition. Their printing technologies will be further discussed in this section.

3.5.1 Sonoplot, Inc.

3.5.1.1 Company

Founded in 2003, Sonoplot, Inc. is a manufacturer of plotting instruments and materials dispensing systems targeted at life sciences and electronics research applications. Sonoplot's patented ultrasonic dispensing technology allows a wide range of materials to be deposited, with viscosity up to 450 cps. Some of these materials include nanoparticle inks, carbon nanotubes, rapheme and polymers. Besides, Sonoplot's systems can also produce fine features of up to 10 to 20 microns. Sonoplot's headquarters is located at 3030 Laura Lane, Suite 120, Middleton, WI 53562, USA.

3.5.1.2 Products

Sonoplot has two main series of products: Microplotter® Proto and Microplotter® II. Both systems share identical software, but offer a different spectrum of feature size, feature types and deposition volumes.

(a) *Microplotter® Proto*
The Sonoplot Microplotter® Proto (see **Figure 3.74**) is an entry-level benchtop picolitre fluid dispensing system which allows non-contact deposition. The Microplotter® Proto can produce droplets and contiguous lines with feature size ranging from 20 μm–200 μm. A wide variety of inks with ink viscosities up to 450 cP can be used by this system. The Microplotter® Proto can achieve deposition variability as low as 10% and 3-axis positioning with 10 μm resolution. The Microplotter® Proto is also equipped with automatic surface height calibration and digital video capturing and recording functions. Its holding platen is interchangeable for different substrate sizes. The provided SonoGuide™ and SonoDraw™ software are used as automation control and CAD layout tools respectively.

Figure 3.74.   Sonoplot Microplotter® Proto benchtop picolitre fluid dispensing system. (Courtesy of Sonoplot, Inc.).

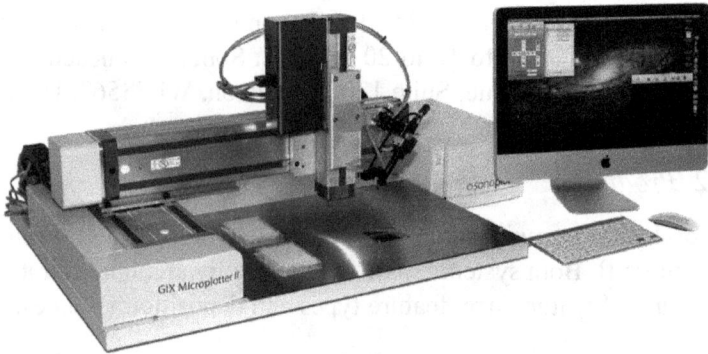

Figure 3.75.   SonoPlot Microplotter® II precision picolitre fluid dispensing system. (Courtesy of Sonoplot, Inc.).

(b) *Microplotter® II*

The Sonoplot Microplotter® II (see **Figure 3.75**) is a precision picolitre fluid dispensing system which allows non-contact deposition. The Microplotter® II can produce droplets, contiguous lines and arcs with feature size ranging from 5 μm–200 μm. A wide variety of inks with ink viscosities up to 450 cP can be used by this system. The Microplotter® II can achieve deposition variability as low as 10% and 3-axis positioning with 5 μm resolution. Similar to the Microplotter® Proto, the Microplotter® II is equipped with automatic surface height calibration and digital video

Table 3.19. Specifications of Microplotter® Proto and Microplotter® II. (Courtesy of Sonoplot, Inc.).

Model	Microplotter® Proto	Microplotter® II
Feature size	20 μm–200 μm	5 μm–200 μm
Feature types	Droplets and contiguous lines	Droplets and contiguous lines and arcs
Deposition volume	≥ 1 pL	≥ 0.6 pL
Deposition variability	As low as 10%	
Viscosity	≤ 450 cP	
Positioning	31 × 31 × 7 cm (X, Y, Z axes)	35 × 30 × 7 cm (X, Y, Z axes)
	10 μm resolution	5 μm resolution
Calibration	Automatic surface height calibration	
Camera	Digital video capture & recording	
Computer	Included iMac	
Software	SonoGuide control & SonoDraw CAD tools included	
Dimensions	58.4 × 59.7 × 61 cm	86.4 × 71.1 × 48.3 cm
Weight	30 kg	91 kg
Power	3.0 A for 100–120 V or 1.5 A for 220–240 V	

capturing and recording functions. Its holding platen is interchangeable for different substrate sizes. The provided SonoGuide™ and SonoDraw™ software are used as automation control and CAD layout tools respectively. Details of the Microplotter® Proto and Microplotter® II are summarised in Table 3.19.

3.5.1.3 *Working principle and process*

Sonoplot's Microplotter systems utilise the ultrasonic dispensing technology for dispensing small precise amount of fluidic materials onto the substrates. This subsection presents the working principle of the ultrasonic dispensing technology and the working process of the Microplotter system.

The ultrasonic dispensing technology is a method that employs ultrasonic pumping for fluid ejection. The fluid dispenser of Sonoplot's Microplotter system comprises a piezoelectric element attached to a hollow glass needle (see **Figure 3.76**). Ultrasonic vibrations are generated

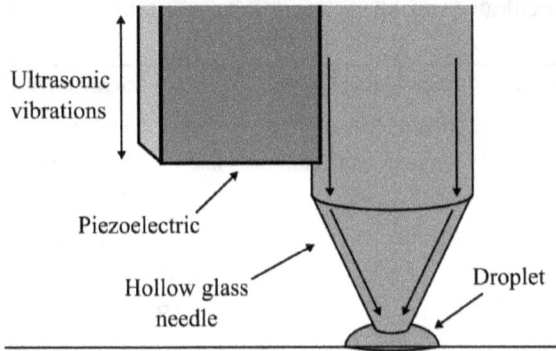

Figure 3.76.   Schematic diagram of the fluid dispenser which demonstrates fluid deposition via ultrasonic pumping. (Courtesy of Sonoplot, Inc.).

when alternating currents are passed through the piezoelectric element. A pumping action is created within the hollow glass needle when the vibration frequencies match the resonant frequencies of the dispenser. This electronically-controlled pumping action allows precise deposition of materials onto the substrate. A spraying action can also be created within the hollow glass needle when the vibration amplitudes are large enough. This spraying action enables quick materials change while preventing cross-contamination.

A substrate material is first loaded onto the Microplotter system's platen. A dispenser cartridge is then loaded and calibrated. The surface height is also calibrated automatically. The hollow glass needle with varying aperture sizes can be chosen to suit different application needs, in which smaller aperture sizes are for printing finer features. The hollow glass needle first lowers itself into the fluid well, and the ink is drawn up through capillary actions. **Figure 3.77** illustrates the process of the fluid dispenser drawing ink from the fluid well.

The dispenser is moved up to a safe distance above the surface and then moved to over the desired position for printing. The fluid-laden dispenser is recalibrated again for surface height until the surface has fluid contact. The pattern file is loaded into the system for printing, in which the printing process is fully automated.

There are two dispensing modes available for Sonoplot's Microplotter systems: (a) dispensing a spot of fluid on a surface; and (b) dispensing continuous lines or arcs.

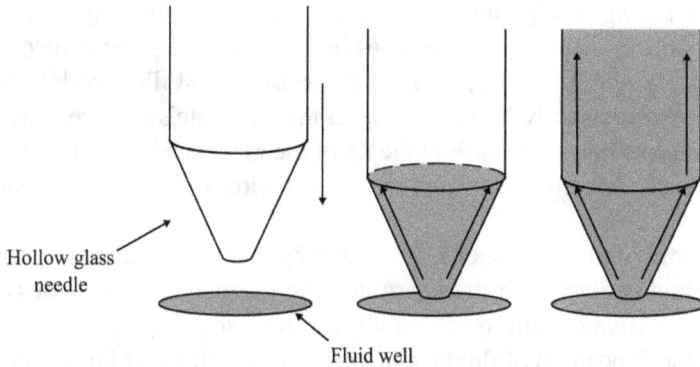

Figure 3.77. Schematic diagram of the fluid dispenser drawing ink from the fluid well. (Courtesy of Sonoplot, Inc.).

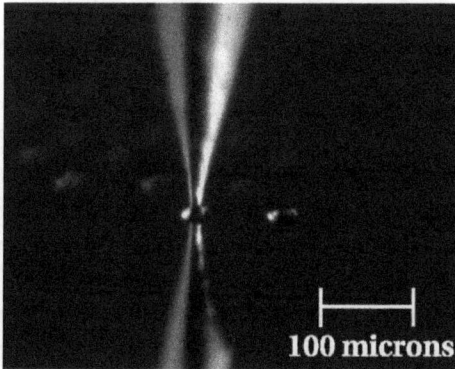

Figure 3.78. Dispensing a 50-micron diameter spot of fluid on a surface. (Courtesy of Sonoplot, Inc.).

(a) *Dispensing a spot of fluid on a surface*: The fluid dispenser is first filled with fluid and then subsequently brought close to the substrate's surface. The fluid meniscus bows outwards at the tip of the glass needle when the piezoelectric element is activated to generate the ultrasonic pumping action. A droplet will be touched off when the tip of the glass needle is brought near to the substrate's surface. The piezoelectric element is deactivated to stop the pumping action and the fluid dispenser is retracted up from the substrate. **Figure 3.78** shows the Sonoplot's Microplotter system dispensing a 50-micron diameter spot of fluid on a surface.

(b) *Dispensing continuous lines or arcs*: The Microplotter system is also capable of dispensing continuous lines and arcs. Similar to dispensing spots on the substrates, the fluid dispenser is first filled with fluid and then subsequently brought close to the substrate's surface. The fluid meniscus bows outwards at the tip of the glass needle when the piezo-electric element is activated to generate ultrasonic pumping actions.

The tip of the glass needle is brought near to the substrate's surface to make fluid contact with the substrate. The piezoelectric element is continuously activated still to maintain the ultrasonic pumping action, which allows the deposition of fluidic material in a smooth arc or line as the fluid dispenser moves on the substrate. The piezoelectric element is deactivated to stop the pumping action and the fluid dispenser is retracted up from the substrate when the arc or line is completed.

3.5.1.4 *Materials*

The ultrasonic liquid dispensing technology in Sonoplot's Microplotter systems can allow deposition of a wide variety of inks, including organic-solvent-based mixtures and aqueous solutions. The recommended viscosity that can be dispensed by these systems ranges between 0–450 cps. Suspensions with particle size smaller than 15 microns can also be dispensed. Some of the printable inks include nanometallic inks (silver, gold, copper), conductive polymers, insulating polymers, graphene, and carbon nanotubes. SonoPlot's ink partners include Applied Nanotech, Cabot, Creative Materials, Dupont, InkTec, Johnson Matthey, Novacentrix, Promethan Particles, UT Dots, Sigma-Aldrich and Xerox.

3.5.1.5 *Strengths and weaknesses*

The strengths of Sonoplot's Microplotter systems are:

(a) *Fine feature printing*: The SonoPlot Microplotter® II can produce droplets, contiguous lines and arcs with feature size ranging from 5 μm to 200 μm. Hence, this printing system is very suitable for fine feature printing applications.
(b) *High accuracy dosing*: The SonoPlot Microplotter® II can deposit volume as small as 0.6 pL and the deposition variability is as small as 10%.

(c) *Wide range of materials*: Sonoplot systems can dispense a wide range of materials, with viscosity ranging from 0–450 cP, on various types of substrates. Suspensions with particle size smaller than 15 microns can also be dispensed.

(d) *Wide variety of substrates*: Sonoplot systems can deposit materials on any substrates with flat planar surfaces.

(e) *Material saving*: As discussed previously, ink is drawn from the fluid well into the hollow glass needle through capillary actions. Hence, only small amount of ink is required to fill the fluid well and at the same time, allowing material saving [104].

By contrast, the weaknesses of Sonoplot's Microplotter systems are:

(a) *Most planar printing*: Sonoplot systems are mostly used for printing on planar substrates. However, they can print on non-planar surfaces, if the surfaces are known and not too abrupt.

(b) *Ink evaporation*: The ink is drawn from an open fluid well and ink dispensing is conducted through an open capillary. Therefore, volatile agents in the ink may evaporate easily and cause changes in the ink concentration during the printing process [105]. Nevertheless, Sonoplot systems do have humidification enclosures that can help prevent evaporation.

3.5.1.6 *Applications and examples*

(a) *Printing conductive traces*: Conductive traces and patterns can be printed directly on a wide variety of substrates. **Figure 3.79**

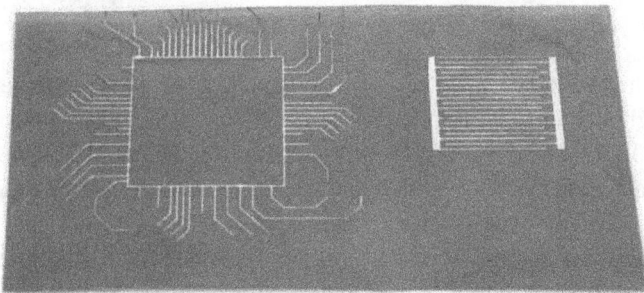

Figure 3.79.   Printed conductive traces on a polyimide substrate. (Courtesy of Sonoplot, Inc.).

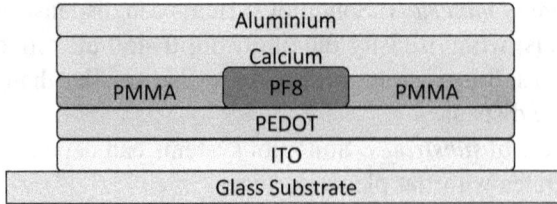

Figure 3.80.   Schematic diagram of the cross-sectional view of the LED structure [104].

shows the conductive traces printed on a polyimide substrate. Silver nanoparticle ink from Cabot was used for printing these conductive traces. The printed samples were sintered at 100°C for one hour to enhance their electrical conductivity.

(b) *Polymer-based LED*: Larson *et al.* [104] fabricated a polymer-based LED on a glass substrate, which comprised of layers that include an indium tin oxide (ITO) electrode, a poly(3,4-ethylenedioxythiophene) (PEDOT) layer, a poly(methylmethacrylate) (PMMA) mask layer and an aluminium-covered calcium top electrode (see **Figure 3.80**). The Sonoplot Microplotter was used to deposit PEDOT directly onto the ITO layer, and then deposit PMMA onto selected areas on the PEDOT layer. The PMMA layer served as insulation masks, in which a small exposed area was left for depositing a blue light-emitting LED polymer material, poly(9,9′-dioctylfluorine) (PF8). A bright blue light was given off with a turn-on voltage of 6V.

3.5.2 *nScrypt, Inc.*

3.5.2.1 *Company*

nScrypt Inc., founded in 2002, is a manufacturer of micro-dispensing and direct digital manufacturing equipment for industrial applications in 3D printing, life science and electronics packaging industries. Their equipment is also commonly used by academic research institutes, defence institutes, national labs and both large and small companies. nScrypt provides various solutions for precision micro-dispensing, 3D printing, electronics and packaging, direct digital manufacturing and life sciences. nScrypt's micro-dispensing system won the Frost and Sullivan product leadership award in 2004. nScrypt's Hybrid 3D Printer also won the top honours at the RAPID's Innovation Auditions competition in 2016. The headquarters of nScrypt Inc. is located at 12151 Research Parkway, Suite 150, Orlando, FL 32826, USA.

3.5.2.2 *Products*

nScrypt has five different products series: 3Dn series, 3Dn-DDM series, SVA series, nRugged, and BioAssembly Tools (BAT) series. The 3Dn series are user-configurable platforms which allow customisations of modular tool head options for individual customer's needs and applications. The 3Dn-DDM series are pre-configured direct digital manufacturing (DDM) platforms for fabricating electronics packaging structures with functionally active circuits and electronics. The SVA series are pre-configured systems specially designed to fill vias and micro-dispense solder pastes and adhesive materials. The nRugged system is a ruggedised digital manufacturing platform designed for printing in harsh conditions and it can be configurable for 3D manufacturing or bioprinting. The BAT series are specially configured systems for bioprinting applications.

nScrypt's systems have high precision linear motor platforms to allow fast and precise positioning of the print heads. The synchronous control of all aspects of motion and printing allows exceptional dispensing and 3D printing results. nScrypt's systems are commonly configured with five process heads which can print, mill, drill, polish and pick-and-place. The systems are equipped with machine vision for full automation, Z scanning and *in-situ* scanning of each layer for a detailed report of layer-by-layer quality. nScrypt's systems can achieve printing speeds up to 1 m/s, in which the printing speed is highly dependent on the material's properties, part size, feature size and the level of detail.

(a) *3Dn DDM Series*
The 3Dn-DDM series (3Dn-DDM Tabletop, 3Dn-DDM and 3Dn-DDM-PF), also known as the "Factory in a Tool" (FiT), are pre-configured direct digital manufacturing (DDM) platforms equipped with micro-dispensing, micro-milling, 3D printing (material extrusion), and pick-and-place functions.

The 3Dn-DDM series can fabricate 3D printed structures together with printed electronics, and pick-and-place electrical components without tool changes. These systems can be customised further with additional post-processing functions such as sintering, heating, laser processing, ultraviolet (UV) curing and post-process inspection. The 3Dn-DDM series are also capable of machine vision and scanning for conformal printing and Z-tracking on a single platform. The 3Dn-DDM series are also capable of micro-dispensing a wide range of conductive metallic and dielectric inks, and 3D printing (by material extrusion) thermoplastics and composites.

The 3Dn-DDM Tabletop model (see **Figure 3.81**) has a X/Y travel range of 300×150 mm and a Z travel range of 100 mm, both with 0.5 μm resolution. The 3Dn-DDM model (see **Figure 3.82**) has a X/Y travel range of 300×300 mm with 100 nm resolution, and a Z travel range of 150 mm with 0.5 μm resolution. The 3Dn-DDM-PF model (see **Figure 3.83**) has a X/Y travel range of 500×500 mm with 10 nm resolution, and a Z travel range of 150 mm with 0.5 μm resolution. The3Dn-DDM Tabletop model is equipped with a SmartPump™100 precision micro-dispensing tool head, while the 3Dn-DDM and 3Dn-DDM-PF models are equipped with two SmartPump™100 precision micro-dispensing tool heads. All the three models come with the *nFD*™ material extrusion tool head, *nPnP* pick-and-place tool head, *nMill*™ milling and polishing tool head, *Real-Time Process View*, *nVision AutoCal Pit*, *Auto Clean Pit* and *nScan* features. The technical specifications of 3Dn-DDM

Figure 3.81.   3Dn-DDM Tabletop. (Courtesy of nScrypt, Inc.).

Figure 3.82. 3Dn-DDM. (Courtesy of nScrypt, Inc.).

Figure 3.83. 3Dn-DDM-PF. (Courtesy of nScrypt, Inc.).

Table 3.20.   Specifications of 3Dn-DDM Tabletop, 3Dn-DDM and 3Dn-DDM-PF. (Courtesy of nScrypt, Inc.).

	3Dn-DDM Tabletop	3Dn-DDM	3Dn-DDM-PF
X/Y accuracy	± 10 μm (± 5μm*)	± 5 μm	± 1.5 μm
X/Y bidirectional repeatability	± 1 μm	± 2 μm	± 0.5 μm
X/Y maximum acceleration	0.5g (no load)	2g (no load)	5g (no load)
X/Y maximum speed	300 mm/s	500 mm/s	1 m/s
X/Y travel range	300 × 150 mm	300 × 300 mm	500 × 500 mm
X/Y resolution	0.5 μm	100 nm	10 nm
Z accuracy	± 6 μm (± 5μm*)	± 5 μm	
Z bidirectional repeatability	± 1 μm	± 0.7 μm	
Z maximum speed	50 mm/s	100 mm/s	
Z travel range	100 mm	150 mm	
Z resolution	0.5 μm		
Dimensions	97 × 92 × 122 cm	112 × 94 × 176 cm	138 × 130 × 213 cm
Weight	185 kg	795 kg	2045 kg
Features	SmartPump™100	2—SmartPump™100 nFD™ nPnP nMill™ Real-Time Process View nVision AutoCal Pit Auto Clean Pit nScan	
Options	Overhead HEPA Filtration, UV Spot Curing Source, Heated Bed, Vacuum Chuck, Rotary Stage, Post-Process Inspection, Laser Integration		

Tabletop, 3Dn-DDM and 3Dn-DDM-PF are summarised in Table 3.20. (Courtesy of nScrypt, Inc.)

(b) *SVA series*
The SVA series (SVA Dev and SVA 3000) are pre-configured systems capable of filling vias and precise micro-dispensing of adhesive

Figure 3.84.   SVA Dev. (Courtesy of nScrypt, Inc.).

materials and solder pastes. These systems can be customised with various functional head options such as micro-dispensing, 3D printing (material extrusion), laser processing, and pick-and-place features. Z tracking for conformal printing and 3D mapping of pre-prints and post-prints can also be performed.

The SVA Dev model (see **Figure 3.84**) has a X/Y travel range of 300 × 150 mm and a Z travel range of 100 mm, both with 0.5 μm resolution. The SVA Dev model is equipped with the SmartPump™100 precision micro-dispensing tool head, and comes with the *Real-Time Process View, nVision AutoCal Pit* and *Auto Clean Pit* features.

The SVA 3000 model (see **Figure 3.85**) is a gantry system equipped with the SmartPump™100 precision micro-dispensing tool head, and also comes with the *Real-Time Process View, nVision AutoCal Pit* and *Auto Clean Pit* features. The SVA 3000 has a X/Y travel range of 300 × 300 mm with 100 nm resolution. and a Z travel range of 150 mm with 0.5 μm resolution. The SVA 3000 can print various fine features (for instance, 30 μm wide conductive traces, 50 μm diameter solder dots, and 20 μm adhesives and dielectrics) and fill 25 μm vias. The SVA 3000 also can dispense up to 3000 solder dots, adhesive dots or via fillings per hour. The technical specifications of the SVA Dev and SVA 3000 are summarised in Table 3.21.

Figure 3.85.   SVA 3000. (Courtesy of nScrypt, Inc.).

Table 3.21.   Specifications of SVA DEV and SVA 3000. (Courtesy of nScrypt, Inc.).

	SVA DEV	**SVA 3000**
X/Y accuracy	± 10 μm (± 5μm)	± 5 μm
X/Y bidirectional repeatability	± 1 μm	± 2 μm
X/Y maximum acceleration	0.5g (no load)	2g (no load)
X/Y maximum speed	300 mm/s	500 mm/s
X/Y travel range	300 × 150 mm	300 × 300 mm
X/Y resolution	0.5 μm	100 nm
Z accuracy	± 6 μm (± 5μm)	± 5 μm
Z bidirectional repeatability	± 1 μm	± 0.7 μm
Z maximum speed	50 mm/s	100 mm/s
Z travel range	100 mm	150 mm
Z resolution	0.5 μm	
Dimensions	97 × 92 × 122 cm	112 × 94 × 176 cm
Weight	185 kg	795 kg
Features	SmartPump™100	Gantry system with SmartPump™100 Tray handling
	Real-time process view	
	nVision AutoCal Pit	
	Auto Clean Pit	
Additional options	Overhead HEPA filtration, post-process inspection	

3.5.2.3 *Working principles and process*

nScrypt's SmartPump™, a positive pressure pump with patented precision dispensing tip and high-precision computer-actuated valve, is the core technology that allows precise micro-dispensing of materials at the pico-litre level. The SmartPump™ controls the material dispensing process with a valve mechanism and air pressure, in which the amount of material dispensed is highly dependent on the applied pressure and dispensing duration [106].

The SmartPump™ can maintain material flow consistently and dispense a wide range of materials with viscosities ranging from 1 – 1,000,000 cP. The dispensing tip is detachable from the SmartPump™. Various dispensing tips are available in different inner diameter sizes, with diameters ranging from 10–125 μm. The dispensing tip diameter can affect the line width of the extruded tracks and the material flow rate [106]. **Figure 3.86** shows the schematic diagram of nScrypt's SmartPump™ valve assembly with the syringe attached.

The operating mechanisms of the SmartPump™ can be described in the steps below [106,107]. The syringe is first filled with ink and then attached to the valve body (see **Figure 3.87(a)**). Pressurised air is connected to the plunger, in which the air pressure can be controlled to apply

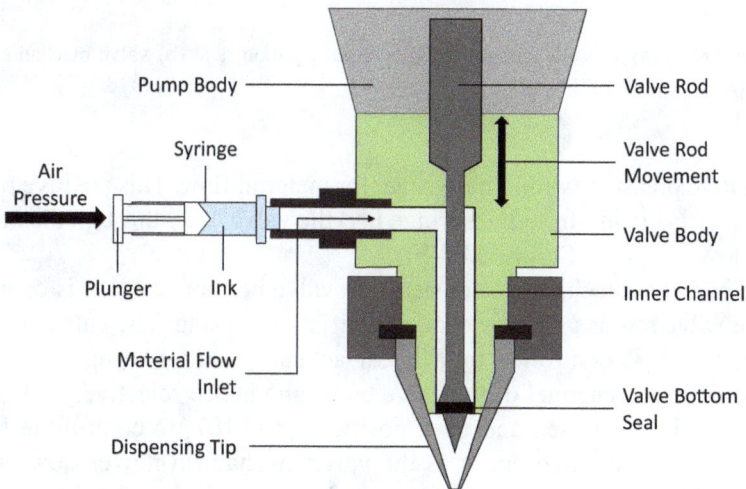

Figure 3.86.   nScrypt SmartPump™ valve assembly [106,107].

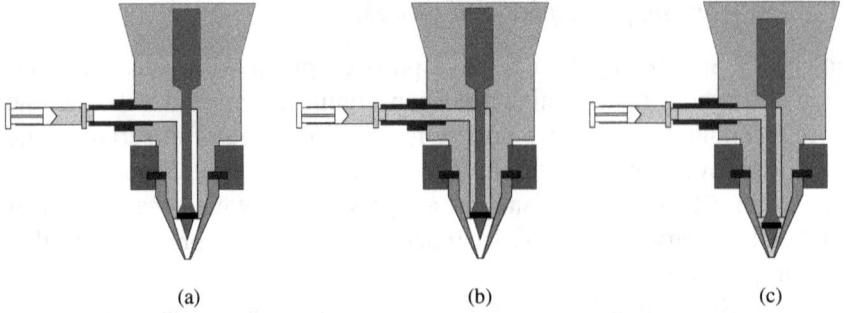

Figure 3.87.   Operating mechanisms of SmartPump™: (a) syringe is filled with ink and attached to the valve body; (b) positive pressure causing ink to flow into the valve body; and (c) the valve is opened and the ink flows into the dispensing tip [106].

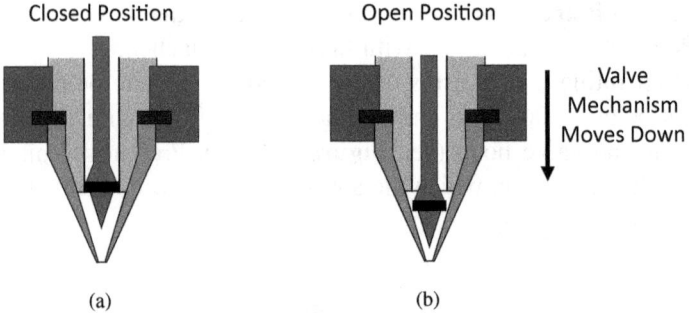

Figure 3.88.   (a) Valve mechanism at the closed position; and (b) valve mechanism at open position [106,107].

a positive pressure within the syringe for material flow. This positive pressure expels the ink from the syringe into the valve body through the material flow inlet (see **Figure 3.87(b)**).

The valve mechanism, consists of a valve bottom seal that is coupled to the valve rod, is used for precise material dispensing. The valve mechanism, which is controlled by a linear actuator, can move up and down along the inner channel of the valve body, and hence selectively toggling the valve in the closed and open positions [106,107] to control the flow of material. The valve opens as the valve mechanism moves down (see **Figure 3.88(b)**), and the valve closes when the valve mechanism moves back up (see **Figure 3.88(a)**).

When the dispensing process is initiated, the valve mechanism moves down. The valve is at the open position to allow material flow from the inner channel into the dispensing tip (see **Figure 3.87(c)**). The ink can only flow out of the dispensing tip if the applied pressure on the syringe surpasses a certain pressure threshold level which is highly dependent on the individual ink's rheological properties [106,107]. The material flow rates can be further controlled by adjusting the applied pressure on the syringe.

To end the dispensing process, the valve mechanism moves back up to the closed position. The valve bottom seal now creates a hermetic seal to the inner channel and stops the ink from flowing out. A small negative pressure is also created within the dispensing tip by this upward movement of the valve mechanism, resulting in the reverse of material flow where remaining materials that are sticking outside are drawn back in the dispensing tip [108,109]. In addition, this negative pressure phenomenon can eliminate material tailings and help in better precision control of material flow, especially at the start and end of every print cycle [106,107]. Hence, material build-up at the tip can be reduced and helps to keep the tip clean for the next dispensing operation.

Any patterns can be printed on the *XY* plane, if necessary conformally in the *Z*-axis, by synchronising the material dispensing process with the *XYZ*-motion control [107]. The users can also vary these process parameters for better control of the material dispensing process: air pressure, valve position, valve opening speed, dispensing height, nozzle diameter, motion delay, feed rate and ink viscosity [107].

3.5.2.4 *Materials*

The SmartPump™ precision micro-dispensing tool head can dispense any type of inks or pastes with a viscosity up to 1,000,000 cPs. There are more than ten thousand commercially available materials that are compatible with the SmartPump™, with materials of varying viscosity such as liquid inks, conductive materials, dielectric materials, pastes and epoxies. Standard conductive materials such as sliver flake or copper nanopaste can be printed with exceptional precision.

The nFD™ material extrusion tool head can extrude any standard filament of 1.75 mm diameter and melting point of up to 400°C. ABS, PLA, polyether ether ketone (PEEK), Polyetherketoneketone (PEKK) and Ultem® (polyetherimide) are some of the materials that can be extruded by the nFD™ material extrusion tool head.

3.5.2.5 *Strengths and weaknesses*

The key strengths of nScrypt's systems are:

(a) *Wide range of materials and substrates*: nScrypt's SmartPump™ precision micro-dispensing tool head can dispense a wide range of materials, with viscosity up to 1,000,000 cP, on various types of substrates. The SmartPump™ can dispense more than 10,000 commercially available materials.

(b) *Fabricating functional embedded electronics*: nScrypt's systems come with micro-dispensing, material extrusion, micro-milling and pick-and-place tool heads, which can allow the fabrication of fully-assembled functional embedded electronics in three-dimensional (3D) printed complex structures all in one build with automated tools changing. This is particularly advantageous for customisation and integration of electronics into product designs.

(c) *Conformal printing*: Conductive patterns can be printed on conformal surfaces by nScrypt systems. **Figure 3.89** shows the conductive patterns printed on a roughly surfaced Kevlar helmet.

(d) *Fine features printing*: Line widths as fine as 20 microns are printable with nScrypt's machines and are also highly dependent on the size of

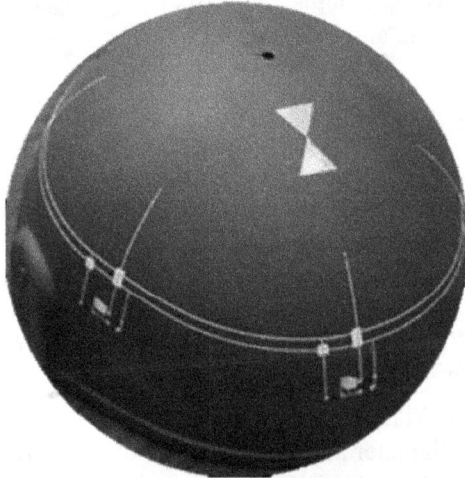

Figure 3.89.   Conductive patterns printed on a roughly surfaced Kevlar helmet. (Courtesy of nScrypt, Inc.).

the dispensing tip used. Various dispensing tips are available in different inner diameter sizes, with diameters ranging from 10–125 μm. The inner diameter of the dispensing tip should generally be at least ten times larger than the materials' particle size for clog-free printing. The key weaknesses of nScrypt's systems are:

(a) *Expensive*: Since nScrypt's systems are heavily focused on high precision and repeatability, their systems are typically more expensive than other non-industrial tools.
(b) *Optimisation of printing parameters needed*: nScrypt's systems can dispense a wide range of materials. Hence, printing parameters for different materials need to be individually optimised.

3.5.2.6 *Applications*

(a) *Printed circuit structures*: Printed circuit structures (PCS) are 3D objects with printed electronics permeating throughout the entire structure, where sensing and other electrical functions can be added in the 3D objects (see **Figure 3.90**). Equipped with material extrusion, micro-dispensing, micro-milling and pick-and-place tool heads, nScrypt 3Dn-DDM Series systems can print packaging structures with fused filament fabrication, then conformally deposit conductive and dielectric materials on them, and pick-and-place electrical components without tool changes.

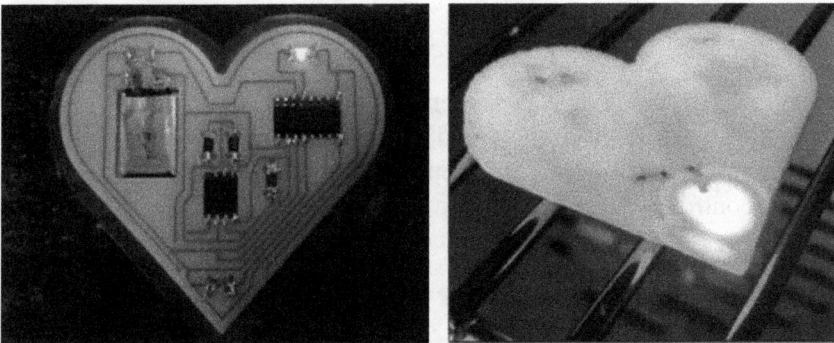

Figure 3.90.   PCS with embedded electronics. (Courtesy of nScrypt, Inc.).

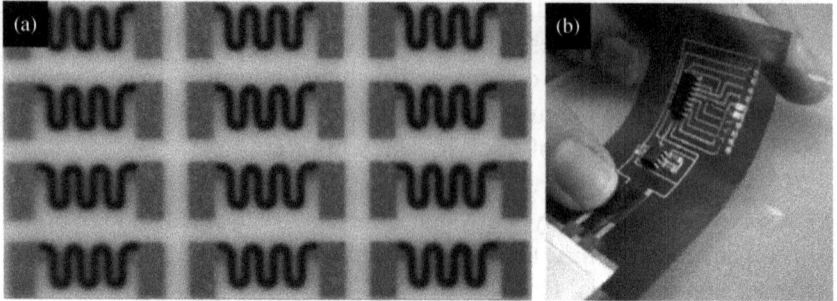

Figure 3.91.   (a) Printed resistors; and (b) printed circuit using lumped elements. (Courtesy of nScrypt, Inc.).

(b) *3D printed electronics and vias filling*: Conductive patterns and electronic devices (such as resistors (see **Figure 3.91(a)**), interconnectors, capacitors, inductors, batteries, sensors and electroluminescent lighting) can be printed directly onto flat or curved substrates which can either be rigid or flexible. **Figure 3.91(b)** shows a printed circuit using lumped elements. Vias as small as 50 microns can also be filled with nScrypt's SVA Series platforms for manufacturing applications.

(c) *Printed antennas*: nScrypt's printing platforms can also print antennas patterns directly onto conformal and complex 3-dimensional surfaces, which include applications such as smartphone antennas. As nScrypt's SmartPump™ micro-dispensing tool head can dispense a wide range of materials, users have a wide choice in selecting suitable materials to obtain the desired radiofrequency (RF) performances for their applications.

3.5.2.7 *Research and development*

nScrypt is advancing its PCS and printed biology capabilities for the next generation printed electronics and regenerative medicine, by allowing wide material range in material deposition and using the most advanced motion control and printing.

References

[1] Lu, B., Lan, H. and Liu, H. (2018). Additive manufacturing frontier: 3D printing electronics, *Opto-Electron. Adv.*, 1, p. 170004.

[2] Tan, H. W., Tran, T. and Chua, C. K. (2016). A review of printed passive electronic components through fully additive manufacturing methods, *Virtual Phys. Prototyp.*, 11, pp. 271–288.

[3] Tan, H. W., Saengchairat, N., Goh, G. L., An, J., Chua, C. K. and Tran, T. (2020). Induction sintering of silver nanoparticle inks on polyimide substrates, *Adv. Mater. Technol.*, 5, p. 1900897.

[4] Tan, H. W., An, J., Chua, C. K. and Tran, T. (2019). Metallic nanoparticle inks for 3D printing of electronics, *Adv. Electron. Mater.*, 5, p. 1800831.

[5] Ian M., H. and Graham D., M. (2012). *Inkjet Technology for Digital Fabrication* (John Wiley & Sons Ltd, United Kingdom).

[6] Fuh, J. (2013). *Handbook of Manufacturing Engineering and Technology*, eds. Andrew Nee, Micro- and Bio-Rapid Prototyping Using Drop-on-Demand 3D Printing (Springer London, London) pp. 1–15.

[7] Junfeng, M., Lovell, M. R. and Mickle, M. H. (2005). Formulation and processing of novel conductive solution inks in continuous inkjet printing of 3-D electric circuits, *IEEE Trans. Electron. Packag. Manuf.*, 28, pp. 265–273.

[8] Tai, J., Gan, H. Y., Liang, Y. N. and Lok, B. K. (2008). Control of Droplet Formation in Inkjet Printing Using Ohnesorge Number Category: Materials and Processes, *presented at the 10th Electronics Packaging Technology Conference*, Singapore, Singapore.

[9] Li, D. (2008). *Encyclopedia of Microfluidics and Nanofluidics*, eds. Dongqing Li, Ohnesorge Number (Springer US, Boston, MA) pp. 1513–1513.

[10] Rapp, B. E. (2017). *Microfluidics: Modelling, Mechanics and Mathematics*, eds. Bastian E. Rapp, Chapter 9: Fluids (Elsevier, Oxford) pp. 243–263.

[11] McKinley, G. H. and Renardy, M. (2011). Wolfgang von Ohnesorge, *Physics of Fluids*, 23, p. 127101.

[12] Chuang, M. Y. (2017). *Inkjet Printing of Ag Nanoparticles using Dimatix Inkjet Printer, No. 2* (Technical Report, University of Pennsylvania).

[13] Brünahl, J. and Grishin, A. M. (2002). Piezoelectric shear mode drop-on-demand inkjet actuator, *Sens. Actuators, A*, 101, pp. 371–382.

[14] Eslamian, M. and Ashgriz, N. (2011). *Handbook of Atomization and Sprays: Theory and Applications*, eds. Nasser Ashgriz, Drop-on-Demand Drop Generators (Springer US, Boston, MA) pp. 581–601.

[15] Haque, R. I., Vié, R., Germainy, M., Valbin, L., Benaben, P. and Boddaert, X. (2015). Inkjet printing of high molecular weight PVDF-TrFE for flexible electronics, *Flexible Printed Electron.*, 1, p. 015001.

[16] Castro, H. F., Correia, V., Sowade, E., Mitra, K. Y., Rocha, J. G., Baumann, R. R. and Lanceros-Méndez, S. (2016). All-inkjet-printed low-pass filters with adjustable cutoff frequency consisting of resistors, inductors and transistors for sensor applications, *Org. Electron.*, 38, pp. 205–212.

[17] Ali, S., Khan, S. and Bermak, A. (2019). Inkjet-printed human body temperature sensor for wearable electronics, *IEEE Access*, 7, pp. 163981–163987.

[18] Konica Minolta Inc. KM512 Series. Retrieved from https://www. konicaminolta.com/inkjet/inkjethead/512/index.html.

[19] Konica Minolta Inc. Technology Overview. Retrieved from https://www. konicaminolta.com/inkjet/technology/technology.html.

[20] Beurer, G. and Kretschmer, J. (1997). Function and performance of a shear mode piezo printhead, *presented at the Proc. IS&T's NIP,*

[21] Takeuchi, Y., Takeuchi, H., Komatsu, K. and Nishi, S. (2005). Improvement of drive energy efficiency in a shear mode piezo-inkjet head, *Hp Company Report.*

[22] Paolella, A. C., Silva-Saez, D., Kozlovski, D. and Even, R. (2019). 3-D Printed RF Amplifier for Wireless Systems, *presented at the IEEE Radio and Wireless Symposium (RWS)*, Orlando, FL, USA.

[23] Fried, S. (2017). 3D printing technologies for electronics, *NIHON GAZO GAKKAISHI (Journal of the Imaging Society of Japan)*, 56, pp. 617–620.

[24] Fima, S. and Elimelech, H. (2019) Double-sided and multilayered printed circuit board fabrication using inkjet printing. U.S. Patent Application 15/527,038, USA.

[25] Choi, C. H., Lin, L. Y., Cheng, C. C. and Chang, C.-H. (2015). Printed oxide thin film transistors: A mini review, *ECS J. Solid State Sci. Technol.*, 4, pp. P3044–P3051.

[26] Kholghi Eshkalak, S., Chinnappan, A., Jayathilaka, W. A. D. M., Khatibzadeh, M., Kowsari, E. and Ramakrishna, S. (2017). A review on inkjet printing of CNT composites for smart applications, *Appl. Mater. Today*, 9, pp. 372–386.

[27] Jung, S. (2011). *Fluid Characterisation and Drop Impact in Inkjet Printing for Organic Semiconductor Devices* (Doctoral Dissertation, University of Cambridge).

[28] Tahernia, M., Mohammadifar, M., Hassett, D. J. and Choi, S. (2019). A fully disposable 64-well papertronic sensing array for screening electro-active microorganisms, *Nano Energy*, 65, p. 104026.

[29] Akhatov, I., Hoey, J., Swenson, O. and Schulz, D. (2008). Aerosol flow through a long micro-capillary: Collimated aerosol beam, *Microfluid. Nanofluid.*, 5, pp. 215–224.

[30] Hoey, J. M., Lutfurakhmanov, A., Schulz, D. L. and Akhatov, I. S. (2012). A review on aerosol-based direct-write and its applications for microelectronics, *J. Nanotechnol.*, 2012, pp. 22.

[31] Hedges, M. and Marin, A. B. (2012). 3D Aerosol Jet® Printing-Adding Electronics Functionality to RP/RM, *presented at the DDMC 2012 Conference*, Berlin, Germany.

[32] Landgraf, M., Reitelshofer, S., Franke, J. and Hedges, M. (2013). Aerosol jet printing and lightweight power electronics for dielectric elastomer actuators, *presented at the 3rd International Electric Drives Production Conference (EDPC)*, Nuremberg. Germany.

[33] Fan, C., Pavlidis, S., Papapolymerou, J., Yung Hang, C., Kan, W., Zhang, C. and Ben, W. (2014). Aerosol jet printing for 3-D multilayer passive microwave circuitry, *presented at the 44th European Microwave Conference (EuMC)*, Rome, Italy.

[34] Wang, X., Kruis, F. E. and McMurry, P. H. (2005). Aerodynamic focusing of nanoparticles: I. guidelines for designing aerodynamic lenses for nanoparticles, *Aerosol Sci. Technol.*, 39, pp. 611–623.

[35] Optomec Inc. (2014). Aerosol Jet® 300 Series Systems. Retrieved from http://www.optomec.com/wp-content/uploads/2014/04/AJ-300-Datasheet_Web.pdf.

[36] Optomec Inc. (n.d.). Aerosol Jet 5X Systems. Retrieved from http://www.optomec.com/printed-electronics/aerosol-jet-printers/aerosol-jet-5x-system/.

[37] Goth, C., Putzo, S. and Franke, J. (2011). Aerosol jet printing on rapid prototyping materials for fine pitch electronic applications, *presented at the IEEE 61st Electronic Components and Technology Conference (ECTC)*, Lake Buena Vista, Florida, USA.

[38] King, B. H., O'Reilly, M. J. and Barnes, S. M. (2009). Characterizing aerosol jet multi-nozzle process parameters for non-contact front side metallization of silicon solar cells, *presented at the 34th IEEE Photovoltaic Specialists Conference (PVSC)*, Philadelphia, Pennsylvania, USA.

[39] Paulsen, J. A., Renn, M., Christenson, K. and Plourde, R. (2012). Printing conformal electronics on 3D structures with aerosol jet technology, *presented at the Future of Instrumentation International Workshop (FIIW)*, Gatlinburg, Tennessee, USA.

[40] Renn, M. J. (2014). Aerosol-Jet Printed Thin Film Transistors *Optomec Aerosol Jet White Paper*.

[41] Tan, H. W., Tran, T. and Chua, C. K. (2016). Aerosol jet printed strain gauge, *presented at the IJIE 2016: The 18th International Conference on Industrial Engineering*, Seoul, South Korea.

[42] Thompson, B. and Yoon, H. S. (2013). Aerosol-printed strain sensor using PEDOT: PSS, *IEEE Sens. J.*, 13, pp. 4256–4263.

[43] Zhao, D., Liu, T., Park, J. G., Zhang, M., Chen, J.-M. and Wang, B. (2012). Conductivity enhancement of aerosol-jet printed electronics by using silver nanoparticles ink with carbon nanotubes, *Microelectron. Eng.*, 96, pp. 71–75.

[44] Fresno, D. D. (2013). Revolutionary "Smart Wing" Created for UAV Model Demonstrates Groundbreaking Technology. Retrieved from https://

www.optomec.com/revolutionary-smart-wing-created-for-uav-model-demonstrates-groundbreaking-technology/.

[45] Salary, R., Lombardi, J. P., Samie Tootooni, M., Donovan, R., Rao, P. K., Borgesen, P. and Poliks, M. D. (2016). Computational fluid dynamics modeling and online monitoring of aerosol jet printing process, *J. Eng. Ind.*, 139, pp. 021015–021021.

[46] Ethan, B. S. (2018). Principles of aerosol jet printing, *Flexible Printed Electron.*, 3, p. 035002.

[47] Chua, C. K. and Leong, K. F. (2017). *3D Printing and Additive Manufacturing — Principles and Applications*, 5th edn. (World Scientific Publishing, Singapore).

[48] Agarwala, S., Goh, G. L. and Yeong, W. Y. (2017). Optimizing aerosol jet printing process of silver ink for printed electronics, *IOP Conf. Ser.: Mater. Sci. Eng.*, 191, p. 012027.

[49] Wilkinson, N. J., Smith, M. A. A., Kay, R. W. and Harris, R. A. (2019). A review of aerosol jet printing — a non-traditional hybrid process for micro-manufacturing, *Int. J. Adv. Manuf. Technol.*, 105, pp. 4599–4619.

[50] Wadhwa, A. (2015). *Run-Time Ink Stability in Pneumatic Aerosol Jet Printing Using a Split Stream Solvent Add Back System* (Master Dissertation Rochester Institute of Technology).

[51] May, K. R. (1973). The collison nebulizer: Description, performance and application, *J. Aerosol Sci.*, 4, pp. 235–243.

[52] Neveille, E. and D'Arezzo, K. (2016). *Aerosol Jet AJ300 System User Manual*.

[53] Mahajan, A., Frisbie, C. D. and Francis, L. F. (2013). Optimization of aerosol jet printing for high-resolution, high-aspect ratio silver lines, *ACS Appl. Mater. Interfaces*, 5, pp. 4856–4864.

[54] Roth, B., Søndergaard, R. R. and Krebs, F. C. (2015). *Handbook of Flexible Organic Electronics* (Woodhead Publishing, Oxford).

[55] Michael, S., Yeon Sik, C., Chess, B. and Sohini, K.-N. (2017). Controlling and assessing the quality of aerosol jet printed features for large area and flexible electronics, *Flexible Printed Electron.*, 2, p. 015004.

[56] Essien, M. (2018). Apparatuses and methods for stable aerosol deposition using an aerodynamic lens system. U.S. Patent No. 10,124,602, U.S. Patent and Trademark Office, Washington, DC, USA.

[57] Nikolic, D., Keicher, D. and Fan, F.-G. (2019). Design of an aerodynamic lens for PM2.5 chemical composition Analysis, *presented at the 49th International Conference on Environmental Systems*, Boston, MA, USA.

[58] Keicher, D. M., Lavin, J. M., Appelhans, L., Whetten, S. R., Essien, M., Mani, S. S., Moore, P. B., Cook, A., Acree, N. A. and Young, N. P. (2016). Process optimization of aerosol based printing of polyimide for capacitor application, *presented at the 26th Annual International Solid Freeform Fabrication Symposium*, Austin, Texas, USA.

[59] Vaezi, M., Zhong, G., Kalami, H. and Yang, S. (2018). *Functional 3D Tissue Engineering Scaffolds*, eds. Ying Deng and Jordan Kuiper, Chapter 10: Extrusion-based 3D printing technologies for 3D scaffold engineering (Woodhead Publishing) pp. 235–254.

[60] Gonzalez-Gutierrez, J., Cano, S., Schuschnigg, S., Kukla, C., Sapkota, J. and Holzer, C. (2018). Additive manufacturing of metallic and ceramic components by the material extrusion of highly-filled polymers: A review and future perspectives, *Materials*, 11, p. 840.

[61] Wlodarczyk-Biegun, M. and Campo, A. (2017). 3D bioprinting of structural proteins, *Biomaterials*, 134, pp. 180–201.

[62] Derakhshanfar, S., Mbeleck, R., Xu, K., Zhang, X., Zhong, W. and Xing, M. (2018). 3D bioprinting for biomedical devices and tissue engineering: A review of recent trends and advances, *Bioact. Mater.*, 3, pp. 144–156.

[63] Mohamed, O. A., Masood, S. H. and Bhowmik, J. L. (2015). Optimization of fused deposition modeling process parameters: A review of current research and future prospects, *Adv. Manuf.*, 3, pp. 42–53.

[64] Horst, D. and Júnior, P. (2019). 3D-printed conductive filaments based on carbon nanostructures embedded in a polymer matrix: A review, *Int. J. Appl. Nanotechnol. Res.*, 4, pp. 26–40.

[65] Leigh, S. J., Bradley, R. J., Purssell, C. P., Billson, D. R. and Hutchins, D. A. (2012). A simple, low-cost conductive composite material for 3D printing of electronic sensors, *PLoS ONE*, 7, p. e49365.

[66] Dul, S., Fambri, L. and Pegoretti, A. (2018). Filaments production and fused deposition modelling of ABS/carbon nanotubes composites, *Nanomaterials*, 8, p. 49.

[67] Dul, S., Fambri, L. and Pegoretti, A. (2016). Fused deposition modelling with ABS–graphene nanocomposites, *Compos. Part A: Appl. Sci. Manuf.*, 85, pp. 181–191.

[68] Wei, X., Li, D., Jiang, W., Gu, Z., Wang, X., Zhang, Z. and Sun, Z. (2015). 3D printable graphene composite, *Sci. Rep.*, 5, p. 11181.

[69] Podsiadły, B., Skalski, A., Wałpuski, B. and Słoma, M. (2019). Heterophase materials for fused filament fabrication of structural electronics, *J. Mater. Sci.: Mater. Electron.*, 30, pp. 1236–1245.

[70] Kwok, S. W., Goh, K. H. H., Tan, Z. D., Tan, S. T. M., Tjiu, W. W., Soh, J. Y., Ng, Z. J. G., Chan, Y. Z., Hui, H. K. and Goh, K. E. J. (2017). Electrically conductive filament for 3D-printed circuits and sensors, *Appl. Mater. Today*, 9, pp. 167–175.

[71] Clingerman, M. L., King, J. A., Schulz, K. H. and Meyers, J. D. (2002). Evaluation of electrical conductivity models for conductive polymer composites, *J. Appl. Polym. Sci.*, 83, pp. 1341–1356.

[72] Li, W., Ghazanfari, A., Leu, M. C. and Landers, R. G. (2017). Extrusion-on-demand methods for high solids loading ceramic paste in freeform extrusion fabrication, *Virtual Phys. Prototyp.*, 12, pp. 193–205.

[73] Li, H., Tan, C. and Li, L. (2018). Review of 3D printable hydrogels and constructs, *Mater. Des.*, 159, pp. 20–38.

[74] Gleadall, A., Visscher, D., Yang, J., Thomas, D. and Segal, J. (2018). Review of additive manufactured tissue engineering scaffolds: Relationship between geometry and performance, *Burns Trauma*, 6(1). Retrieved from https://doi.org/10.1186/s41038-018-0121-4.

[75] Katarina, I., MacGregor, J. A. Z., Almeida, A. A., Pickard, J. D. M. and Ewertowski, M. (2019) Interchangeable fabrication head assembly. U.S. Patent 10,414,092, USA.

[76] Katarina, I., Almeida, A., Pickard, J., Zozaya, J. and Ewertowski, M. (2018) Apparatus and method for printing circuitry. U.S. Patent 10,091,891.

[77] Abera, B. D., Falco, A., Ibba, P., Cantarella, G., Petti, L. and Lugli, P. (2019). Development of flexible dispense-printed electrochemical immunosensor for aflatoxin M1 detection in milk, *Sensors* 19(18), p. 3912.

[78] Can La. Piezo Jet Valve Dispensing. Retrieved from https://info.tech consystems.com/hubfs/TS9800/Piezo%20Jet%20Valve%20Dispensing%20Article-V3.pdf.

[79] Zhou, C., Duan, J., Deng, G. and Li, J. (2017). A novel high-speed jet dispenser driven by double piezoelectric stacks, *IEEE Trans. Indus. Electron.*, 64, pp. 412–419.

[80] Trimzi, M. A., Ham, Y. B., An, B. C., Choi, Y. M., Park, J. H. and Yun, S. N. (2020). Development of a piezo-driven liquid jet dispenser with hinge-lever amplification mechanism, *Micromachines*, 11, p. 117.

[81] Nordson EFD. PICO Pulse Series Valves Operating Manual. Retrieved from https://www.nordson.com/en/divisions/efd/products/jet-dispensers/pico-pulse-valve.

[82] Deng, G., Wang, N., Zhou, C. and Li, J. (2018). A simplified analysis method for the piezo jet dispenser with a diamond amplifier, *Sensors*, 18, p. 2115.

[83] Zhou, C., Li, J. H., Duan, J. A. and Deng, G. L. (2015). The principle and physical models of novel jetting dispenser with giant magnetostrictive and a magnifier, *Sci. Rep.*, 5, p. 18294.

[84] ViscoTec. Endless Piston Principle. Retrieved from https://www.viscotec.de/en/technology/endless-piston-principle/.

[85] Ghazanfari, A., Li, W., Leu, M. C. and Hilmas, G. E. (2017). A novel free-form extrusion fabrication process for producing solid ceramic components with uniform layered radiation drying, *Addit. Manuf.*, 15, pp. 102–112.

[86] Swanson, P. The "Endless Piston" Pump Technology for Precision Dispensing. Retrieved from https://www.intertronics.co.uk/wp-content/uploads/2017/05/wp12-1The-Endless-Piston-Pump-Technology-for-Precision-Dispensing.pdf.

[87] Li, W., Armani, A., Leu, M. and Landers, R. (2017). Extrusion-on-demand methods for high solids loading ceramic paste in freeform extrusion

fabrication, *Virtual Phys. Prototyp.* pp. 1–13. doi: 10.1080/17452759. 2017.1312735.

[88] Park, J.-U., Hardy, M., Kang, S. J., Barton, K., Adair, K., Mukhopadhyay, D. k., Lee, C. Y., Strano, M. S., Alleyne, A. G., Georgiadis, J. G., Ferreira, P. M. and Rogers, J. A. (2007). High-resolution electrohydrodynamic jet printing, *Nat. Mater.*, 6, pp. 782–789.

[89] Huang, Q. and Zhu, Y. (2019). Printing conductive nanomaterials for flexible and stretchable electronics: A review of materials, processes, and applications, *Adv. Mater. Technol.*, 4, p. 1800546.

[90] An, B. W., Kim, K., Lee, H., Kim, S.-Y., Shim, Y., Lee, D.-Y., Song, J. Y. and Park, J.-U. (2015). High-resolution printing of 3D structures using an electrohydrodynamic inkjet with multiple functional inks, *Adv. Mater.*, 27, pp. 4322–4328.

[91] Raje, P. V. and Murmu, N. C. (2014). A review on electrohydrodynamic-inkjet printing technology, *Int. J. Emerg. Technol. Adv. Eng.*, 4, pp. 174–183.

[92] Yin, Z., Huang, Y., Duan, Y. and Zhang, H. (2018). *Electrohydrodynamic Direct-Writing for Flexible Electronic Manufacturing* (Springer, Singapore).

[93] Liu, W. C. and Watt, A. A. R. (2019). Solvodynamic printing as a high resolution printing method, *Sci. Rep.*, 9, p. 10766.

[94] Onses, M. S., Sutanto, E., Ferreira, P. M., Alleyne, A. G. and Rogers, J. A. (2015). Mechanisms, capabilities, and applications of high-resolution electrohydrodynamic jet printing, *Small*, 11, pp. 4237–4266.

[95] Can, T. T. T., Nguyen, T. C. and Choi, W.-S. (2020). High-viscosity copper paste patterning and application to thin-film transistors using electrohydrodynamic jet printing, *Adv. Eng. Mater.*, 22, p. 1901384.

[96] Qin, H., Wei, C., Dong, J. and Lee, Y.-S. (2014). AC-pulse modulated electrohydrodynamic (EHD) direct printing of conductive micro silver tracks for micro-manufacturing, *presented at the 24th International FAIM Conference*, San Antonio, Texas, USA.

[97] Barton, K., Mishra, S., Alleyne, A., Ferreira, P. and Rogers, J. (2011). Control of high-resolution electrohydrodynamic jet printing, *Control Eng. Prac.*, 19, pp. 1266–1273.

[98] Barton, K., Mishra, S., Alex Shorter, K., Alleyne, A., Ferreira, P. and Rogers, J. (2010). A desktop electrohydrodynamic jet printing system, *Mechatronics*, 20, pp. 611–616.

[99] Sutanto, E., Shigeta, K., Kim, Y. K., Graf, P. G., Hoelzle, D. J., Barton, K. L., Alleyne, A. G., Ferreira, P. M. and Rogers, J. A. (2012). A multimaterial electrohydrodynamic jet (E-jet) printing system, *J. Micromech. Microeng.*, 22, p. 045008.

[100] Cui, B. (2011). *Recent Advances in Nanofabrication Techniques and Applications* (BoD–Books on Demand).

[101] Silva, F. (2014). Smart Sensors and MEMS: Intelligent Devices and Microsystems for Industrial Applications, *IEEE Ind. Electron. Mag.*, 8, pp. 74–74.

[102] Jaworek, A. and Krupa, A. (1999). Classification of the Modes of EHD Spraying, *J Aerosol Sci.*, 30, pp. 873–893.

[103] Bober, D. B. and Chen, C.-H. (2011). Pulsating electrohydrodynamic cone-jets: From choked jet to oscillating cone, *J. Fluid Mechan.*, 689, pp. 552–563.

[104] Larson, B. J. (2005). *New Technologies for Fabricating Biological Microarrays* (Doctoral Dissertation, University of Wisconsin-Madison).

[105] Allanurov, A. M., Zdrok, A. Y., Loschilov, A. G. and Malyutin, N. D. (2014). Problem of ink evaporation while using plotter systems to manufacture printed electronic products, *Procedia Technol.*, 18, pp. 19–24.

[106] Datar, A. (2012). *Micro-Extrusion Process Parameter Modeling* (Master Dissertation, Rochester Institute of Technology).

[107] Deffenbaugh, P., Church, K., Goldfarb, J. and Chen, X. (2013). Fully 3D printed 2.4 GHz bluetooth/Wi-Fi antenna, *presented at the 46th International Symposium on Microelectronics, IMAPS 2013*, Orlando, Florida, USA.

[108] Li, B., Clark, P. A. and Hail, K. (2007). Robust printing and dispensing solutions with three sigma volumetric control for 21st century manufacturing and packaging, *MRS Proc.*, 1002, pp. 1002-N1005-1008.

[109] Church, K. H., Clark, P. A., Chen, X., Owens, M. W. and Stone, K. M. (2010) Dispensing patterns including lines and dots at high speeds. US 2010/0055299 A1 USA.

Problems

1. List the advantages and disadvantages of inkjet printing.
2. Using a sketch to illustrate your answer, describe the process for a bend-mode piezoelectric DoD inkjet printhead.
3. Describe the process for Optomec's Aerosol Jet® system with the ultrasonic atomiser.
4. List some possible applications with the availability of the Aerosol Jet® technology.
5. Briefly describe three different types of extrusion-based printing.
6. Describe the five sub-systems of the electro hydrodynamic (EHD) jet printing.
7. Describe nScrypt's SmartPump™ dispensing process.
8. Among the systems discussed in this chapter, which system do you think is most suitable to fabricate antennas for smartphones and why?

Chapter 4

Materials and Inks for
3D Printed Electronics

3D printing approach can be used to fabricate electronic components by depositing a wide variety of functional materials onto substrates. These materials include metallic nanoparticle inks, metal-organic decomposition (MOD) inks, conductive polymers, carbon nanotubes (CNTs), graphene, semiconductor inks and dielectric inks. This chapter provides an overview of the types of inks used in 3D printing of electronics, their physical properties, as well as their advantages and disadvantages.

4.1 Metallic Nanoparticle Inks

Metallic nanoparticle inks are usually the most preferred for fabricating conductive tracks and patterns because of their relatively higher electrical conductivity [1]. As the name suggests, metallic nanoparticle inks are suspensions of electrically conductive metallic nanoparticles in liquid mediums. Metallic nanoparticles can be synthesised by either top-down or bottom-up approaches. The former approach breaks down bulk materials into small nanoparticles through using grinding, ball-milling, laser ablation or etching techniques, whereas the latter synthesises nanoparticles chemically from atomic-level precursors. This approach is more preferred as it provides better control over the nanoparticle's shape, average size, size distribution and particle stability [2,3]. The size of nanoparticles in this type of inks usually varies between 1–100 nm, and they are also encapsulated in organic additives and stabilising agents to prevent agglomerations. The encapsulated metallic nanoparticles are then dispersed in a liquid medium to form the metallic nanoparticle inks. This section discusses the critical parameters of a metallic nanoparticle ink, including material compositions, particle concentration, particle size and shape, encapsulating organic additives, stabilising agents and liquid medium.

4.1.1 *Materials Compositions*

Many metallic nanoparticle inks have been developed for 3D printed electronics applications over the years. They are typically categorised according to the material compositions of their metallic nanoparticles [1]. The bulk of the organic substances and solvents are decomposed away from the deposited metallic nanoparticle inks after the sintering process, leaving mostly metallic nanoparticles behind. Therefore, the material composition of the metallic nanoparticles largely determines the electrical, material and mechanical properties or any functional features of the printed patterns in 3D printed electronics. The material composition of metallic nanoparticles can also determine the sintering temperatures required. Key deciding factors for material compositions of metallic nanoparticle inks are largely dependent on the costs, oxidation stability, electrical conductivity and requirements for electrical and magnetic properties for the applications. Table 4.1 tabulates and categorises various types of metallic nanoparticle inks into single element metallic nanoparticle inks, metallic oxides nanoparticle inks, alloy metallic nanoparticle inks and core-shell bimetallic nanoparticle inks [1].

Single element metallic nanoparticle inks (for instance, silver, gold and copper nanoparticle inks) are more commonly used as they are widely available in the market [1]. Among the single element metallic nanoparticle inks, silver nanoparticle inks are predominantly used for 3D printed electronics applications due to their excellent electrical conductivity and oxidation stability characteristics [4–7]. The cost of silver, although significantly lower than gold, is still much higher compared to common metals like copper and aluminium. In recent years, copper nanoparticle inks have been explored as a substitute for silver nanoparticle inks due to their comparable electrical conductivity [8] and lower cost [9]. However, many existing challenges are limiting the feasibility of copper nanoparticles in 3D printed electronics. Copper's higher melting point

Table 4.1.   Various types of metallic nanoparticle inks [1].

Single Element Metallic Nanoparticle Inks	Metallic Oxides Nanoparticle Inks	Alloy Metallic Nanoparticle Inks	Core-Shell Bimetallic Nanoparticle Inks
• Silver nanoparticle inks • Gold nanoparticle inks • Copper nanoparticle inks • Platinum nanoparticle inks • Aluminium nanoparticle inks • Palladium nanoparticle inks • Nickel nanoparticle inks • Cobalt nanoparticle inks • Tin nanoparticle inks	• Copper oxide-based nanoparticle inks • Iron oxide nanoparticle inks • Zinc oxide (ZnO) nanoparticle inks • Indium tin oxide (ITO) nanoparticle inks	• Copper-nickel alloy nanoparticle inks	• Copper-silver core-shell bimetallic nanoparticle inks • Copper-nickel core-shell bimetallic nanoparticle Inks

implies that a higher sintering temperature is required for sintering copper nanoparticle inks. In addition, copper nanoparticles tend to oxidise easily in non-inert environment during high temperature sintering process [6,8,10–12]; copper oxides are considerably less conductive than pure copper [13]. Development of a cost effective and efficient process to overcome challenges in formulating copper nanoparticle inks is still an active research area.

There are also other single element metallic nanoparticle inks that possess diverse functional properties. For example, cobalt nanoparticle inks have ferromagnetic properties which are favourable for applications involving interactions with high frequencies and electromagnetic waves [14,15]. Due to nickel's high corrosion and oxidation resistance properties, nickel nanoparticle inks may be used to cover and protect underlying materials [16]. Palladium has electrocatalytic properties, thus palladium nanoparticle inks can be used for fabricating electrochemical sensors [17].

Metallic oxides, alloy and core-shell bimetallic nanoparticle inks mostly are still under research and development and most are not available in the market for consumer use. Some of these inks have unique electrical properties and interesting features that potentially allow them to revolutionise the future of 3D printing of electronics. For instance, metallic oxides nanoparticle inks can be sintered in an ambient environment without oxidising due to its high oxidation stability [18]. Inks made of alloy metallic nanoparticles, such as copper-nickel alloy, have strain sensitivity, good fatigue life and relatively high elongation capability properties, all of which makes them ideal for sensing applications like thermocouples and strain gauges [19].

A core-shell bimetallic nanoparticle is made up of two different metals; the outer shell is made of a different material from the inner core. The materials for the outer shell are usually those with high oxidation stability, such as silver and gold, to protect the inner core from oxidations. This structure offers electrical, material, optical, catalytic and photocatalytic properties from two different metallic materials [20,21]. Copper-silver (Cu-Ag) core-shell bimetallic nanoparticle inks are currently explored as potential alternative inks to silver nanoparticle inks [8]. The use of Cu-Ag core-shell bimetallic nanoparticle inks is a solution to overcome the oxidation limitations of copper while maintaining high electrical conductivity. The Cu-Ag core-shell configuration also reduces the silver load in nanoparticles by substituting the inner core with cheaper copper, thus reducing the material costs [22]. As core-shell bimetallic nanoparticle inks are still under research and development, the current synthesis process is still time-consuming and complicated [1].

4.1.2 *Solid Loading of Metallic Nanoparticle Inks*

The amount of suspended metallic nanoparticles in the ink is measured by the ink's solid loading [1]. Inks with higher solid loading have higher particle concentration of metallic nanoparticles and lesser organic additives. For metallic nanoparticle inks with different solid loadings, printed layers of ink having high solid loading tend to have lower electrical resistance compared to those printed by low solid loading inks. This is because more metallic nanoparticles are present in a high solid loading ink layer, thus generating printed layers with higher cross-sectional areas and subsequently lower electrical resistance. Lower organic additives present in the inks can also reduce the effects of contact resistance between nanoparticles [23]. In the market, metallic nanoparticle inks typically have solid loadings ranging from 20–80% [24].

4.1.3 *Particle Shape*

The shape of metallic nanoparticles can influence their optical, electrical, magnetic and catalytic properties [7]. Apart from spherical nanoparticles, those with unconventional geometries [25] such as nanowires [26–28], nanocubes [29], nanoprisms [30], nanorods [31–33], nanoplatelets [34] and multifaceted nanoparticles are also used in many innovative applications (see **Figure 4.1**). Most metallic nanoparticle inks available in the market are formulated with spherical metallic nanoparticles. This is largely because spherical nanoparticles are simpler and more cost-efficient to synthesise. Spherical shapes also have the least energy shape of all geometries, thus requiring minimal dispersants for ensuring dispersion stability. In addition, shear thickening and thixotropy are least influenced by spherical shapes. However, the spherical geometry may not offer the best packing efficiency and high-density packing as the maximum achievable packing density for identical spherical particles is only 74%. One way to increase

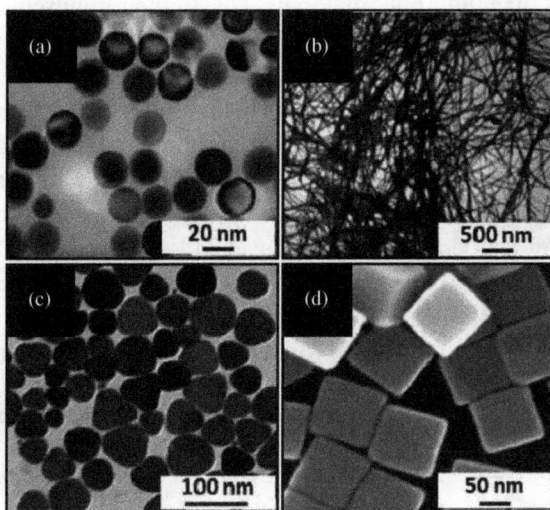

Figure 4.1. Transmission electron microscopy (TEM) of (a) spherical silver nanoparticles; (b) silver nanowires; (c) silver nanoprisms; and (d) silver nanocubes. Adapted from Ref. [25], Copyright (2019).

the packing density is to use polydispersed nanoparticles or particles with non-spherical geometries. The spherical geometry also may not be optimal for migration of electrons over long linear distances, and limiting the maximum achievable electrical conductance [7,35].

In recent years, silver nanowires have gained increasing interests for their high optical transparency (up to 90% transparency), flexibility and excellent electrical conductivity [36]. These properties make silver nanowires very suitable for the fabrication of flexible and transparent electrodes [7], except that they tend to easily clog the printing nozzles because of their elongated shapes [27,37]. Nanoplatelets, due to their flat structure, is also a potential candidate as they allow extremely dense microstructures for better electrical conductance [7,38].

4.1.4 *Particle Size*

Interestingly, the melting temperatures are significantly lower for metallic nanoparticles as compared to their bulk materials. This can be attributed to the surface effects phenomenon [39–41] that correlates melting temperatures to the particle size [42]. The surface-to-volume ratio increases as the particle size decreases, and so nanoparticles have much higher surface-to-volume ratios than their bulk materials. Due to their minuscule size, nanoparticles mainly compose of surface atoms (high percentage of surface atoms due to high surface-to-volume ratios) [41]. Surface atoms have a lack of chemical bonding between them, resulting in lesser stability. Hence, requiring lesser energy to break down the chemical bonds, which implies lower temperatures for sintering or melting nanoparticles [39,40,43]. For instance, bulk gold has a melting temperature of 1064°C and its melting temperature may be lower to less than 300°C by breaking down the bulk material into particles smaller than 5nm in diameter [44,45]. Studies also indicate a linear relationship between melting temperature and particle size [39]. It can therefore be inferred that lower sintering temperatures are generally required for inks with smaller metallic nanoparticles as compared with those with larger particle size [39,40,43].

4.1.5 *Organic Additives and Stabilising Agents*

The strength of van der Waals forces between particles increases as the size of metallic nanoparticles get smaller. This results in agglomeration of metallic nanoparticles that usually cause nozzles clogging and inhomogeneous dispersion. Organic additives and stabilising agents are added to the formulation of the metallic nanoparticle inks to inhibit agglomeration of nanoparticles and facilitating dispersion stability and printability [7,35,45,46]. The encapsulating organic additives and stabilising agents around each nanoparticle create steric repulsion forces between individual particles, thereby decreasing the Van der Waals forces between nanoparticles to discourage agglomeration [45]. Usually, organic additives such as surfactants, humectants and adhesion promoters, are added to promote good printability, substrate wettability and adhesion [7,35].

Although smaller nanoparticles may substantially reduce the sintering temperature required, the addition of organic additives and stabilising agents in ink formulations can cause an increase of sintering temperature [47]. Hence, the inks must be appropriately optimised with adequate amounts of organic additives and stabilising agents to achieve a balance between good printability and low sintering temperatures. An additional sintering process is required to decompose these electrically insulating organic additives and stabilising agents away, allowing the metallic nanoparticles to form contact points among each other to increase the electrical conductivity. The decomposition of these organic substances in the sintered inks forms pores, which can negatively impact their densification and electrical conductivity [3].

4.1.6 *Liquid Medium*

The volumetric bulk of metallic nanoparticle inks are made up of liquid mediums which serve as carriers for carrying, encompassing and stabilising the metallic nanoparticles. Liquid mediums with high evaporation rates are usually preferred for faster drying when deposited onto the substrates. They also should not leave behind any residues after evaporating [48]. Water, alcohols, hydrocarbons, esters, amides and aromatics are some of the commonly used liquid mediums for these inks [7,35].

4.2 Metal-Organic Decomposition Inks

Metal-organic decomposition (MOD) inks are high concentration metal salts or metal-organic complexes dissolved in either aqueous solutions or organic solvents [49–51]. They are also commonly referred to as precursor-type inks [52], organometallic inks [48] or even nanomaterials-free inks [3]. MOD inks are particle-free, as opposed to metallic nanoparticle inks that are particulate suspension-based [53]. Hence, MOD inks do not require the addition of colloidal stabilisers or pose any major concerns for agglomeration [18,51]. MOD inks thereby have a lower probability of nozzle clogging, better printability and tend to have longer higher shelf life than metallic nanoparticle inks [50]. In 3D printed electronics, MOD inks are conductive inks that are also used for fabricating conductive tracks and patterns. However, MOD inks usually have lower electrical conductivity as compare to metallic nanoparticle inks due to the limited amount of metal salts that can be dissolved in the aqueous solutions [53].

MOD inks are not as widely available in the commercial market as metallic nanoparticle inks due to higher complexity in the organic salt formulations and synthesis procedures [49]. Nonetheless, there are still many different types of MOD inks that can be used for 3D printed electronics applications, including silver [54], gold [55,56], copper [57,58], platinum [59] and aluminium [60] MOD inks. The choice of metal-organic complexes for formulating MOD inks is also dependent on factors such as cost, conversion ratio, electrical resistivity, solubility and stability in solvents, oxidation stability, decomposition temperature and ease of handling [18,52,53]. Among the different types of MOD inks, silver MOD inks are more widely utilised as silver has better electrical conductivity, oxidation stability and high redox potential, even though it has low electromigration resistance and high cost price [61]. Silver MOD inks are usually formulated with silver salts, such as silver acetate [62], silver carboxylate [63], silver neodecanoate [63,64] and silver oxalate [65], that are dissolvable in compatible solvents like toluene, xylene, methanol, ethanol and ethylene glycol [1,50].

MOD inks are not electrically conductive initially when they are first deposited onto the substrates. To obtain electrical conductivity, the metal-organic complexes in MOD inks need to be decomposed back into their conductive elemental metals mainly through thermal sintering [18,49–52]. During the thermal sintering process, the metal cations in the inks are reduced into elemental metal while the displaced ligands are oxidised. Ligands are electrically non-conductive and their presence can decrease the electrical conductivity in the sintered inks. If the sintering temperatures are higher than their decomposition temperatures, the displaced ligands can be removed from the inks [51]. MOD inks can lose more than 80% volume during solvent evaporation and metal-organic complexes decomposition. Printing multiple passes can help to prevent missing voids in the printed patterns and improve the electrical conductance of the printed patterns [52]. Sintering temperatures ranging from 150–200 °C are typically required for thermal sintering of MOD inks [66], but these temperatures are too damaging for low cost polymer substrates [3,52]. Alternative sintering techniques, including laser sintering [67], intense pulse light (IPL) sintering [58] and plasma sintering [58,68], are explored as new methods for sintering MOD inks to reduce the sintering temperatures.

4.3 Conductive Polymers

Conductive polymers can be further categorised into either structural conductive polymers or composite conductive polymers [53,69].

4.3.1 *Structural Conductive Polymers*

Structural conductive polymers are intrinsically conductive organic polymers which have conjugated polymer backbones to allow electron flow [48,53,70]. Most polymers are electrically insulative. This is due to the immobility of electrons along with the polymer networks since the electrons are bound to the polymers' atoms [71]. However, conjugated polymers have alternating carbon–carbon single and multiple bonds to connect p-orbitals with delocalised electrons in the main polymer chain. To transmit electrical currents by their own charge carriers, the conjugated polymers must possess molecular orbits that are strongly delocalised and overlapped. The number of π electrons and their activation energy in the conjugated polymer chain can determine how easily the electrons escape the molecular orbit. Longer conjugated polymer chains tend to have more π electrons and also require lower activation energy for electrons. Hence, these electrons are easier to be delocalised and better electrical conductivity can be achieved [53]. Some of the structural conductive polymers include poly(3,4-ethylenedioxy-thiophene) (PEDOT), polythiophene, polyaniline, polyacetylene, polyphenylene vinylene (PPV) and polypyrrole.

The electrical conductivity of conjugated polymers is still very low to be effectively used for 3D printed electronics applications. Therefore, conjugated polymers have to be doped, where electron acceptors or donors can be added to improve the electrical conductivity by increasing the carrier concentration [53,72]. An electron is removed when the conjugated polymer is oxidised by an electron acceptor. This creates a positively charged hole site which moves through the conjugated polymer chain which facilitates the movement of electrons. In contrast, an electron is added when the conjugated polymer is reduced by an electron donor. As opposed to inorganic semiconductor doping, conjugated polymers can have high doping concentration, up to 0.1 dopant molecule per node of a polymer chain. The conjugated polymers can be doped through either physical or chemical doping techniques. Physical doping mainly includes ion implantation, whereas various chemical doping techniques include electrochemical doping, liquid phase doping, gas-phase doping and photon-induced doping [53].

Although structural conductive polymers may not have high electrical conductivity like metallic inks, structural conductive polymers are soft, highly flexible, lightweight, resistant to corrosion, low cost, good compatibility with aqueous and organic solvents, and have unique optical properties [71,73]. They also have good adhesion and mechanical stability on flexible polymer substrates due to their organic nature [49]. Postprocessing of the structural conductive polymers is usually not necessary after printing, thereby saving time and labour [48]. Hence, structural conductive polymers are suitable for fabricating electrochromic devices [74], light-emitting diodes (LEDs) [75], supercapacitors [76], solar cells [77], fuel cells [78], batteries [78,79], sensors [80], transparent electrodes, transparent conductive films and anti-static layers [81]. However, structural conductive polymers have few drawbacks to address. They typically have poor thermal stability, low electrical conductivity and poor solubility in solvents [53,82]. Furthermore, they are usually unstable in air and humidity, and their electrical conductivity will also degrade over time. The non-Newtonian behaviour of structural conductive polymers may also cause printing problems, especially with inkjet printing [48].

Poly(3,4-ethylenedioxythiophene)-poly(styrenesulfonate) (PEDOT: PSS), with its unique characteristics and advantages, is one of the most popular and widely used structural conductive polymers for 3D printed electronics applications. PEDOT:PSS is a type of intrinsically conductive polymer which is made up of two ionomers, namely PEDOT and polystyrene sulfonate (PSS). PEDOT is a conjugated polymer which carries positive charges, whereas PSS contains parts of the sulfonyl groups that are deprotonated and carries negative charges [53]. PEDOT has poor solubility in solvents when existing as its own. However, in the presence of PSS,

PEDOT becomes soluble as a colloidal dispersion in aqueous solutions [71] and thereby giving PEDOT:PSS better processability and printability [49]. PEDOT:PSS can attain electrical conductivity up to 1000 S/cm, which is almost comparable to indium tin oxide (ITO) [71]. The electrical conductivity of PEDOT:PSS can also be further enhanced 2 to 3 orders of magnitude by adding organic additives and surfactants to its ink formulation [53,69,77]. PEDOT:PSS also has low resource barrier, low cost, low processing temperature, high flexibility and high transparency [69]. However, PEDOT:PSS is susceptible to degradation upon exposure to high temperatures, humidity and ultraviolet irradiations [53,71].

4.3.2 *Composite Conductive Polymers*

Composite conductive polymers are extrinsically conductive polymers that incorporate conductive fillers (e.g. metallic powders, carbon black, carbon nanotubes or graphene) in an insulating polymeric matrix [69,83]. The insulating polymeric matrix mainly serves as an adhesive for the conductive fillers, which includes acrylic resin, epoxy resin, polyethylene, polypropylene etc. The minimum concentration of conductive fillers required for composite conductive polymers to have electrical conductivity is also known as the percolation threshold [84]. The percolation threshold is reached when the conductive fillers make contacting points with each other throughout the entire materials. The percolation threshold is also dependent on the particle morphology, particle size distribution, particle dispersion and the dielectric properties of the polymeric matrix. Composite conductive polymers are relatively cheaper and have a wide range of material choices [83], but their electrical conductivity is highly dependent on the type of conductive fillers materials used.

4.4 Carbon Nanotubes

Carbon nanotubes (CNTs) are graphene sheets which are rolled up into hollow cylinders, with open or closed ends [48,85]. CNTs can be further categorised by the number of concentric shells of rolled-up graphene sheets that they have. Single-walled CNT (SWCNT) only has a single concentric shell of the rolled-up graphene sheet, whereas multi-walled CNTs (MWCNTs) have multiple concentric shells of rolled-up graphene sheets (see **Figure 4.2**). SWCNTs typically have diameters ranging from

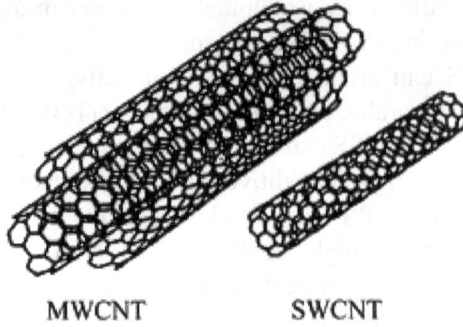

MWCNT   SWCNT

Figure 4.2.   Structures of MWCNT and SWCNT. Adapted from Ref. [88], Copyright (2011), with permission from RSC Publishing.

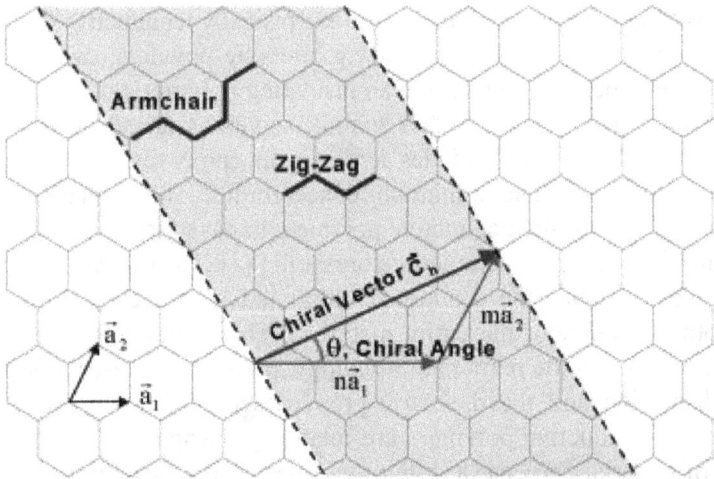

Figure 4.3.   Schematic diagram of the chiral vector and chiral vector on a graphene sheet. Reprinted from Ref. [89], Copyright (2001), with permission from Elsevier.

0.4 nm to 4 nm and lengths in the micrometre range. MWCNTs can have tens to hundreds of concentric shells of rolled-up graphene sheets, and these concentric shells are held together by weak van der Waals forces [86]. Hence, their diameters can vary from several nanometres to tens of nanometres, depending on the number of layers that they have [87].

The term "tube chirality" is used to describe how a sheet of graphene is rolled to form a SWCNT, and it is defined by the chiral vector, $\vec{C_h}$ and chiral angle, θ (see **Figure 4.3**) [89]. The chiral vector determines the

SWCNT's diameter and it can be described as $\vec{C}_h = m\vec{a}_1 + n\vec{a}_2$, where the integers (n, m) are the number of steps along the zigzag carbon bonds of the hexagonal lattice and both $\vec{a}_1$ and $\vec{a}_2$ are unit vectors [89]. The chiral angle is the angle that is defined between the vectors $\vec{C}_h$ and $\vec{a}_1$ (see **Figure 4.3**). The chiral angle determines the degree of twisting in the SWCNT and only ranges between 0°–30° [90]. The two limiting cases, known as the armchair and zigzag configurations, have chiral angles of 30° and 0° respectively (see **Figures 4.4(a) and (c)**). SWCNTs with chiral angles in between 0°–30° are called chiral SWCNTs (see **Figure 4.4(b)**).

The tube chirality affects the atomic structure of SWCNTs, and thus directly influence their material and electrical properties [89]. SWCNTs can be either metallic conductive or semiconducting depending on their chirality. Armchair SWCNTs ($\theta = 30°$) are always metallic, whereas zigzag SWCNTs ($\theta = 0°$) are mostly semiconducting [91,92]. Zigzag SWCNTs will only be metallic when n is a multiple of 3. Similarly, chiral SWCNTs ($0° < \theta < 30°$) are metallic when $(2n + m)/3$ is an integer, otherwise semiconducting [93]. Metallic SWNTs exhibit high electrical conductivity and able to withstand larger current densities, while semiconducting SWNTs display equal hole and electron mobilities due to their intrinsically ambipolar nature [94]. The atomic structure, length, diameter

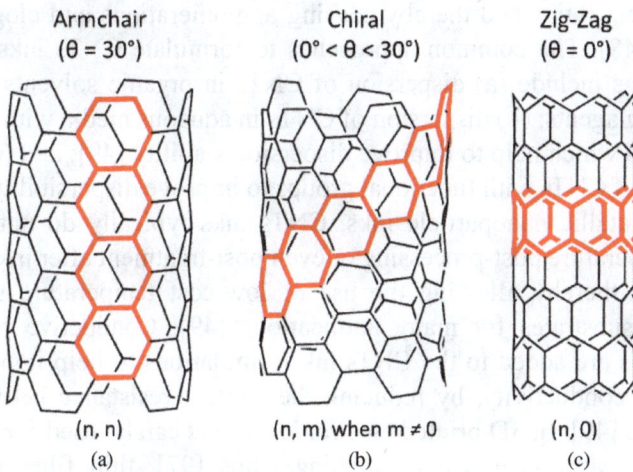

Armchair ($\theta = 30°$)	Chiral ($0° < \theta < 30°$)	Zig-Zag ($\theta = 0°$)
(n, n)	(n, m) when m ≠ 0	(n, 0)
(a)	(b)	(c)

Figure 4.4. Different types of CNT configurations which are dependent on the chiral angles: (a) armchair; (b) chiral; and (c) zigzag. Adapted from Ref. [90], Copyright (2016), with permission from The Royal Society of Chemistry.

and morphology of SWCNTs can also affect their mechanical, electrical, optical and thermal properties [87,95]. Apart from their high current-carrying capacities and high carrier mobility [96], SWCNTs also exhibit many other favourable properties for fabricating 3D printed flexible and stretchable electronic devices, such as high mechanical flexibility, high tensile strength, high optical transparency, good stability and durability, and low temperatures dispersions and processing [86]. It is also key to have good control over chirality distributions and diameters of SWCNTs during synthesis, to achieve SWCNTs with better purity and uniformity. This will allow formulation of inks with stable and reproducible electrical properties. SWCNTs are usually synthesised using selective growth methods such as chemical vapor deposition (CVD), laser ablation and arc discharge [94].

It is also interesting to note that each concentric shell of rolled-up graphene sheets in MWCNTs can also have different tube chirality [89]. As MWCNTs are made up of multiple concentric shells, it can result in infinite forms of MWCNTS that arise from different possible combinations of the shells [86]. Hence, MWCNTs are generally metallic in nature due to their rolled-up directions [48]. As compared to SWCNTs, MWCNTs tend to have better chemical and thermal stability, lower production cost and higher ease of mass production.

CNTs are more susceptible to large van der Waals forces due to their large aspect ratios and thereby causing agglomerations and clogging of nozzles [48]. The common approaches to formulate CNTs inks for ink depositions include (a) dispersion of CNTs in organic solvents without dispersing agents; (b) dispersion of CNTs in aqueous media with dispersing agents which help to improve dispersion stability [49]; and (c) modifications of CNTs with functional groups to improve dispersibility [2,24]. Unlike metallic nanoparticle inks, CNTs inks typically do not require high temperature post-processing or even post-treatment after ink depositions and thereby allowing the use of low-cost temperature sensitive polymer substrates for many applications [49]. Conductive polymers sometimes are added to the CNTs ink formulations to help improve the electrical conductivity, by reducing the contact resistance between the nanotubes [49]. In 3D printed electronics, CNTs can be used for applications such as transparent conducting films [97], thin film transistor (TFTs) [98,99], sensors [100], supercapacitors [101,102], touch panels [103] and RFID tags [104].

4.5 Graphene

Graphene, an allotrope of carbon, is a single atomic layer of graphite, in which the atoms are arranged in two-dimensional (2D) honeycomb lattice structures [3,105,106]. Graphene has been gaining increasing attention for 3D printed electronics applications in the recent years for its plethora of unique properties, such as excellent intrinsic electrical conductivity, mechanical flexibility, thermal stability, chemical stability, scalable production capability and compatibility with other materials [107,108]. Some of the 3D printed electronics applications include flexible conductive circuits [109], electrochemical sensors [110,111], flexible thin film transistors (TFTs) [112], flexible all carbon-based field effect transistor (FET) [113], supercapacitors [114] and photovoltaic devices [115].

Chemical vapour deposition (CVD), liquid-phase exfoliation of graphite, epitaxial growth, sublimation of silicon (Si) from silicon carbide (SiC) and chemical reduction of graphene oxide (GO) are some of the primary techniques that are used for mass production of graphene [108,116,117]. Among all, the liquid-phase exfoliation of graphite technique is considered as one of the most effective ways to produce pristine graphene inks. In bulk graphite, parallel layers of graphene are held together by weak van der Waals forces. The liquid-phase exfoliation of graphite technique uses ultrasonication to induce high shear forces for delaminating adjacent layers from the bulk graphite [118]. However, the exfoliated graphene sheets are prone to re-aggregation due to their high surface areas [119]. To obtain a pristine graphene ink with good stability and printability, the exfoliated graphene sheets need to be dispersed either in organic solvents without additives or using common solvents with stabilising surfactants and polymeric additives [106,120]. However, these surfactants and additives are electrically insulating and can significantly increase the contact resistance between overlapped graphene sheets. Furthermore, it is also challenging to fully remove the surfactants and additives from the deposited pristine graphene inks. Therefore, the printed patterns tend to have significantly lower electrical conductivity as compared to polycrystalline graphite. Note that the lateral size of the graphene sheets in pristine graphene inks must also be well-controlled to prevent nozzle clogging, especially in inkjet printing. However, printed patterns that are made up of smaller graphene sheet size tend to have higher electrical resistivity, due to the increased number of sheet-to-sheet junctions [106,108]. Surfactants-free and additives-free pristine graphene inks generally can attain higher electrical conductivity while allowing for low-temperature processing, due to the absence of surfactants and additives residue in the printed patterns [107]. However, without binder additives, these pristine graphene inks usually have weak adhesions to the substrates [106].

Apart from pristine graphene inks, there are also many other types of graphene-based inks such as chemically derived graphene inks and graphene hybrid inks. Chemically derived graphene inks, usually known as graphene oxide (GO) ink formulations, can cost-effectively be produced in an industrially accessible scale [120]. GO can be reduced into reduced graphene oxide (rGO), also known as chemically derived graphene, by thermal, chemical, photonic and other methods [121]. However, rGO has substantially different properties, or even inferior electrical properties [108], as compared to pristine graphene due to the presence of defects and residual functional groups of GO that changes the carbon plane structures [117]. Graphene hybrid inks are graphene inks that are mixed with either metallic particles or conductive polymers to help improve their electrical conductivity while tailoring the ink properties for various applications [120].

4.6 Semiconductor Inks

Semiconductor inks are essential building blocks for 3D printed active devices and their ink properties (such as mobility, on/off ratio and bandgap) can significantly affect the devices' performances.

Organic-based semiconductor inks exist in either polymer or small-molecules varieties [122]. For successful implementations in 3D printed electronics applications, an ideal organic-based semiconductor material should have a π-conjugated backbone structure that composes of linked unsaturated units, in which the extended π orbitals help to achieve its characteristic optical and charge-transport properties. In addition, it should also have core functionalisation with solubilising substituents to allow better solubility and solid-state core interactions [123].

Small-molecules semiconductor inks, including anthracene, tetracene, pentacene and rubrene, have strong intermolecular interactions and display high-carrier mobility [124]. However, they tend to have poorer solubility and low environmental stability, and hence limiting their uses in 3D printed electronics applications [125]. To solve these limitations, functionalised groups can be introduced to these small-molecules organic-based semiconductor inks. For instance, pentacene can be functionalised into 6,13-bis(tri isopropyl silylethynyl)pentacene (TIPS-pentacene) [126] to gain better electrical performances and environmental stability. On the other hand, polymer semiconductor inks generally have low viscosity and good solubility in organic solvents due to appropriate side chains, and therefore they are more compatible with 3D electronics printing techniques for ink depositions [124,125]. In addition, they are highly compatible with flexible substrates and require low temperature processability. As compared to small-molecules semiconductor inks, polymer semiconductor inks typically have inferior electrical performances but better environmental stability. Commonly-used small-molecule organic semiconductor inks include 6,13-bis(tri isopropyl silylethynyl)pentacene (TIPS-pentacene) [127–129] and 2,7-Dioctyl[1]benzothieno[3,2-b][1]benzothiophene (C8-BTBT) [130], whereas polymer organic semiconductor inks include poly(3-hexylthiophene) (P3HT) [131,132] and regioregular poly(3-hexaylthiophene) (RR-P3HT) [133,134]. Some of the 3D printed electronics applications with organic-based semiconductor inks include organic thin-film transistor (OTFTs) [127–130], organic field-effect transistor (OFETs) [135], and organic solar cells (OSCs) [131,134,136].

Inorganic metal oxide semiconductor materials are also gaining increasing interests over organic semiconductor materials due to their superior electronic transport properties, electrical and environmental stability [137,138]. Metal oxide semiconductor materials usually require high processing temperatures after ink deposition to convert chemical precursors into metal-oxide-metal (M-O-M) bonds, and thus restricting the use of temperature-sensitive polymeric substrates for 3D printed electronics applications [124]. Hence, tremendous research efforts in the recent years are focusing on formulating metal oxide semiconductor inks that require lower processing temperatures (i.e. below 250°C), while maintaining good electrical performances [124].

Due to their chemical bonding, metal oxide semiconductor materials have a high degree of ionicity. The metal (M) ns-orbitals and oxygen (O) 2p-orbitals help to form a highly dispersive conduction band minimum (CBM) and a localising valence band maximum (VBM) respectively [139]. The interactions between these two orbitals give better electron transport as compared to hole transport. Therefore, the majority of metal oxide semiconductors is predominantly n-type conductivity. Inorganic oxide semiconductor materials can be further classified as crystalline oxides and amorphous oxides [140]. Electron conducting (n-type) oxides and hole conducting (p-type) oxides are generally crystalline. The commonly used n-type oxides include zinc oxide (ZnO), indium oxide (In_2O_3), whereas p-type oxides include cuprous oxide (Cu_2O), cupric oxide (CuO) and tin monoxide (SnO). Amorphous oxides semiconductor materials include indium-gallium-zinc-oxide (IGZO), indium-zinc-oxide (IZO), zine-tin-oxide (ZTO) and amorphous indium-tin-oxide (a-ITO) [140]. 3D printed electronics applications with metal oxide semiconductors inks include thin-film transistor (TFTs), field-effect transistor (FETs), diodes and CMOS logic. Other inorganic materials such as single-walled carbon nanotubes (SWCNTs), graphene and 2D semiconducting materials (for instance, molybdenum disulfide (MoS_2), tungsten diselenide (WSe_2) and tungsten disulfide (WS_2)) can also be used as semiconducting materials in 3D printed electronics applications [124].

4.7 Dielectric Inks

Dielectric inks also play a significant role in 3D printed electronics. They are primarily used for multilayer insulation against electrical conduction, circuit protection from the environment, or making capacitors and transistors [141,142]. Insulating organic polymers or inorganic nanocomposites suspensions are usually used as dielectric inks.

Electrically insulative dielectric materials have a large bandgap, implying that there is a large energy difference between the valence band and conduction band. Hence, increasing the difficulty for electrons to move from the valence band to the conduction band to create an electrical current flow [142,143]. Electrically insulative dielectric materials are especially important for making insulative layers in multilayer circuitry. Multilayer circuitry comprises of many conductive and insulative layers that are patterned and combined in a layered structure with a layer-by-layer manner. Electrically insulative dielectric materials are printed directly on electrically conductive traces to help with insulating, so that crossing conductive traces can be printed on the former layers without short-circuiting (see **Figure 4.5**). It is also critical to ensure that the dielectric layers withstand multiple post-processing treatments (such as ultraviolet (UV) curing and heat treatments). The dielectric layers must be sufficiently thick to avoid electrical leakages, and they are also required to be smooth and defect-free to provide good electrical insulation and

1. Cleaning and chemical treatment
of substrate

4. Printing dielectric

2. Printing of conductive material

Printhead

Ink

5. Curing

UV+Heat

3. Conductor sintering

Heat

N repeating
steps

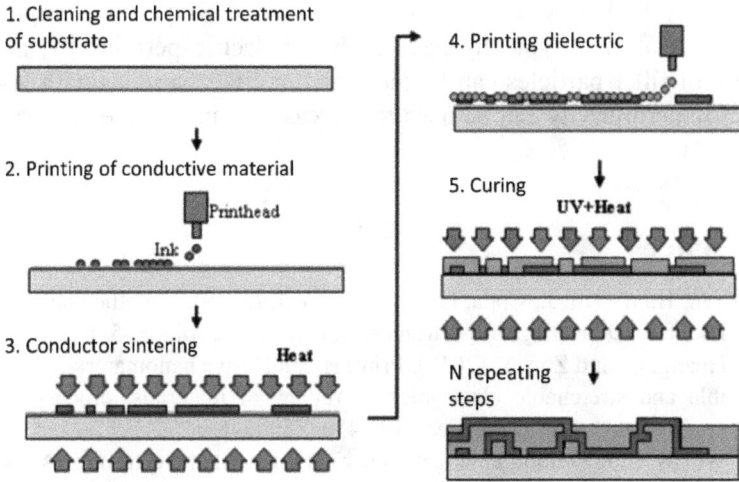

Figure 4.5. Schematic diagrams of inkjet printing of multilayer circuitry. Adapted from Ref. [144]. Copyright (2010), with permission from Elsevier.

printability for the subsequent layers [144]. Organic polymers, such as polyimide (PI), polymethyl methacrylate (PMMA), polyvinyl chloride (PVC), polystyrene (PS), polyvinyl alcohol (PVA), polyvinylpyrrolidone (PVP) and polyvinylidene fluoride (PVDF), are commonly used as electrically insulating dielectric materials in 3D printed electronics [145]. They usually have high optical transparency, low surface roughness, high dielectric constant, good compatibility with flexible substrates, low temperature processing and low cost [53,145]. However, dielectric materials can lose their insulating properties when a sufficiently large voltage, also known as the breakdown voltage, is applied. The electrons in the valence band will gain sufficient energy to be excited to the conductance band [142].

Apart from multilayer insulation and circuit protection, dielectric inks are also commonly used for fabricating essential electrical components such as capacitors [146,147] and transistors [148,149]. The chemical and physical properties of the dielectric layers in these electronic components can significantly affect their electrical performances. For instance, the dielectric permittivity of dielectric inks can directly affect the capacitance of capacitors. Dielectric inks with high dielectric permittivity and low loss are preferred for fabricating capacitors with high capacitance density [150]. Organic polymeric dielectric inks usually have lower dielectric permittivity, higher hysteresis and lower device stability. Nanocomposites dielectric materials typically have higher dielectric permittivity, as the loading of filler particles can be increased or filler particles with higher dielectric permittivity can be used to increase the inks' dielectric permittivity [53].

References

[1] Tan, H. W., An, J., Chua, C. K. and Tran, T. (2019). Metallic nanoparticle inks for 3D printing of electronics, *Adv. Electron. Mater.*, 5, p. 1800831.

[2] Huang, Q. and Zhu, Y. (2019). Printing conductive nanomaterials for flexible and stretchable electronics: A review of materials, processes, and applications, *Adv. Mater. Technol.*, 4, p. 1800546.

[3] Wu, W. (2017). Inorganic nanomaterials for printed electronics: A review, *Nanoscale*, 9, pp. 7342–7372.

[4] Tan, H. W., Tran, T. and Chua, C. K. (2016). A review of printed passive electronic components through fully additive manufacturing methods, *Virtual Phys. Prototyp.*, 11, pp. 271–288.

[5] Hwang, Y.-T., Chung, W.-H., Jang, Y.-R. and Kim, H.-S. (2016). Intensive plasmonic flash light sintering of copper nanoinks using a band-pass light filter for highly electrically conductive electrodes in printed electronics, *ACS Appl. Mater. Interfaces*, 8, pp. 8591–8599.

[6] Park, B. K., Kim, D., Jeong, S., Moon, J. and Kim, J. S. (2007). Direct writing of copper conductive patterns by ink-jet printing, *Thin Solid Films*, 515, pp. 7706–7711.

[7] Rajan, K., Roppolo, I., Chiappone, A., Bocchini, S., Perrone, D. and Chiolerio, A. (2016). Silver nanoparticle ink technology: State of the art, *Nanotechnol. Sci. Appl.*, 9, pp. 1–13.

[8] Magdassi, S., Grouchko, M. and Kamyshny, A. (2010). Copper nanoparticles for printed electronics: Routes towards achieving oxidation stability, *Materials*, 3, pp. 4626–4638.

[9] Ryu, J., Kim, H.-S. and Hahn, H. T. (2011). Reactive sintering of copper nanoparticles using intense pulsed light for printed electronics, *J. Electron. Mater.*, 40, pp. 42–50.

[10] Dharmadasa, R., Jha, M., Amos, D. A. and Druffel, T. (2013). Room temperature synthesis of a copper ink for the intense pulsed light sintering of conductive copper films, *ACS Appl. Mater. Interfaces*, 5, pp. 13227–13234.

[11] Kang, J. S., Ryu, J., Kim, H. S. and Hahn, H. T. (2011). Sintering of inkjet-printed silver nanoparticles at room temperature using intense pulsed light, *J. Electron. Mater.*, 40, pp. 2268–2277.

[12] Kim, Y. J., Ryu, C.-H., Park, S.-H. and Kim, H.-S. (2014). The effect of poly (N-vinylpyrrolidone) molecular weight on flash light sintering of copper nanopaste, *Thin Solid Films*, 570, pp. 114–122.

[13] Li, W., Li, W., Wei, J., Tan, J. and Chen, M. (2014). Preparation of conductive Cu patterns by directly writing using nano-Cu ink, *Mater. Chem. Phys.*, 146, pp. 82–87.

[14] Nelo, M., Sowpati, A., Palukuru, V. K., Juuti, J. and Jantunen, H. (2010). Formulation of screen printable cobalt nanoparticle ink for high frequency applications, *Prog. Electromagn. Res.*, 110, pp. 253–266.

[15] Sowpati, A. K., Nelo, M., Palukuru, V. K., Juuti, J. and Jantunen, H. (2013). Miniaturisation of dual band monopole antennas loaded with screen printed cobalt nanoparticle ink, *IET Microw. Antenna P.*, 7, pp. 180–186.

[16] (2016). Nano Dimension teams with Tel Aviv University to 3D print sensors using nickel nanoparticles. Retrieved from https://www.3ders.org/articles/20160223-nano-dimension-teams-with-tel-aviv-university-to-3d-print-nickel-nanoparticle-sensors.html.

[17] Qin, Y., Alam, A. U., Howlader, M. M. R., Hu, N.-X. and Deen, M. J. (2016). Inkjet printing of a highly loaded palladium ink for integrated, low-cost pH sensors, *Adv. Funct. Mater.*, 26, pp. 4923–4933.

[18] Perelaer, J., Smith, P. J., Mager, D., Soltman, D., Volkman, S. K., Subramanian, V., Korvink, J. G. and Schubert, U. S. (2010). Printed electronics: The challenges involved in printing devices, interconnects, and contacts based on inorganic materials, *J. Mater. Chem.*, 20, pp. 8446–8453.

[19] Applied Nanotech Inc. Copper-Nickel Alloy Ink: CuNi-OC5050 Retrieved from http://www.appliednanotech.net/wp-content/uploads/2018/02/ANI-CuNi-OC5050.pdf.

[20] Zaleska-Medynska, A., Marchelek, M., Diak, M. and Grabowska, E. (2016). Noble metal-based bimetallic nanoparticles: The effect of the structure on the optical, catalytic and photocatalytic properties, *Adv. Colloid Interface Sci.*, 229, pp. 80–107.

[21] Sharma, G., Kumar, A., Sharma, S., Naushad, M., Prakash Dwivedi, R., Alothman, Z. A. and Mola, G. T. (2017). Novel development of nanoparticles to bimetallic nanoparticles and their composites: A review, *J. King Saud Univ. Sci.* https://doi.org/10.1016/j.jksus.2017.06.012.

[22] Changsoo, L., Na Rae, K., Jahyun, K., Yung Jong, L. and Hyuck Mo, L. (2015). Cu-Ag core–shell nanoparticles with enhanced oxidation stability for printed electronics, *Nanotechnology*, 26, p. 455601.

[23] Wang, F., Mao, P. and He, H. (2016). Dispensing of high concentration Ag nano-particles ink for ultra-low resistivity paper-based writing electronics, *Sci. Rep.*, 6, p. 21398.

[24] Kamyshny, A. and Magdassi, S. (2014). Conductive nanomaterials for printed electronics, *Small*, 10, pp. 3515–3535.

[25] Loiseau, A., Asila, V., Boitel-Aullen, G., Lam, M., Salmain, M. and Boujday, S. (2019). Silver-based plasmonic nanoparticles for and their use in biosensing, *Biosensors*, 9, p. 78.

[26] Fan, Z., Ho, J. C., Takahashi, T., Yerushalmi, R., Takei, K., Ford, A. C., Chueh, Y.-L. and Javey, A. (2009). Toward the development of printable nanowire electronics and sensors, *Adv. Mater.*, 21, pp. 3730–3743.

[27] Finn, D. J., Lotya, M. and Coleman, J. N. (2015). Inkjet printing of silver nanowire networks, *ACS Appl. Mater. Interfaces*, 7, pp. 9254–9261.

[28] Williams, N. X., Noyce, S., Cardenas, J. A., Catenacci, M., Wiley, B. J. and Franklin, A. D. (2019). Silver nanowire inks for direct-write electronic tattoo applications, *Nanoscale*, 11, pp. 14294–14302.

[29] Siddiqui, G. U., Rehman, M. M. and Choi, K. H. (2017). Resistive switching phenomena induced by the heterostructure composite of $ZnSnO_3$ nanocubes interspersed ZnO nanowires, *J. Mater. Chem. C*, 5, pp. 5528–5537.

[30] Sinar, D., Knopf, G. K. and Nikumb, S. (Year). Graphene and silver-nanoprism dispersion for printing optically-transparent electrodes. *In Organic Photonic Materials and Devices XIX*, International Society for Optics and Photonics, p. 101010L.

[31] Reiser, B., González-García, L., Kanelidis, I., Maurer, J. and Kraus, T. (2016). Gold nanorods with conjugated polymer ligands: Sintering-free conductive inks for printed electronics, *Chem. Sci.*, 7, pp. 4190–4196.

[32] Manjakkal, L., Sakthivel, B., Gopalakrishnan, N. and Dahiya, R. (2018). Printed flexible electrochemical pH sensors based on CuO nanorods, *Sens. Actuators, B*, 263, pp. 50–58.

[33] Amin, G., Sandberg, M. O., Zainelabdin, A., Zaman, S., Nur, O. and Willander, M. (2012). Scale-up synthesis of ZnO nanorods for printing inexpensive ZnO/polymer white light-emitting diode, *J. Mater. Sci.*, 47, pp. 4726–4731.

[34] Lee, Y.-I., Kim, S., Jung, S.-B., Myung, N. V. and Choa, Y.-H. (2013). Enhanced electrical and mechanical properties of silver nanoplatelet-based conductive features direct printed on a flexible substrate, *ACS Appl. Mater. Interfaces*, 5, pp. 5908–5913.

[35] Ian M., H. and Graham D., M. (2012). *Inkjet Technology for Digital Fabrication* (John Wiley & Sons Ltd, United Kingdom).

[36] Shi, Y., He, L., Deng, Q., Liu, Q., Li, L., Wang, W., Xin, Z. and Liu, R. (2019). Synthesis and applications of silver nanowires for transparent conductive films, *Micromachines*, 10, p. 330.

[37] Jeevika, A. and Ravi Shankaran, D. (2015). Seed-free synthesis of 1D silver nanowires ink using clove oil (Syzygium Aromaticum) at room temperature, *J. Colloid Interface Sci.*, 458, pp. 155–159.

[38] Lee, C.-L., Chang, K.-C. and Syu, C.-M. (2011). Silver nanoplates as inkjet ink particles for metallization at a low baking temperature of 100°C, *Colloids Surf., A*, 381, pp. 85–91.

[39] Allen, G. L., Bayles, R. A., Gile, W. W. and Jesser, W. A. (1986). Small particle melting of pure metals, *Thin Solid Films*, 144, pp. 297–308.

[40] Perelaer, J., Abbel, R., Wünscher, S., Jani, R., van Lammeren, T. and Schubert, U. S. (2012). Roll-to-roll compatible sintering of inkjet printed features by photonic and microwave exposure: From non-conductive ink to 40% bulk silver conductivity in less than 15 seconds, *Adv. Mater.*, 24, pp. 2620–2625.

[41] Schmidt, M., Kusche, R., von Issendorff, B. and Haberland, H. (1998). Irregular variations in the melting point of size-selected atomic clusters, *Nature*, 393, pp. 238–240.

[42] Luo, W., Hu, W. and Xiao, S. (2008). Size effect on the thermodynamic properties of silver nanoparticles, *J. Phys. Chem. C*, 112, pp. 2359–2369.

[43] Roduner, E. (2006). Size matters: Why nanomaterials are different, *Chem. Soc. Rev.*, 35, pp. 583–592.

[44] Buffat, P. and Borel, J. P. (1976). Size effect on the melting temperature of gold particles, *Phys. Rev. A*, 13, pp. 2287–2298.

[45] Perelaer, J. and Schubert, U. S. (2013). Novel approaches for low temperature sintering of inkjet-printed inorganic nanoparticles for roll-to-roll (R2R) applications, *J. Mater. Res.*, 28, pp. 564–573.

[46] Galagan, Y., Coenen, E. W. C., Abbel, R., van Lammeren, T. J., Sabik, S., Barink, M., Meinders, E. R., Andriessen, R. and Blom, P. W. M. (2013). Photonic sintering of inkjet printed current collecting grids for organic solar cell applications, *Org. Electron.*, 14, pp. 38–46.

[47] Perelaer, J., Jani, R., Grouchko, M., Kamyshny, A., Magdassi, S. and Schubert, U. S. (2012). Plasma and microwave flash sintering of a tailored silver nanoparticle ink, yielding 60% bulk conductivity on cost-effective polymer foils, *Adv. Mater.*, 24, pp. 3993–3998.

[48] Cummins, G. and Desmulliez Marc, P. Y. (2012). Inkjet printing of conductive materials: A review, *Circuit World*, 38, pp. 193–213.

[49] Chen, S. P., Chiu, H. L., Wang, P. H. and Liao, Y. C. (2015). Inkjet printed conductive tracks for printed electronics, *ECS J. Solid State Sci. Technol.*, 4, pp. P3026–P3033.

[50] Meda, M. (2019). *Effect of Process Conditions on Feature Size of Inkjet Printed Silver MOD Ink* (Master Dissertation, Rochester Institute of Technology, Rochester, NY, USA).

[51] Wunscher, S., Abbel, R., Perelaer, J. and Schubert, U. S. (2014). Progress of alternative sintering approaches of inkjet-printed metal inks and their application for manufacturing of flexible electronic devices, *J. Mater. Chem. C*, 2, pp. 10232–10261.

[52] Choi, Y., Seong, K.-D. and Piao, Y. (2019). Metal–organic decomposition ink for printed electronics, *Adv. Mater. Interfaces*, 6, p. 1901002.

[53] Cui, Z. (2016). *Printed Electronics: Materials, Technologies and Applications* (John Wiley & Sons, Singapore).

[54] Smith, P. J., Shin, D. Y., Stringer, J. E., Derby, B. and Reis, N. (2006). Direct ink-jet printing and low temperature conversion of conductive silver patterns, *J. Mater. Sci.*, 41, pp. 4153–4158.

[55] Nur, H., Song, J., Evans, J. and Edirisinghe, M. (2002). Ink-jet printing of gold conductive tracks, *J. Mater. Sci.: Mater. Electron.*, 13, pp. 213–219.

[56] Schoner, C., Tuchscherer, A., Blaudeck, T., Jahn, S. F., Baumann, R. R. and Lang, H. (2013). Particle-free gold metal–organic decomposition ink for inkjet printing of gold structures, *Thin Solid Films*, 531, pp. 147–151.

[57] Rozenberg, G. G., Bresler, E., Speakman, S. P., Jeynes, C. and Steinke, J. H. G. (2002). Patterned low temperature copper-rich deposits using inkjet printing, *Appl. Phys. Lett.*, 81, pp. 5249–5251.

[58] Araki, T., Sugahara, T., Jiu, J., Nagao, S., Nogi, M., Koga, H., Uchida, H., Shinozaki, K. and Suganuma, K. (2013). Cu salt ink formulation for printed electronics using photonic sintering, *Langmuir*, 29, pp. 11192–11197.

[59] Cummins, G., Kay, R., Terry, J., Desmulliez, M. P. and Walton, A. J. (Year). Optimization and characterization of drop-on-demand inkjet printing process for platinum organometallic inks. *In 2011 IEEE 13th Electronics Packaging Technology Conference*, IEEE, pp. 256–261.

[60] Curtis, C. J., Miedaner, A., Van Hest, M. F. A. M. and Ginley, D. S. (2010) Printing aluminum films and patterned contacts using organometallic precursor inks. U.S. Patent Application 12/678,647, USA.

[61] Shin, D.-H., Woo, S., Yem, H., Cha, M., Cho, S., Kang, M., Jeong, S., Kim, Y., Kang, K. and Piao, Y. (2014). A self-reducible and alcohol-soluble copper-based metal–organic decomposition ink for printed electronics, *ACS Appl. Mater. Interfaces*, 6, pp. 3312–3319.

[62] Vaseem, M., Lee, S.-K., Kim, J.-G. and Hahn, Y.-B. (2016). Silver-ethanolamine-formate complex based transparent and stable ink: Electrical assessment with microwave plasma vs thermal sintering, *Chem. Eng. J.*, 306, pp. 796–805.

[63] Dearden, A. L., Smith, P. J., Shin, D.-Y., Reis, N., Derby, B. and O'Brien, P. (2005). A low curing temperature silver ink for use in ink-jet printing and subsequent production of conductive tracks, *Macromol. Rapid Commun.*, 26, pp. 315–318.

[64] Teng, K. F. and Vest, R. W. (1988). Metallization of solar cells with ink jet printing and silver metallo-organic inks, *IEEE Trans. Components, Hybrids, and Manuf. Technol.*, 11, pp. 291–297.

[65] Dong, Y., Li, X., Liu, S., Zhu, Q., Li, J.-G. and Sun, X. (2015). Facile synthesis of high silver content MOD ink by using silver oxalate precursor for inkjet printing applications, *Thin Solid Films*, 589, pp. 381–387.

[66] Mou, Y., Zhang, Y., Cheng, H., Peng, Y. and Chen, M. (2018). Fabrication of highly conductive and flexible printed electronics by low temperature sintering reactive silver ink, *Appl. Surf. Sci.*, 459, pp. 249–256.

[67] Min, H., Lee, B., Jeong, S. and Lee, M. (2016). Laser-direct process of Cu nano-ink to coat highly conductive and adhesive metallization patterns on plastic substrate, *Opt. Laser. Eng.*, 80, pp. 12–16.

[68] Bromberg, V., Ma, S., Egitto, F. D. and Singler, T. J. (2013). Highly conductive lines by plasma-induced conversion of inkjet-printed silver nitrate traces, *J. Mater. Chem. C*, 1, pp. 6842–6849.

[69] Cruz, S. M. F., Rocha, L. A. and Viana, J. C. (2018). *Flexible Electronics*, Chapter 2: Printing technologies on flexible substrates for printed electronics (IntechOpen).

[70] Kitto, T., Bodart-Le Guen, C., Rossetti, N. and Cicoira, F. (2019). *Handbook of Organic Materials for Electronic and Photonic Devices (Second Edition)*, eds. Oksana Ostroverkhova, Chapter 25: Processing and patterning of conducting polymers for flexible, stretchable, and biomedical electronics (Woodhead Publishing, Cambridge, UK) pp. 817–842.

[71] Suganuma, K. (2014). *Introduction to Printed Electronics* (Springer Science+Business Media, New York).

[72] Ziadan, K. M. (2012). Conducting polymers application, *New Polymers for Special Applications*, pp. 3–24.

[73] Kitto, T., Bodart-Le Guen, C., Rossetti, N. and Cicoira, F. (2019). *Handbook of Organic Materials for Electronic and Photonic Devices (Second Edition)*, eds. Oksana Ostroverkhova, Chapter 25: Processing and patterning of conducting polymers for flexible, stretchable, and biomedical electronics (Woodhead Publishing, UK) pp. 817–842.

[74] Mortimer, R. J., Dyer, A. L. and Reynolds, J. R. (2006). Electrochromic organic and polymeric materials for display applications, *Displays*, 27, pp. 2–18.

[75] White, M. S., Kaltenbrunner, M., Głowacki, E. D., Gutnichenko, K., Kettlgruber, G., Graz, I., Aazou, S., Ulbricht, C., Egbe, D. A. M., Miron, M. C., Major, Z., Scharber, M. C., Sekitani, T., Someya, T., Bauer, S. and Sariciftci, N. S. (2013). Ultrathin, highly flexible and stretchable PLEDs, *Nat. Photon.*, 7, pp. 811–816.

[76] Han, Y. and Dai, L. (2019). Conducting polymers for flexible supercapacitors, *Macromol. Chemi. Phys.*, 220, p. 1800355.

[77] Eom, S. H., Senthilarasu, S., Uthirakumar, P., Yoon, S. C., Lim, J., Lee, C., Lim, H. S., Lee, J. and Lee, S.-H. (2009). Polymer solar cells based on inkjet-printed PEDOT: PSS layer, *Org. Electron.*, 10, pp. 536–542.

[78] Wegner, G. (2006). Polymers as functional components in batteries and fuel cells, *Polym. Adv. Technol.*, 17, pp. 705–708.

[79] Ryu, K. S., Jeong, S. K., Joo, J. and Kim, K. M. (2007). Polyaniline doped with dimethyl sulfate as a nucleophilic dopant and its electrochemical properties as an electrode in a lithium secondary battery and a redox super-capacitor, *J. Phys. Chem. B*, 111, pp. 731–739.

[80] Wang, Y., Liu, A., Han, Y. and Li, T. (2020). Sensors based on conductive polymers and their composites: A review, *Polymer International*, 69, pp. 7–17.

[81] Angelopoulos, M. (2001). Conducting polymers in microelectronics, *IBM J. Res. Dev.*, 45, pp. 57–75.

[82] Li, D., Lai, W. Y., Zhang, Y. Z. and Huang, W. (2018). Printable transparent conductive films for flexible electronics, *Adv. Mater.*, 30, p. 1704738.

[83] Wang, Y., Zhu, C., Pfattner, R., Yan, H., Jin, L., Chen, S., Molina-Lopez, F., Lissel, F., Liu, J. and Rabiah, N. I. (2017). A highly stretchable, transparent, and conductive polymer, *Sci. Adv.*, 3, p. e1602076.

[84] Shah, S., Shiblee, M. D. N. I., Rahman, J. M. H., Basher, S., Mir, S. H., Kawakami, M., Furukawa, H. and Khosla, A. (2018). 3D printing of electrically conductive hybrid organic–inorganic composite materials, *Microsyst. Technol.*, 24, pp. 4341–4345.

[85] De Volder, M. F. L., Tawfick, S. H., Baughman, R. H. and Hart, A. J. (2013). Carbon nanotubes: Present and future commercial applications, *Science*, 339, p. 535.

[86] Pitroda, J., Jethwa, B. and Dave, S. (2016). A critical review on carbon nanotubes, *Int. J. Constr. Res. Civ. Eng*, 2, pp. 36–42.

[87] Eatemadi, A., Daraee, H., Karimkhanloo, H., Kouhi, M., Zarghami, N., Akbarzadeh, A., Abasi, M., Hanifehpour, Y. and Joo, S. W. (2014). Carbon nanotubes: Properties, synthesis, purification, and medical applications, *Nanoscale Res. Lett.*, 9, p. 393.

[88] Zhao, G., Wang, C., Wu, Q. and Wang, Z. (2011). Determination of carbamate pesticides in water and fruit samples using carbon nanotube reinforced hollow fiber liquid-phase microextraction followed by high performance liquid chromatography, *Anal. Methods*, 3, pp. 1410–1417.

[89] Thostenson, E. T., Ren, Z. and Chou, T.-W. (2001). Advances in the science and technology of carbon nanotubes and their composites: a review, *Compos. Sci. Technol.*, 61, pp. 1899–1912.

[90] Sisto, T. J., Zakharov, L. N., White, B. M. and Jasti, R. (2016). Towards pi-extended cycloparaphenylenes as seeds for CNT growth: Investigating strain relieving ring-openings and rearrangements, *Chem. Sci.*, 7, pp. 3681–3688.

[91] Wilder, J. W. G., Venema, L. C., Rinzler, A. G., Smalley, R. E. and Dekker, C. (1998). Electronic structure of atomically resolved carbon nanotubes, *Nature*, 391, pp. 59–62.

[92] Zulhairun, A. K., Abdullah, M. S., Ismail, A. F. and Goh, P. S. (2019). *Current Trends and Future Developments on (Bio-) Membranes*, eds. Angelo Basile, Efrem Curcio and Inamuddin, Chapter 1: Graphene and CNT technology (Elsevier) pp. 3–26.

[93] Odom, T. W., Huang, J.-L., Kim, P. and Lieber, C. M. (1998). Atomic structure and electronic properties of single-walled carbon nanotubes, *Nature*, 391, pp. 62–64.

[94] Zaumseil, J. (2015). Single-walled carbon nanotube networks for flexible and printed electronics, *Semicond. Sci. Technol.*, 30, p. 074001.

[95] Saifuddin, N., Raziah, A. and Junizah, A. (2013). Carbon nanotubes: A review on structure and their interaction with proteins, *J. Chem.*, 2013. Retrieved from https://doi.org/10.1155/2013/676815.

[96] Xiang, L., Zhang, H., Hu, Y. and Peng, L.-M. (2018). Carbon nanotube-based flexible electronics, *J. Mater. Chem. C*, 6, pp. 7714–7727.

[97] Lee, Y.-I., Kim, S., Lee, K.-J., Myung, N. V. and Choa, Y.-H. (2013). Inkjet printed transparent conductive films using water-dispersible single-walled carbon nanotubes treated by UV/ozone irradiation, *Thin Solid Films*, 536, pp. 160–165.

[98] Vaillancourt, J., Zhang, H., Vasinajindakaw, P., Xia, H., Lu, X., Han, X., Janzen, D. C., Shih, W.-S., Jones, C. S., Stroder, M., Chen, M. Y., Subbaraman, H., Chen, R. T., Berger, U. and Renn, M. (2008). All ink-jet-printed carbon nanotube thin-film transistor on a polyimide substrate with an ultrahigh operating frequency of over 5 GHz, *Appl. Phys. Lett.*, 93, p. 243301.

[99] Xiao, K., Liu, Y., Hu, P. a., Yu, G., Wang, X. and Zhu, D. (2003). High-mobility thin-film transistors based on aligned carbon nanotubes, *Appl. Phys. Lett.*, 83, pp. 150–152.

[100] Qin, Y., Kwon, H.-J., Subrahmanyam, A., Howlader, M. M. R., Selvaganapathy, P. R., Adronov, A. and Deen, M. J. (2016). Inkjet-printed bifunctional carbon nanotubes for pH sensing, *Mater. Lett.*, 176, pp. 68–70.

[101] Niu, Z., Dong, H., Zhu, B., Li, J., Hng, H. H., Zhou, W., Chen, X. and Xie, S. (2013). Highly stretchable, integrated supercapacitors based on single-walled carbon nanotube films with continuous reticulate architecture, *Adv. Mater.*, 25, pp. 1058–1064.

[102] Yu, C., Masarapu, C., Rong, J., Wei, B. and Jiang, H. (2009). Stretchable supercapacitors based on buckled single-walled carbon-nanotube macrofilms, *Adv. Mater.*, 21, pp. 4793–4797.

[103] Feng, C., Liu, K., Wu, J.-S., Liu, L., Cheng, J.-S., Zhang, Y., Sun, Y., Li, Q., Fan, S. and Jiang, K. (2010). Flexible, stretchable, transparent conducting films made from superaligned carbon nanotubes, *Adv. Funct. Mater.*, 20, pp. 885–891.

[104] Occhiuzzi, C., Rida, A., Marrocco, G. and Tentzeris, M. M. (Year). CNT-based RFID passive gas sensor. *In 2011 IEEE MTT-S International Microwave Symposium*, pp. 1–4.

[105] Andersson, M., Lloyd Spetz, A. and Pearce, R. (2013). *Semiconductor Gas Sensors*, eds. Raivo Jaaniso and Ooi Kiang Tan, Chapter 4: Recent trends in silicon carbide (SiC) and graphene-based gas sensors (Woodhead Publishing, Cambridge, UK) pp. 117–158.

[106] He, P., Cao, J., Ding, H., Liu, C., Neilson, J., Li, Z., Kinloch, I. A. and Derby, B. (2019). Screen-printing of a highly conductive graphene ink for flexible printed electronics, *ACS Appl. Mater. Interfaces*, 11, pp. 32225–32234.

[107] Secor, E. B. and Hersam, M. C. (2015). Emerging carbon and post-carbon nanomaterial inks for printed electronics, *J. Phys. Chem. Lett.*, 6, pp. 620–626.

[108] Secor, E. B., Prabhumirashi, P. L., Puntambekar, K., Geier, M. L. and Hersam, M. C. (2013). Inkjet printing of high conductivity, flexible graphene patterns, *J. Phys. Chem. Lett.*, 4, pp. 1347–1351.

[109] Gao, Y., Shi, W., Wang, W., Leng, Y. and Zhao, Y. (2014). Inkjet Printing Patterns of Highly Conductive Pristine Graphene on Flexible Substrates, *Ind. Eng. Chem. Res.*, 53, pp. 16777–16784.

[110] Kanso, H., González García, M. B., Llano, L. F., Ma, S., Ludwig, R., Fanjul Bolado, P. and Santos, D. H. (2017). Novel thin layer flow-cell screen-printed graphene electrode for enzymatic sensors, *Biosens. Bioelectron.*, 93, pp. 298–304.

[111] Seekaew, Y., Lokavee, S., Phokharatkul, D., Wisitsoraat, A., Kerdcharoen, T. and Wongchoosuk, C. (2014). Low-cost and flexible printed graphene–PEDOT: PSS gas sensor for ammonia detection, *Org. Electron.*, 15, pp. 2971–2981.

[112] Sire, C., Ardiaca, F., Lepilliet, S., Seo, J.-W. T., Hersam, M. C., Dambrine, G., Happy, H. and Derycke, V. (2012). Flexible gigahertz transistors derived from solution-based single-layer graphene, *Nano Lett.*, 12, pp. 1184–1188.

[113] Liu, R., Shen, F., Ding, H., Lin, J., Gu, W., Cui, Z. and Zhang, T. (2013). All-carbon-based field effect transistors fabricated by aerosol jet printing on flexible substrates, *J. Micromech. Microeng.*, 23, p. 065027.

[114] Xu, Y., Schwab, M. G., Strudwick, A. J., Hennig, I., Feng, X., Wu, Z. and Müllen, K. (2013). Screen-printable thin film supercapacitor device utilizing graphene/polyaniline inks, *Adv. Energy Mater.*, 3, pp. 1035–1040.

[115] Dodoo-Arhin, D., Howe, R. C. T., Hu, G., Zhang, Y., Hiralal, P., Bello, A., Amaratunga, G. and Hasan, T. (2016). Inkjet-printed graphene electrodes for dye-sensitized solar cells, *Carbon*, 105, pp. 33–41.

[116] Nayak, L., Mohanty, S., Nayak, S. K. and Ramadoss, A. (2019). A review on inkjet printing of nanoparticle inks for flexible electronics, *J. Mater. Chem. C*, 7, pp. 8771–8795.

[117] Pei, S. and Cheng, H.-M. (2012). The reduction of graphene oxide, *Carbon*, 50, pp. 3210–3228.

[118] Eredia, M., Ciesielski, A. and Samorì, P. (2016). Graphene via molecule-assisted ultrasound-induced liquid-phase exfoliation: A supramolecular approach, *Phys. Sci. Rev.*, 1. Retrieved from https://doi.org/10.1515/psr-2016-0101.

[119] Barwich, S. (2015). *A Study of Liquid Phase Exfoliation and Properties of 2D Nanomaterials* (Doctoral Dissertation, Trinity College Dublin).

[120] Tran, T. S., Dutta, N. K. and Choudhury, N. R. (2018). Graphene inks for printed flexible electronics: Graphene dispersions, ink formulations, printing techniques and applications, *Adv. Colloid Interface Sci.*, 261, pp. 41–61.

[121] Papageorgiou, D. G., Kinloch, I. A. and Young, R. J. (2015). Graphene/elastomer nanocomposites, *Carbon*, 95, pp. 460–484.

[122] Matsui, H., Takeda, Y. and Tokito, S. (2019). Flexible and printed organic transistors: From materials to integrated circuits, *Org. Electron.*, 75, p. 105432.

[123] Usta, H. and Facchetti, A. (2015). Polymeric and Small-Molecule Semiconductors for Organic Field-Effect Transistors, *Large Area and Flexible Electronics*, pp. 1–100.

[124] Chung, S., Cho, K. and Lee, T. (2019). Recent progress in inkjet-printed thin-film transistors, *Adv. Sci.*, 6, p. 1801445.

[125] Ahmad, S. (2014). Organic semiconductors for device applications: Current trends and future prospects, *J. Polym. Eng.*, 34, p. 279.

[126] Anthony, J. E., Brooks, J. S., Eaton, D. L. and Parkin, S. R. (2001). Functionalized Pentacene: Improved Electronic Properties from Control of Solid-State Order, *J. Amer. Chem. Soc.*, 123, pp. 9482–9483.

[127] Cho, S. Y., Ko, J. M., Lim, J., Lee, J. Y. and Lee, C. (2013). Inkjet-printed organic thin film transistors based on TIPS pentacene with insulating polymers, *J. Mater. Chem. C*, 1, pp. 914–923.

[128] Lee, S. H., Choi, M. H., Han, S. H., Choo, D. J., Jang, J. and Kwon, S. K. (2008). High-performance thin-film transistor with 6,13-bis(triisopropylsilyle thynyl) pentacene by inkjet printing, *Org. Electron.*, 9, pp. 721–726.

[129] Cho, S. Y., Ko, J. M., Jung, J.-Y., Lee, J. Y., Choi, D. H. and Lee, C. (2012). High-performance organic thin film transistors based on inkjet-printed polymer/TIPS pentacene blends, *Org. Electron.*, 13, pp. 1329–1339.

[130] Jiang, C. (2020). All-inkjet printed organic thin-film transistors with and without photo-sensitivity to visible lights, *Crystals*, 10, p. 727.

[131] Eom, S. H., Park, H., Mujawar, S. H., Yoon, S. C., Kim, S.-S., Na, S.-I., Kang, S.-J., Khim, D., Kim, D.-Y. and Lee, S.-H. (2010). High efficiency

polymer solar cells via sequential inkjet-printing of PEDOT:PSS and P3HT: PCBM inks with additives, *Org. Electron.*, 11, pp. 1516–1522.

[132] Xue, F., Liu, Z., Su, Y. and Varahramyan, K. (2006). Inkjet printed silver source/drain electrodes for low-cost polymer thin film transistors, *Microelectron. Eng.*, 83, pp. 298–302.

[133] Speakman, S. P., Rozenberg, G. G., Clay, K. J., Milne, W. I., Ille, A., Gardner, I. A., Bresler, E. and Steinke, J. H. G. (2001). High performance organic semiconducting thin films: Ink jet printed polythiophene [rr-P3HT], *Org. Electron.*, 2, pp. 65–73.

[134] Hoth, C. N., Choulis, S. A., Schilinsky, P. and Brabec, C. J. (2009). On the effect of poly(3-hexylthiophene) regioregularity on inkjet printed organic solar cells, *J. Mater. Chem.*, 19, pp. 5398–5404.

[135] Gamerith, S., Klug, A., Scheiber, H., Scherf, U., Moderegger, E. and List, E. J. W. (2007). Direct ink-jet printing of Ag–Cu nanoparticle and Ag-precursor based electrodes for OFET applications, *Adv. Funct. Mater.*, 17, pp. 3111–3118.

[136] Jung, J., Kim, D., Lim, J., Lee, C. and Yoon, S. C. (2010). Highly efficient inkjet-printed organic photovoltaic cells, *Jpn. J. Appl. Phys.*, 49, p. 05EB03.

[137] Yu, X., Marks, T. J. and Facchetti, A. (2016). Metal oxides for optoelectronic applications, *Nat. Mater.*, 15, pp. 383–396.

[138] Yu, K. J., Yan, Z., Han, M. and Rogers, J. A. (2017). Inorganic semiconducting materials for flexible and stretchable electronics, *Npj Flex. Electron.*, 1, p. 4.

[139] Thomas, S. R., Pattanasattayavong, P. and Anthopoulos, T. D. (2013). Solution-processable metal oxide semiconductors for thin-film transistor applications, *Chem. Soc. Rev.*, 42, pp. 6910–6923.

[140] Garlapati, S. K., Divya, M., Breitung, B., Kruk, R., Hahn, H. and Dasgupta, S. (2018). Printed Electronics Based on Inorganic Semiconductors: From Processes and Materials to Devices, *Adv. Mater.*, 30, p. 1707600.

[141] Varghese, J. and Sebastian, M. T. (2017). *Microwave Materials and Applications 2V Set*, eds. M.T. Sebastian, H. Jantunen and R. Ubic, Chapter 10: Dielectric inks (John Wiley & Sons, USA) pp. 457–480.

[142] Zhou, Y., Han, S.-T. and Roy, V (2014). *Nanocrystalline Materials*, Nanocomposite dielectric materials for organic flexible electronics (Elsevier) pp. 195–220.

[143] Serway, R. A., Beichner, R. J. and Jewett, J. W. (2000). *Physics for Scientists and Engineers with Modern Physics* (Cengage Learning, Boston, MA, USA).

[144] Kaija, K., Pekkanen, V., Mäntysalo, M., Koskinen, S., Niittynen, J., Halonen, E. and Mansikkamäki, P. (2010). Inkjetting dielectric layer for electronic applications, *Microelectron. Eng.*, 87, pp. 1984–1991.

[145] Khan, Y., Thielens, A., Muin, S., Ting, J., Baumbauer, C. and Arias, A. C. (2020). A new frontier of printed electronics: Flexible hybrid electronics, *Adv. Mater.*, 32, p. 1905279.

[146] Chang, J., Zhang, X., Ge, T. and Zhou, J. (2014). Fully printed electronics on flexible substrates: High gain amplifiers and DAC, *Org. Electron.*, 15, pp. 701–710.

[147] Kang, B. J., Lee, C. K. and Oh, J. H. (2012). All-inkjet-printed electrical components and circuit fabrication on a plastic substrate, *Microelectron. Eng.*, 97, pp. 251–254.

[148] Li, J., Tang, W., Wang, Q., Sun, W., Zhang, Q., Guo, X., Wang, X. and Yan, F. (2018). Solution-processable organic and hybrid gate dielectrics for printed electronics, *Mater. Sci. Eng.: R: Reports*, 127, pp. 1–36.

[149] Gaikwad, A. M., Khan, Y., Ostfeld, A. E., Pandya, S., Abraham, S. and Arias, A. C. (2016). Identifying orthogonal solvents for solution processed organic transistors, *Org. Electron.*, 30, pp. 18–29.

[150] Zhang, Z., Zheng, J., Premasiri, K., Kwok, M.-H., Li, Q., Li, R., Zhang, S., Litt, M. H., Gao, X. P. A. and Zhu, L. (2020). High-κ polymers of intrinsic microporosity: A new class of high temperature and low loss dielectrics for printed electronics, *Mater. Horiz.*, 7, pp. 592–597.

Problems

1. List the key parameters of a metallic nanoparticle ink.
2. Why the melting temperatures are significantly lower for metallic nanoparticles as compared to their bulk materials?
3. Why are organic additives and stabilising agents added to the formulation of the metallic nanoparticle inks? How can the addition of organic additives and stabilising agents in the ink formulations affect the ink printability and the final sintered printed patterns?
4. Compare and contrast the differences between metallic nanoparticle inks and MOD inks.
5. How are structural conductive polymers different from the composite conductive polymers?
6. What are the properties of SWCNTs that are favourable other for fabricating 3D printed flexible and stretchable electronic devices?
7. Why graphene has been gaining increasing attention for 3D printed electronics applications in the recent years? What are some of the 3D printed electronics applications that involve the use of graphene?
8. Why dielectric inks are important in 3D printed electronics and what are their primary usages?

Chapter 5

Substrates and Processing for 3D Printed Electronics

Conventionally, electronics are fabricated on rigid, brittle and planar substrates (for instance, FR-4, glass and silicon wafers). However, these rigid substrates lack conformability, bendability and stretchability that may be required for more innovative applications such as conformal electronics, flexible electronics, and stretchable electronics [1–4]. Therefore, substrates play a vital role in 3D printed electronics since they are the base materials for the deposition of functional inks and materials through various 3D electronics printing techniques. Most 3D electronics printing techniques have the capability to print directly onto planar and conformal surfaces and thus, giving the users a wide range of different type of substrates to choose from. The choice of substrates for 3D printed electronics is highly dependent on the ink properties, the required sintering techniques, and the intended application. Some of the critical properties of the substrates that influence the printing and device fabrication processes include surface wettability, surface smoothness, cost, optical transparency, heat resistance, thermal expansion, film thickness, mechanical properties such as, hardness, conformability, bendability and stretchability [5].

5.1 Compatibility Between Inks and Substrates

It is particularly vital to ensure good compatibility between inks and substrates to have good printing quality. Surface cleanliness, surface wettability, adhesion, porosity and surface roughness are some of the critical parameters to consider [6].

5.1.1 *Surface Cleanliness*

Surface cleanliness of substrates is one of the most important considerations to ensure good printing quality. Substrates may contain organic substances (such as greases and oils) and dust particles on their surfaces. Organic substances can decrease the wettability of the inks on the substrates and dust particles entrapped within the deposited inks may cause an insulating effect and degrade the electrical properties of the printed electrical components. Therefore, substrates must be thoroughly cleaned before ink depositions.

Literature shows that there are many different methods for cleaning different substrates. For instance, Werner *et al.* [7] demonstrated that glass substrates can be cleaned by treating the surfaces with acetone first to remove any organic contaminants. Isopropyl alcohol (IPA) is then used to rinse off the contaminated acetone from the surfaces. For polymer substrates, they are usually cleaned in ultrasonic baths using ethanol or IPA for 15 minutes to remove the bulk of the organic contaminants [8,9]. They are then cleaned in ultrasonic baths again with distilled water for another 15 minutes to remove any contaminated ethanol or IPA. The polymer substrates are air-dried before printing.

5.1.2 *Surface Wettability*

Surface wettability is the ability of fluid to spread over a surface [10] and it can be determined by the fluid contact angle [6]. A surface is said to have good wettability when the fluid spreads evenly over the surface without beading up [6], and the fluid contact angle is less than 90°. Good surface wettability can help achieve more uniform printing resolutions and features [11]. Both surface energy of substrate and surface tension of the ink can affect proper wetting of substrates. The surface wettability is also highly dependent on the substrates' surface morphology and chemical compositions [5]. Plasma treatments can be utilised to improve surface wettability on certain substrates, such as polymer substrates, by chemically and physically modifying their surfaces.

5.1.2.1 *Surface energy of substrates*

Surface energy is defined as "the work necessary to form unit area of surface by a process of division" [12]. The surface energy of a substrate describes "the degree of energy with which the molecules of the surface of a solid draw and allow adherence of a fluid" [6]. Therefore, surface energy also determines the substrates' strength of attraction and is a measure of surface wettability.

High surface energy materials, such as ceramics, glass, metals and oxides [13], have atoms that are strongly bonded to each other by ionic, covalent or metallic bonds [14]. Fluids that are deposited on these surfaces are attracted to the atoms by their strong attractive forces. Hence, high surface energy substrates tend to have better wettability and adhesion. Most polymers [13] have medium to low surface energy, as the molecules in these materials are bonded to each other by weak hydrogen bonds or *van der Waals* forces [14]. There is very little attraction for fluids that are deposited on these surfaces, and thus resulting in poor wettability.

5.1.2.2 *Surface tension of inks*

Surface tension is defined as "the work required to increase the area of a surface isothermally and reversibly by unit amount" [13] and also refers to as "the amount of cohesive forces between liquid molecules" [6]. Surface tension can affect the inks' printability and wettability on the substrates and it is also highly dependent on the ink's type and composition [5]. The surface energy of the substrate and the surface tension of the ink are closely interrelated, and therefore they must be properly matched for ensuring good surface wettability [15]. As a rule of thumb, the surface tension of the ink must be lower than the surface energy of the substrate for good wettability. The ink will bead up on the substrate's surface if the ink's surface tension is comparatively higher and may cause discontinuous lines and missing dots. If the surface tension of the ink is too much lower than the surface energy of substrate, the ink may be difficult to control or bleed and thereby limiting the printing resolution [16].

5.1.3 *Adhesion*

Adhesion is a very important factor for depositing inks onto the substrates and it is also affected by the surface energy between inks and substrates [16]. To have a strong adhesion between the ink and substrate, the substrate's surface must be free of contaminants. The ink must also properly wet the substrate to maintain a stable solid-liquid interface, and the surface energy of the substrate needs to exceed the surface tension of the ink [5,6]. A non-wetting or beading ink droplet on a substrate has a little surface area for adhesion and thus, having poorer adhesion [16]. There are also few techniques to improve ink adhesion on substrates, including surface treatments (e.g. plasma, chemical and flame treatments), mechanical or chemical surface roughening of surfaces, and addition of adhesion promoter to the ink formulations [16].

5.1.4 *Porosity*

The porosity of the substrate, which is dependent on the size and number of pores, can directly affect its permeability [5] and result in ink penetrating or absorbing into the substrate. Studies have shown that high ink penetration into the substrates generally causes lower attainable electrically conductivity [17]. Therefore, macroporous substrates like uncoated papers and textiles are not favourable to be used as substrates in printed electronics applications. A layer of coating is typically applied on top of these macroporous substrates to turn them into non-porous substrates. Note that the coating material must also be compatible with the deposited ink and sintering process [6]. For non-absorbent characteristics, non-porous or slightly microporous substrates such as polymer films and glass are usually preferred [5].

5.1.5 *Surface Roughness*

Surface roughness of the substrates can negatively affect ink deposition and printing quality, hence smooth substrate surfaces are highly preferred. Substrates with high surface roughness have a high presence of irregularities on them and usually can cause many serious printing problems such as disconnected lines and missing dots [18,19]. This is particularly evident when the deposited ink layer is very thin, as insufficient materials are available to fill up the surface valleys of the substrate [20,21]. However, substrates with high surface roughness tend to have better ink adhesion due to the increase of effective interface areas between inks and substrates [20].

5.2 Compatibility between Sintering Processes and Substrates

It is also critical to ensure that there is good compatibility between the sintering processes and substrates. In sintering processes, especially thermal sintering, substrates need to endure high temperatures which can range from 100°C to 300°C [22,23]. Therefore, the selected substrates should be dimensionally and mechanical stable against high temperatures to avoid substrate deformations and deteriorations [18].

5.3 Other Considerations for Substrates

The mechanical properties of the chosen substrates must suit the intended application so that the printed electronics devices can be as reliable as possible [24]. For instance, the chosen substrate intended for stretchable electronics applications must be able to withstand the imposed mechanical stresses and retain its geometric dimensions after numerous cycles of repeated stretching.

The material properties of substrates may also affect the electrical performances of 3D printed electronics, especially high-frequency radiofrequency (RF) devices, mainly because of dispersion and dielectric losses in the printed conductive traces. The dielectric properties of substrate materials are essential in designing antennas and filters to minimise signal losses [25,26].

5.4 Commonly Used 2D Substrates

There are countless substrate choices available in the market and it can be overwhelming to select a suitable and cost-effective substrate. This section generally discusses the unique features of some commonly used 2D substrates, including glass, polymer, paper, metallic foils and textiles. After evaluating the individual features of the different substrates, users can make a better choice to suit their intended application and as well as the ink and sintering technique used.

5.4.1 *Glass*

Glass is a very attractive transparent substrate material that has long been used in the electronics industry [27]. Glass has many appealing characteristics that are suitable for use in electronics applications, including high surface quality, high chemical resistance, very high thermo-mechanical stability, high optical transparency, and an excellent barrier to moisture and oxygen. Glass is a good substrate choice for photovoltaics, displays and lightings applications for its high optical transparency. However, the high rigidity, high brittleness, high cost and heavy weight of glass substrates make it unfavourable, especially in flexible electronics applications [28]. Although ultrathin glass is commercially available as flexible substrates, they are costly and have limited mechanical durability.

5.4.2 *Polymer Substrates*

A wide variety of polymer substrates have been used for printed electronic applications including polycarbonate (PC), polyethylene naphthalate (PEN), polyethylene terephthalate (PET) and polyimide (PI) [6]. Polymer substrates are highly applicable for flexible electronics applications due to their lightweight, flexibility, bendability, rollability, optical transparency and low cost. The glass transition and melting temperatures of these substrates are tabulated in Table 5.1.

Polymer substrates, like PC, PEN and PET, have lower temperature stabilities compared to glass, with glass transition temperatures around 150°C or even lower [29]. Hence, these substrates are not suitable for exposure to high processing temperatures during the sintering process as deformations and deteriorations of these substrates will occur. Therefore, sintering techniques with selective and low temperature sintering are highly preferred for these substrates. Most polymer substrates also have

Table 5.1.   Glass transition temperatures and melting temperatures of the various polymer substrates [6].

Polymer	Abbreviation	Glass Transition Temperature, T_g (°C)	Melting Temperature, T_m (°C)
Polycarbonate	PC	145	115–160
Polyethylene naphthalate	PEN	120–155	269
Polyethylene terephthalate	PET	70–110	115–258
Polyimide	PI	155–270	250–452

low surface energy and require surface pre-treatments to improve ink wettability and adhesion [6].

Polyimide is a high performance, high temperature engineering polymer which is known for its excellent mechanical properties, high thermal dimensional stability, good chemical resistance and its characteristic yellow colour [30,31]. Therefore, polyimide is widely used in printed electronics applications as it can withstand the high temperatures in sintering processes. However, polyimide is considerably expensive to be used as substrates in low-cost electronics applications.

5.4.3 *Paper Substrates*

Paper is a widely available, low cost, organic-based substrate. In recent years, paper substrates have gained increasing attention for low cost disposable electronic devices due to their environmentally friendly nature and sustainability [32–35]. Paper substrates are also flexible, foldable and bendable, and thereby leading to many other interesting applications including foldable circuits [36] and flexible displays [37]. Transparent nanocellulose papers have excellent optical properties and can also be used as display windows for paper electronics [28,38]. Applications of coatings or passivation layers [37] may be needed to help resolve high ink permeability, high surface roughness and high moisture sensitivity issues from which uncoated paper substrates face [32,39]. However, the additional step of applying coatings and passivation layers can increase cost and processing time.

5.4.4 *Metallic Foils*

Metallic foils can be used as substrates for printed electronics applications (for instance, organic light-emitting diode (OLED) top-emitting-based displays [40], OLED lightings [41], organic thin-film transistors (OTFTs) [42], TFT display backplanes [43] and organic photovoltaics [44]). Metallic foils have favourable characteristics such as low moisture and air permeability, good dimensional stability, good chemical resistance, high heat resistance and good flexibility [6,28,31,45]. Metallic foils may also function as a common voltage terminal for shielding or grounding, owing to their electrical conductive nature [45].

Since the metallic foils are electrically conductive, it is infeasible to deposit functional inks directly on their surfaces. Electrically insulating passivation layers are usually required to be applied on their surfaces to prevent the printed patterns from short circuiting [28]. These passivation layers can also help to reduce surface roughness and cover minor morphological defects on the metallic foils' surfaces [41]. Similarly, application of passivation layers on metallic foils can increase cost and processing time. Metallic foils are opaque and they usually more expensive and heavier than the other types of substrates too [28]. Hence, limiting their applications and affecting their cost effectiveness.

5.4.5 *Textiles*

Textiles, also known as fabrics, are made of natural or synthetic fibres. Their soft, flexible, conformable, lightweight and breathable characteristics make them ideal substrate choices for wearable electronics and sensors [46–48]. Textiles can also be twisted, stretch, shear, fold or bent while maintaining their mechanical properties. Due to textiles' high porosity and surface roughness, it may be challenging to deposit inks onto their surfaces and more print passes may be required [49]. Hence, densely packed fabric structures are preferred for better printing performance. Some of the other interesting applications also include conformable RFIDs [50], pressure sensors [51] and flexible circuits [52].

5.5 3D Substrates

As mentioned before, 3D electronics printing techniques have the capabilities to print directly onto most conformal surfaces. Hence inspiring many innovative applications that were previously unachievable by using the common 2D substrates. Many industries have anticipated the integration of 3D electronics printing techniques with 3D substrates for more significant weight reduction and better space utilisation. **Figure 5.1** shows that electrically conductive antennas can be fabricated by depositing metallic inks directly onto convex and concave hemispherical surfaces [53] and thereby demonstrates the full optimisations of available spaces in electronic devices.

Similar to 2D substrates, the surface cleanliness, surface wettability, adhesion, porosity, surface roughness and material properties of 3D substrates can affect the final printing quality. The 3D substrates are either moulded through conventional techniques or be 3D printed by various

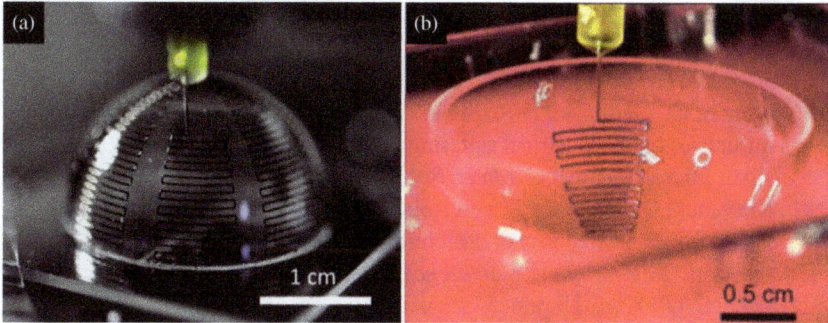

Figure 5.1. Electrically conductive antennas fabricated by depositing metallic inks directly onto: (a) convex; and (b) concave hemispherical surfaces. Reprinted with permission from Ref. [53]. Copyright (2011) from John Wiley & Sons, Inc.

3D printing technologies such as fused filament fabrication (FFF) or stereolithography (STL It is important to point out that the geometries of the 3D substrate can have a substantial effect on printability and compatibility with inks. For instance, surface wetting properties are dependent on both macroscopic curvature and microscopic surface topographies [54]. Printing of electrically conductive ink on curved surfaces is a relatively new technology and further research is needed to fully establish the printability of such substrates.

References

[1] Huang, Y., Wu, H., Xiao, L., Duan, Y., Zhu, H., Bian, J., Ye, D. and Yin, Z. (2019). Assembly and applications of 3D conformal electronics on curvilinear surfaces, *Mater. Horiz.*, 6, pp. 642–683.

[2] Tan, H. W., An, J., Chua, C. K. and Tran, T. (2019). Metallic nanoparticle inks for 3D printing of electronics, *Adv. Electron. Mater.*, 5, p. 1800831.

[3] Tan, H. W., Saengchairat, N., Goh, G. L., An, J., Chua, C. K. and Tran, T. (2020). Induction sintering of silver nanoparticle inks on polyimide substrates, *Adv. Mater. Technol.*, 5, p. 1900897.

[4] Tan, H. W., Tran, T. and Chua, C. K. (2018). Review of 3D printed electronics: Metallic nanoparticles inks, *presented at the 3rd International Conference on Progress in Additive Manufacturing (Pro-AM 2018)*, Singapore.

[5] Izdebska-Podsiadły, J. and Thomas, S. (2015). *Printing on Polymers: Fundamentals and Applications* (William Andrew, USA).

[6]   Cruz, S. M. F., Rocha, L. A. and Viana, J. C. (2018). *Flexible Electronics*, Chapter 2: Printing technologies on flexible substrates for printed electronics (IntechOpen).

[7]   Werner, C., Godlinski, D., Zöllmer, V. and Busse, M. (2013). Morphological influences on the electrical sintering process of aerosol jet and ink jet printed silver microstructures, *J. Mater. Sci.: Mater. Electron.*, 24, pp. 4367–4377.

[8]   Kravchuk, O., Bobitski, Y. and Reichenberger, M. (2016). Electrical sintering of inkjet printed sensor structures on polyimide substrate, *presented at the 36th International Conference on Electronics and Nanotechnology (ELNANO)*, Kiev, Ukraine.

[9]   Chung, W.-H., Hwang, H.-J. and Kim, H.-S. (2015). Flash light sintered copper precursor/nanoparticle pattern with high electrical conductivity and low porosity for printed electronics, *Thin Solid Films*, 580, pp. 61–70.

[10] Praveen, K. M., Pious, C. V., Thomas, S. and Grohens, Y. (2019). *Non-Thermal Plasma Technology for Polymeric Materials*, eds. Sabu Thomas, Miran Mozetič, Uroš Cvelbar, Petr Špatenka and Praveen K.M, Chapter 1: Relevance of plasma processing on polymeric materials and interfaces (Elsevier) pp. 1–21.

[11] Dong Jun, L., Sung Hyeon, P., Shin, J., Hak Sung, K., Je Hoon, O. and Yong Won, S. (2011). Pulsed light sintering characteristics of inkjet-printed nanosilver films on a polymer substrate, *J. Micromech. Microeng.*, 21, p. 125023.

[12] Shuttleworth, R. (1950). The surface tension of solids, *Proceed. Phys. Soc. Section A*, 63, pp. 444–457.

[13] Ebnesajjad, S. (2011). *Handbook of Adhesives and Surface Preparation*, eds. Sina Ebnesajjad, Chapter 3: Surface tension and its measurement (William Andrew Publishing, Oxford) pp. 21–30.

[14] Gözen, I., Dommersnes, P. and Jesorka, A. (2015). *Surface Energy*, eds. Aliofkhazraei Mahmood, Chapter 12: Lipid self-spreading on solid substrates (InTech, Croatia) p. 337.

[15] Janule Victor, P. (1995). On-site surface and wetting tension measurements of water-based coatings and substrates, *Pigment & Resin Technology*, 24, pp. 7–12.

[16] Clem, P. G., Bell, N. S., Brennecka, G. L., Dlmos, D. B. and King, B. H. (2002). *Direct-Write Technologies for Rapid Prototyping*, eds. Alberto Piqué, Chapter 8: Micropen printing of electronic components (Academic Press, San Diego) pp. 229–259.

[17] Hrehorova, E., Pekarovicova, A., Bliznyuk, V. and Fleming, P. D. (2007). Polymeric materials for printed electronics and their interactions with paper substrates, *presented at the NIP & Digital Fabrication Conference, 2007 International Conference on Digital Printing Technologies*, Anchorage, AK, USA.

[18] Torvinen, K., Sievänen, J., Hjelt, T. and Hellén, E. (2012). Smooth and flexible filler-nanocellulose composite structure for printed electronics applications, *Cellulose*, 19, pp. 821–829.
[19] Aijazi, A. T. (2014). *Printing Functional Electronic Circuits and Components* (Doctoral dissertation, Western Michigan University).
[20] Ryan, A. and Lewis, H. (2012). Effect of surface roughness on paper substrate circuit board, *IEEE Transactions on Components, Packaging and Manufacturing Technology*, 2, pp. 1202–1208.
[21] Thorman, S. (2015). *Absorption Non-uniformity Characterisation and Its Impact on Flexographic Ink Distribution of Coated Packaging Boards* (Licentiate Thesis, Karlstad University).
[22] Hsien-Hsueh, L., Kan-Sen, C. and Kuo-Cheng, H. (2005). Inkjet printing of nanosized silver colloids, *Nanotechnology*, 16, p. 2436.
[23] Gaspar, C., Passoja, S., Olkkonen, J. and Smolander, M. (2016). IR-sintering efficiency on inkjet-printed conductive structures on paper substrates, *Microelectron. Eng.*, 149, pp. 135–140.
[24] Kim, D., Kwak, Y. and Park, J. (2018). Effect of substrates on the dynamic properties of inkjet-printed Ag thin films, *Applied Sciences*, 8, p. 195.
[25] Lim, Y. Y. (2015). *Printing Conductive Traces to Enable High Frequency Wearable Electronics Applications* (Doctoral Thesis Loughborough University).
[26] Rosker, E. S., Sandhu, R., Hester, J., Goorsky, M. S. and Tice, J. (2018). Printable materials for the realization of high performance RF components: Challenges and opportunities, *Int. J. Antenn. Propag.*, 2018, p. 9359528.
[27] Hrehorova, E., Rebros, M., Pekarovicova, A., Bazuin, B., Ranganathan, A., Garner, S., Merz, G., Tosch, J. and Boudreau, R. (2011). Gravure printing of conductive inks on glass substrates for applications in printed electronics, *J. Disp. Technol.*, 7, pp. 318–324.
[28] Suganuma, K. (2014). *Introduction to Printed Electronics* (Springer Science+Business Media, New York).
[29] Wunscher, S., Abbel, R., Perelaer, J. and Schubert, U. S. (2014). Progress of alternative sintering approaches of inkjet-printed metal inks and their application for manufacturing of flexible electronic devices, *J. Mater. Chem. C*, 2, pp. 10232–10261.
[30] McKeen, L. W. (2014). *The Effect of Long Term Thermal Exposure on Plastics and Elastomers*, eds. Laurence W. McKeen, Chapter 6: Polyimides (William Andrew Publishing, Oxford) pp. 117–137.
[31] Lu, Q.-H. and Zheng, F. (2018). *Advanced Polyimide Materials*, eds. Shi-Yong Yang, Chapter 5: Polyimides for electronic applications (Elsevier) pp. 195–255.
[32] Quddious, A., Yang, S., Khan, M. M., Tahir, F. A., Shamim, A., Salama, K. N. and Cheema, H. M. (2016). Disposable, paper-based, inkjet-printed

humidity and H2S gas sensor for passive sensing applications, *Sensors*, 16, p. 2073.

[33] Rida, A., Yang, L., Vyas, R. and Tentzeris, M. M. (2009). Conductive inkjet-printed antennas on flexible low-cost paper-based substrates for RFID and WSN applications, *IEEE Antennas and Propagation Magazine*, 51, pp. 13–23.

[34] Orecchini, G., Alimenti, F., Palazzari, V., Rida, A., Tentzeris, M. M. and Roselli, L. (2011). Design and fabrication of ultra-low cost radio frequency identification antennas and tags exploiting paper substrates and inkjet printing technology, *IET Microw. Antenna P.*, 5, pp. 993–1001.

[35] Li, S. and Lee, P. S. (2017). Development and applications of transparent conductive nanocellulose paper, *Sci. Technol. Adv. Mater.*, 18, pp. 620–633.

[36] Siegel, A. C., Phillips, S. T., Dickey, M. D., Lu, N., Suo, Z. and Whitesides, G. M. (2010). Foldable printed circuit boards on paper substrates, *Adv. Funct. Mater.*, 20, pp. 28–35.

[37] Kim, J., Park, S. H., Jeong, T., Bae, M. J., Song, S., Lee, J., Han, I. T., Jung, D. and Yu, S. (2010). Paper as a substrate for inorganic powder electroluminescence devices, *IEEE Trans. Electron Devices*, 57, pp. 1470–1474.

[38] Hoeng, F., Denneulin, A. and Bras, J. (2016). Use of nanocellulose in printed electronics: A review, *Nanoscale*, 8, pp. 13131–13154.

[39] Xie, L., Mäntysalo, M., Cabezas, A. L., Feng, Y., Jonsson, F. and Zheng, L.-R. (2012). Electrical performance and reliability evaluation of inkjet-printed Ag interconnections on paper substrates, *Mater. Lett.*, 88, pp. 68–72.

[40] Afentakis, T., Hatalis, M., Voutsas, A. and Hartzell, J. (2003). Poly-silicon TFT AM-OLED on thin flexible metal substrates, *presented at the Electronic Imaging 2003*, Santa Clara, CA, United States.

[41] Guaino, P., Maseri, F., Schutz, R., Hofmann, M., Birnstock, J., Avril, L. L., Pireaux, J.-J., Viville, P., Kanaan, H. and Lazzaroni, R. (2011). Large white organic light-emitting diode lighting panel on metal foils, *J. Photonics. Energy*, 1, p. 011015.

[42] Chen, F.-C., Chen, T.-D., Zeng, B.-R. and Chung, Y.-W. (2011). Influence of mechanical strain on the electrical properties of flexible organic thin-film transistors, *Semicond. Sci. Technol.*, 26, p. 034005.

[43] Kattamis, A. Z., Giebink, N., Cheng, I.-C., Wagner, S., Forrest, S. R., Hong, Y. and Cannella, V. (2007). Active-matrix organic light-emitting displays employing two thin-film-transistor a-Si:H pixels on flexible stainless-steel foil, *J. Soc. Inf. Disp.*, 15, pp. 433–437.

[44] Gaynor, W., Lee, J.-Y. and Peumans, P. (2009). Fully solution-processed organic solar cells on metal foil substrates, *presented at the SPIE Photonic Devices + Applications,*, San Diego, California, United States.

[45] D'Andrade, B. W., Kattamis, A. Z. and Murphy, P. F. (2015). *Handbook of Flexible Organic Electronics*. Flexible organic electronic devices on metal foil substrates for lighting, photovoltaic, and other applications, pp. 315–341.

[46] Zeng, W., Shu, L., Li, Q., Chen, S., Wang, F. and Tao, X.-M. (2014). Fiber-based wearable electronics: A review of materials, fabrication, devices, and applications, *Adv. Mater.*, 26, pp. 5310–5336.

[47] Castano, L. M. and Flatau, A. B. (2014). Smart fabric sensors and e-textile technologies: A review, *Smart Mater. Struct.*, 23, p. 053001.

[48] Acar, G., Ozturk, O., Golparvar, A. J., Elboshra, T. A., Böhringer, K. and Yapici, M. K. (2019). Wearable and flexible textile electrodes for biopotential signal monitoring: A review, *Electronics*, 8, p. 479.

[49] Shahariar, H., Kim, I., Soewardiman, H. and Jur, J. S. (2019). Inkjet printing of reactive silver ink on textiles, *ACS Appl. Mater. Interfaces*, 11, pp. 6208–6216.

[50] Rao, S., Llombart, N., Moradi, E., Koski, K., Bjorninen, T., Sydanheimo, L., Rabaey, J. M., Carmena, J. M., Rahmat-Samii, Y. and Ukkonen, L. (2014). Miniature implantable and wearable on-body antennas: Towards the new era of wireless body-centric systems [antenna applications corner], *IEEE Antennas and Propagation Magazine*, 56, pp. 271–291.

[51] Li, Z. and Wang, Z. L. (2011). Air/liquid-pressure and heartbeat-driven flexible fiber nanogenerators as a micro/nano-power source or diagnostic sensor, *Adv. Mater.*, 23, pp. 84–89.

[52] Kim, D.-H., Kim, Y.-S., Wu, J., Liu, Z., Song, J., Kim, H.-S., Huang, Y. Y., Hwang, K.-C. and Rogers, J. A. (2009). Ultrathin silicon circuits with strain-isolation layers and mesh layouts for high-performance electronics on fabric, vinyl, leather, and paper, *Adv. Mater.*, 21, pp. 3703–3707.

[53] Adams, J. J., Duoss, E. B., Malkowski, T. F., Motala, M. J., Ahn, B. Y., Nuzzo, R. G., Bernhard, J. T. and Lewis, J. A. (2011). Conformal printing of electrically small antennas on three-dimensional surfaces, *Adv. Mater.*, 23, pp. 1335–1340.

[54] Lee, W. L., Wang, D., Wu, J., Ge, Q. and Low, H. Y. (2019). Injection molding of superhydrophobic submicrometer surface topography on macroscopically curved objects: Experimental and simulation studies, *ACS Appl. Polym. Mater.*, 1, pp. 1547–1558.

Problems

1. Why is it important to have good compatibility between the inks and substrates? If so, which are some of the critical parameters to consider?

2. Define surface wettability and explain why it is an important factor to consider for achieving good printability.
3. Why is it important to have good compatibility between the sintering processes and substrates?
4. In your opinion, discuss and explain which substrate material is most suitable for fabricating wearable heart rate monitoring devices.

Chapter 6

Sintering Techniques for Metallic Nanoparticle Inks

Metallic nanoparticle inks [1–3] are commercially available in the market and widely used for printed electronics applications. They are typically favoured over metal-organic decomposition (MOD) inks as they have better electrical conductance [4] and omit the need for additional chemical reduction processes [5]. However, undesirable coffee ring effects are prominent and inevitable on deposited metallic nanoparticle inks, in which metallic nanoparticles tend to concentrate on the perimeters [6].

As discussed in the earlier chapter, metallic nanoparticle inks are suspensions of metallic nanoparticles in liquid mediums (see **Figure 6.1(a)**), in which organic additives and stabilising agents encapsulate each metallic nanoparticle to prevent agglomeration. Although the bulk of the liquid medium is removed through evaporation after the inks are deposited onto the substrates, the metallic nanoparticles are still encapsulated in organic additives and stabilising agents. The organic additives and stabilising agents prevent the contact of neighbouring nanoparticles with each other (see **Figure 6.1(b)**), thus preventing the flow of electrons. Consequently, it causes the as-deposited metallic nanoparticle inks to be initially electrically non-conductive. These organic additives and stabilising agents typically can be decomposed at operating temperatures ranging from 200–350°C in a time frame of 10–60 minutes [7]. Therefore, an irreversible [8] and essential post-processing process, known as the sintering process, is required to introduce electrical conductance within the printed patterns. The sintering process is critical in the 3D printing of electronics

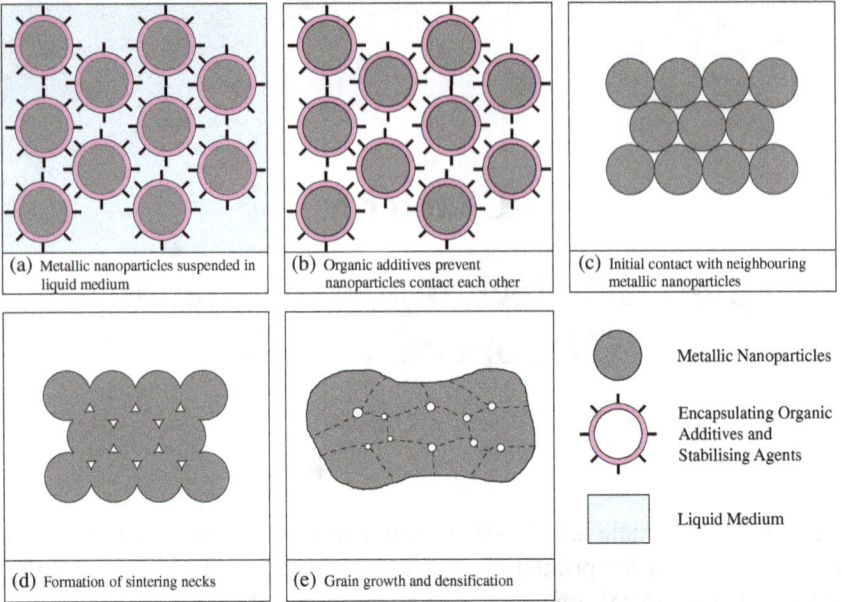

Figure 6.1.   Sintering processes of the metallic nanoparticle inks [1].

as it can influence the final electrical conductivity of the printed patterns.

This chapter will cover some of the most commonly used sintering techniques for sintering metallic nanoparticle inks, namely thermal sintering, laser sintering, infrared (IR) sintering, intense pulse light (IPL) sintering, ultraviolet (UV) sintering, microwave sintering, electrical sintering, low-pressure plasma sintering, localised atmospheric plasma sintering and chemical sintering (see **Figure 6.2**). This chapter particularly emphasises on each sintering technique's working principle, and its strengths and weaknesses.

6.1 Physical Principles of Sintering

The sintering process is a process whereby energy is introduced to the as-deposited metallic nanoparticle inks and heat is generated to fuse neighbouring metallic nanoparticles into a single continuous entity. The introduction of energy to the as-deposited metallic nanoparticle inks can be done through many means, such as thermal conduction, convection and radiation [9], laser [10–13], IR irradiations [14], high intensity white light

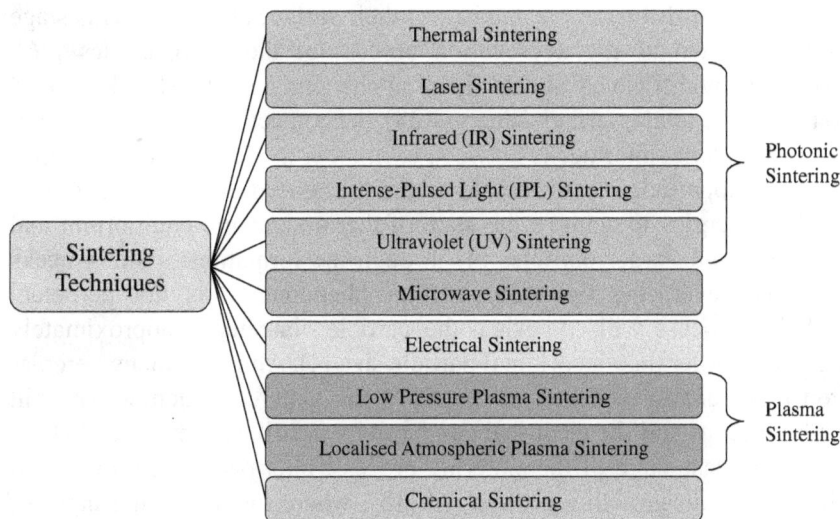

Figure 6.2.   Sintering techniques for metallic nanoparticle inks.

[15], UV irradiations, microwaves [16], electricity [17], plasma [18–23], and chemical reactions. As each nanoparticle absorbs the introduced energy, it will bring about a rise in the temperature. The sintering process takes place at the temperature above the glass transition temperature but below the melting temperature of the metallic nanoparticle [24].

To achieve the optimal electrical conductance, the as-deposited metallic nanoparticle inks should experience these three different sequential stages in an ideal sintering process: initial contact with neighbouring metallic nanoparticles [4], the formation of sintering necks, and grain growth and densification.

In this first stage, initial contacts with neighbouring metallic nanoparticles are formed [13,25] (see **Figure 6.1(c)**) after decomposition of the enveloping organic additives and stabilising agents in the as-deposited metallic nanoparticle inks [4]. Electrons can now flow freely throughout the entire printed patterns [5,13]. However, the electrical resistivity of the printed patterns is still very high [4]. Note that shrinkage in the volume of the as-deposited metallic nanoparticle inks can also be observed during the first stage of the sintering process [13].

At the second stage of the sintering process, sinter necks are formed between adjacent nanoparticles (see **Figure 6.1(d)**). These initial neck formations between neighbouring nanoparticles are induced by the enhanced self-diffusion mechanism of their surface atoms [7]. This stage is also known as the aggregation process of the nanoparticles [26]. Triggered by diffusive Ostwald ripening phenomenon [4,27] and the need to reduce the total interfacial energy [28], the contacting metallic nanoparticles can form sinter necks between each other at elevated temperatures. These nanoparticles, with the aid of the introduced energy, release high surface energies to achieve the state of thermodynamic equilibrium and aggregate into larger particles [4] through the formation of sinter necks [29]. However, the Ostwald ripening phenomenon is not perpetual and this process will end when the particle diameter is approximately 1.5 times larger than its original particle size [4]. Although many percolation paths are formed at this stage [7], the sintered structures are still highly porous and the electrical conductivity is still not optimised [4].

In the last stage of the sintering process, inter-particle atomic diffusion and grain growth are observed [13], where the additional input of energy and sintering time promote grain growth and densification (see **Figure 6.1(e)**) [28,30]. Besides, the densification process is also affected by the grain growth kinetics and initial grain size [28]. Ingham *et al.* [26] observed that the grain growth process took substantially longer than the nanoparticles aggregation process and the grain boundaries within the sintered structures do affect the electrical conductance. Larger grain sizes can help to facilitate electrons movements by widening electrons flow paths and reduce the electrical resistivity of the sintered patterns. Therefore, highly dense microstructures with larger grain sizes and low porosities can significantly reduce the electrical resistivity and improve the electrical performance of the sintered patterns. Hence, electrons can flow throughout the entire printed patterns to conduct electricity, where every individual particle is integrated into a single continuous entity to form a continuous path, having electrical conductivity comparable to the bulk material.

6.2 Thermal Sintering

Thermal sintering, also known as conventional oven sintering [9], is one of the most widely used and simplest sintering techniques for sintering metallic nanoparticle inks. Thermal energy can be transferred from the heating elements to the printed patterns via conduction, convection and radiation [31]. The required operating temperatures of the thermal

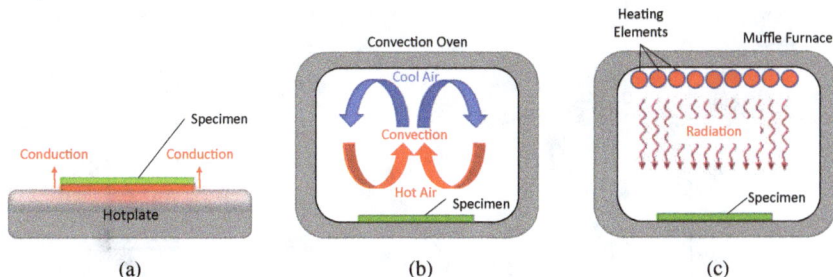

Figure 6.3.   Schematic diagrams of thermal sintering equipment: (a) hotplate, (b) convection oven and (c) muffle furnace.

sintering process usually range from 100°C to 300°C [32,33] and the printed patterns can be simply sintered through the use of hotplates, convection ovens and muffle furnaces (see **Figure 6.3**). Hotplates transmit thermal heat through conduction, while convection ovens and muffle furnaces transmit through convection and radiation respectively.

During the thermal sintering process, thermal energy first decomposes the organic additives and stabilising agents that envelop the metallic nanoparticles. The organic additives and stabilising agents typically can be decomposed at temperatures ranging from 200–350°C in a time frame of 10–60 minutes [7]. Without the enveloping organic additives, the exposed metallic nanoparticles can form sintering necks between adjacent nanoparticles and coalescences take place. Electrons are now able to flow through the entire printed pattern as a continuous path to conduct electricity [32,34]. Grain growth and densification of the metallic nanoparticles film further take place with an additional input of thermal energy [34].

As a rule of thumb, longer sintering time and higher sintering temperatures usually lead to an improvement of electrical conductivity of the printed metallic nanoparticles film. Greer and Street [35] revealed a reduction in the electrical resistivity of silver nanoparticles films with increasing isothermal sintering time(see **Figure 6.4(a)**). The evolution of the morphology of a printed silver nanoparticle film after sintering for various sintering time at 150°C [35] is shown in **Figures 6.4(b)–(d)**. From these figures, it can be observed that the silver nanoparticle film densifies as the sintering time increases, which is also associated with the decrease in electrical resistivity.

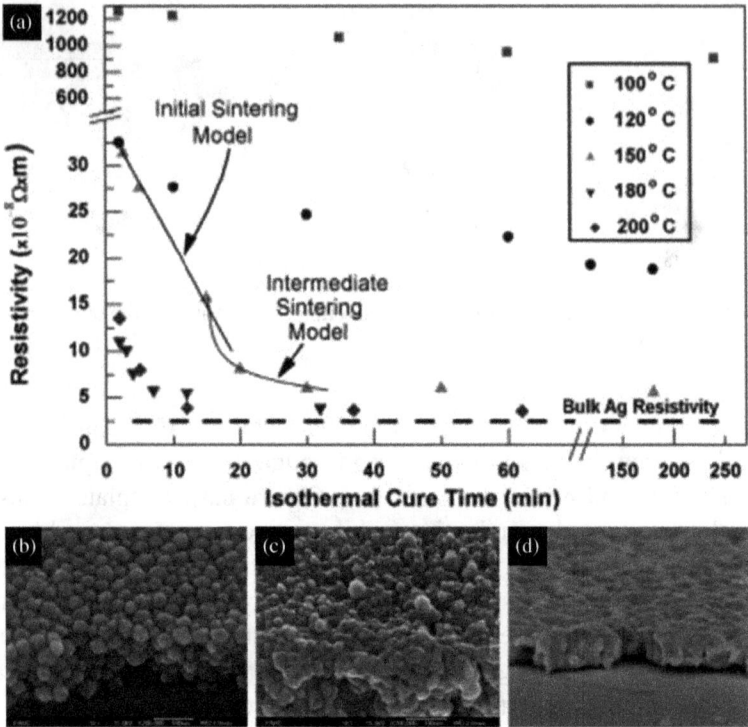

Figure 6.4.   Electrical resistivity of 100 nm silver nanoparticle film as a function of sintering time at various sintering temperatures as compared to bulk silver resistivity; and scanning electron microscope (SEM) images of silver nanoparticle film at various sintering conditions: (a) as dried; (b) 150°C for 5 minutes; and (c) 150°C for 3 hours. Adapted from Ref. [35], Copyright (2007), with permission from Elsevier.

From **Figure 6.4(a)**, it is also prominent that the electrical resistivity of the silver nanoparticles films decreases considerably at higher sintering temperatures [35] while achieving near bulk silver resistivity. The rates of densification and grain growth are accelerated with increasing sintering temperatures, which in turn improves the electrical conductivity of the printed films [30]. Hence, achieving better electrical conductivity by increasing the sintering temperatures [10]. The key variables that decide the sintering time and temperature are the material, particle size and particle shape of the metallic nanoparticles, and as well as the decomposition temperatures of the organic additives and stabilising agents present in the ink [13].

6.2.1 *Strengths*

The key strengths of thermal sintering are:

(a) *Uncomplicated and straightforward*: Thermal sintering is an uncomplicated and very straightforward sintering process, which only involves optimisation of the sintering time and temperature.
(b) *Simple sintering equipment*: Thermal sintering requires very simple sintering equipment only, such as hotplates, convection ovens and muffle furnaces.
(c) *Reliable sintering process*: Thermal sintering, relative to other sintering techniques, is one of the most reliable techniques for sintering metallic nanoparticle inks [9] as thermal sintering can offer a high degree of predictability and control [36].

6.2.2 *Weaknesses*

The key weaknesses of thermal sintering are:

(a) *Time-consuming*: Thermal sintering is a very time-consuming process and can take up to several hours [33], hence lengthening the total fabrication time [37].
(b) *Not suitable for temperature sensitive substrates*: Both the printed patterns and substrates are equally exposed to heat during the thermal sintering process [36]. Hence, the mechanical properties of the substrates restrict the maximum thermal sintering temperature. Heat sensitive substrates, such as coated papers and cheap polymer films substrates, are very reactive to the applied heat and not suitable to be heated at high temperatures as deformation and deterioration of the substrates will occur [23].
(c) *Deformation and shrinkage*: Thermal sintering over a long period at high temperatures can often induce thermal stresses onto the substrates, causing deformations and shrinkages that result in microcracks within the printed patterns [33]. These microcracks can disrupt the electrical conductivity within the printed patterns. Thus, the thermal sintering process must be coupled with the appropriate substrates with extra care.
(d) *Not suitable for multi-layered, multi-materials printed components*: Multi-layered, multi-materials printed components require the

previous layer to be sintered first before the next subsequent layer of another material type can be printed directly on it [12]. Therefore, it is time-consuming to use the thermal sintering technique for sintering multi-layered multi-materials printed components. In addition, thermal ageing [33] and deterioration of temperature sensitive substrates may also occur after many repetitions of this thermal sintering process under prolonged exposures at elevated temperatures.

6.3 Photonic Sintering

Electromagnetic (EM) irradiations specifically from the near-infrared (NIR), visible and ultraviolet regions of the EM spectrum are used in photonic sintering processes for sintering metallic nanoparticle inks. The photonic sintering processes comprise of laser sintering, IR sintering, IPL sintering and UV sintering, where each sintering process is categorised by its corresponding range of EM irradiation excitation frequencies [23].

Metallic nanoparticles, such as silver, gold and copper, are known to exhibit interesting optical phenomenon especially when they interact with EM irradiations from the visible light region [23,38]. The scattering and absorption rates of the metallic nanoparticles are significantly intensified [39] when exposed to the required optical power density and optimal irradiation excitation frequency [23]. This phenomenon is known as the surface plasmon resonance (SPR), whereby free electrons are 'oscillated under resonant conditions' throughout the entire nanoparticle in the presence of light [23,38–41]. SPR is triggered when the wavelengths of the EM irradiations are significantly greater than the nanoparticles' diameters [38]. The light energy absorbed by the metallic nanoparticles is rapidly transformed into thermal energy under the SPR effect and is used for localised heating [23,39,41]. The thermal energy decomposes the encompassing organic additives and stabilising agents and promotes coalescence of adjacent metallic nanoparticles to form conductive tracks.

Many studies have shown that the SPR effect is often determined by the nanoparticles' sizes, shapes, structures and materials, and the dielectric properties of the organic additives and stabilising agents that surround the nanoparticles [23,38,39]. Therefore, different metallic nanoparticle ink formulations require different irradiation excitation frequency [23,41–45].

Since the metallic nanoparticle inks and substrates have different optical properties and absorption characteristics, it is possible to couple metallic nanoparticle inks with suitable substrates to exploit the possibility of selective heating. The metallic nanoparticle inks should be highly absorptive in the employed wavelength range, whereas the selected substrates should be as reflective or transparent as possible in minimising EM irradiations absorption to prevent substrate degradation [23].

6.3.1 *Laser Sintering*

The laser sintering process involves the use of a focused laser beam at a tuned wavelength [23], that is integrated into a scanning system, to sinter the metallic nanoparticle inks. The laser transfers and directs the energy directly onto a small localised area of metallic nanoparticles encompassed by the laser beam [46]. The metallic nanoparticles absorb the laser irradiation and get rapidly heated up due to the photothermal effect [23], thus giving a rapid sintering effect [13]. The critical parameters of laser sintering include writing speed, type of metallic nanoparticle inks, number of treatment cycles and the laser type, wavelength and power [23].

There are two approaches to use the laser sintering technique for sintering metallic nanoparticle inks and creating conductive structures for 3D printed electronics. The first approach creates conductive structures through patterning by laser sintering [47]. A layer of metallic nanoparticle ink is first deposited onto the substrate. The laser beam selectively scans over the ink, and the exposed ink gets sintered and bonded to the substrate. The unsintered ink is then washed away with solvents to reveal the conductive patterns [48]. The laser spot size primarily defines the resolution of conductive structures [23]. The second approach is to first create the printed patterns using 3D electronics printing techniques and then the laser selectively scans over the printed patterns to sinter them [46]. The second approach can minimise material wastage and eliminate the washing step that is required in the first approach.

Pulse and continuous-wave (CW) lasers are generally used for the laser sintering process [12,46], with each having its advantages. Pulse lasers usually produce pulses with a higher intensity that can allow more material depth penetration, whereas CW lasers can achieve more uniform heating [12]. The use of short laser pulses in pulse lasers can also help in minimising heat dissipations and reduce thermal stress to the substrates. Note that the penetrating depth is also affected by the laser's wavelength and the metallic nanoparticles' absorptivity.

6.3.1.1 *Strengths*

The key strengths of laser sintering are:

(a) *Fast heating*: The laser sintering technique can sinter the metallic nanoparticles in a relatively short time locally [49]. With its fast sintering speed, laser sintering is also ideal for sintering metallic nanoparticles with very fast oxidation rates, such as copper nanoparticles [13,37], under ambient conditions to prevent oxidation.

(b) *Good control of sintering parameters*: The laser wavelength can be matched to the absorption maximum of the metallic nanoparticles to maximise energy transfer efficiency of metallic nanoparticles [47] for optimal sintering speed, electrical conductivity and structure morphology [23]. The position and power of the laser beam can also be digitally controlled for sintering the printed patterns [37,49]. For instance, more laser power can be used to sinter thicker traces to better optimise the sintering process. It is also possible to integrate the laser sintering system into 3D electronics printing systems, such as inkjet and aerosol jet printing systems, to expedite the fabrication process.

(c) *Selective sintering*: The laser sintering technique allows for "selective sintering" [37], in which the metallic nanoparticle ink is only sintered when exposed to the laser beam. The laser selectively scans over the printed patterns to sinter them. The laser wavelength can also be selected for minimising energy absorption by substrates to reduce thermal stress and prevent substrate degradation [50].

　　Thermal energy can still be transferred to the surrounding substrates through thermal conduction during the laser sintering process [50]. Hence, temperature sensitive substrates, such as paper, textiles and polymer films, may only be utilised with careful tuning and optimisations of the laser sintering process to ensure the generated temperature kept to a minimum in preventing heat dissipations into the substrates [37].

(d) *High resolution patterning:* Laser sintering can also produce low-micrometre, high-resolution patterns which are advantageous for applications such as 3D printed transistors [23].

6.3.1.2 *Weaknesses*

The key weaknesses of laser sintering are:

(a) *Complex and expensive equipment*: The equipment used in the laser sintering process is complex and expensive [47].

(b) *Not suitable for sintering large areas*: Instead of having the entire printed pattern heated globally [51], laser sintering is a local heating process in which only a small localised area of metallic nanoparticles encompassed by the laser beam is sintered. Furthermore, the optimal writing speed for laser sintering is approximately around 0.2 mm/s, which can be too time-consuming for sintering large areas [47].

6.3.2 *Infrared (IR) Sintering*

IR sintering employs EM irradiations from the NIR [52] to mid-infrared (MIR) spectrum for sintering metallic nanoparticles [23,47]. In general, a simple IR sintering system includes an IR lamp and a reflector to focus the IR irradiations onto the printed patterns [53]. The critical process parameters of IR sintering include sintering time, power output from the IR lamp, the distance between the IR lamp and printed patterns, and the presence of reflector to focus the IR irradiations onto the printed patterns [53].

As metallic nanoparticles aggregate into larger particles during the IR sintering process, their colour changes into reflective silvery, and as a result, reflecting away the IR irradiations. Hence, slowing down the light absorption rates and the sintering process. This negative feedback mechanism in IR sintering can also help prevent metallic nanoparticles from overheating and damaging the underlying substrates [47,52].

6.3.2.1 *Strengths*

The key strengths of IR sintering are:

(a) *Low cost and simple to operate*: Among all photonic sintering techniques, the IR sintering technique has the cheapest system setup costs [53] as cheap incandescent lamps can also be used as the source of IR irradiations [47].
(b) *Highly efficient*: The IR sintering technique is more efficient than the conventional thermal sintering technique in sintering metallic nanoparticle inks, in which IR sintering can achieve better electrical conductivity in a shorter time. Gaspar *et al.* [33] demonstrated that IR sintering was able to sinter silver nanoparticle inks on Lumi Silk substrates within 10 minutes and achieved 40% bulk conductivity of silver,

whereas conventional thermal sintering took 60 minutes of sintering time and only achieved 20% bulk conductivity of silver. The IR sintering technique utilises concentrated localised heating and thus explains its excellent heating efficiency and reduced sintering time [33].

(c) *Compatible with paper substrates*: Most paper substrates are suitable for IR sintering because they have low thermal conductivity, high diffuse reflectance and high thermal stability properties [33,47,53]. As compared to temperature-sensitive polymer substrates, paper substrates are highly likely to preserve their mechanical properties under exposures to stronger intensities of IR irradiations. Therefore, it is possible to fabricate low-cost electronics, such as disposable radio-frequency identification (RFID) tags [54], on paper substrates.

Despite the good compatibility of IR sintering with paper substrates, prolonged exposures of IR irradiations can cause thermal ageing that deteriorates the paper substrates' original mechanical properties [33]. However, this issue can be resolved by reducing either the sintering time or IR irradiations power.

6.3.2.2 *Weaknesses*

The key weaknesses of IR sintering are:

(a) *Not selective sintering*: The metallic nanoparticle inks are heated up together with substrates during the IR sintering process.
(b) *Not suitable for temperature-sensitive polymer substrates*: Temperature-sensitive polymer substrates, such as polyethylene terephthalate (PET), tend to have very high absorbance rates in the IR regions of the electromagnetic spectrum [55], low glass transition temperatures, low melting points and high thermal expansion coefficients [47]. Hence, these substrates tend to heat up rapidly and experience damages and deformations, such as warping, shrinkage or deformations [23,33,55], when they are exposed to high-intensity IR irradiations during IR sintering.

Nonetheless, temperature-sensitive polymer substrates usually have substantially decreased absorbance rates in the NIR electromagnetic spectrum region. Thus, irradiations in this region can be used for sintering metallic nanoparticle inks [52,55]. Gu *et al.* [52] had demonstrated successful IR sintering of silver nanoparticle inks on PET substrates using irradiations in the NIR spectrum, with a peak wavelength of 1100 nm.

6.3.3 *Intense Pulse Light Sintering*

Intense pulse light (IPL) sintering, also known as photonic flash sintering, uses short pulses of high intensity white light from high power xenon flash lamps to sinter metallic nanoparticle inks [23]. As metallic nanoparticles have plasmon resonance bands in the visible spectrum, they can effectively absorb and convert the light energy into thermal energy under the SPR phenomenon during the IPL sintering process [56]. The thermal energy decomposes the encompassing organic additives around each nanoparticle and at the same time, heats the metallic nanoparticles to the sintering temperature for coalescence to take place to form a continuous conductive pattern [57].

A typical IPL sintering system comprises of these essential components: aluminium reflector, xenon flash lamp, light filter, capacitors, power supply and triggering pulse controller [15,23,56,58–60]. The xenon flash lamp generates short pulses of high intensity white light through the arc plasma phenomenon. The white light covers the entire spectrum of visible light with a small portion of NIR and UV irradiations [61] and has wavelengths ranging from 380 nm to 950 nm [61]. The printed patterns are usually placed at 3–14 mm under the xenon flash lamp [56,57], and aluminium reflector focuses the light onto them to help achieve high energy densities [15,59]. A light filter is used to block out wavelengths from the ultraviolet light spectrum to prevent causing severe damages to the substrates [57,59]. The critical process parameters of IPL sintering include pulse shape, pulse duration, time intervals between pulses, flashing frequency, flashing intensity and distance between the lamp and printed patterns [23].

6.3.3.1 *Strengths*

The key strengths of IPL sintering are:

(a) *Fast sintering*: IPL sintering is one of the fastest sintering techniques, in which a single pass of the IPL sintering process can be completed within milliseconds [56,60].

(b) *No protective atmosphere required*: With its fast sintering speed, IPL sintering can sinter metallic nanoparticles with very fast oxidation rates, such as copper nanoparticles, under ambient conditions without oxidising while maintaining their desired final electrical properties.

(c) *Selective sintering*: IPL sintering can allow selective sintering of different metallic nanoparticle inks and prevent unnecessary heating of the substrates. It is possible to couple the emission spectrum of the light source to the absorption spectrum of the metallic nanoparticles so that the metallic nanoparticles can efficiently absorb the light energy in the least amount of time. Temperature sensitive polymer substrates are usually transparent and can allow visible light to pass through. Thus, preventing the substrates from heating up due to the reduced absorption of light energy. However, the heat generated from the sintering of metallic nanoparticles may be thermally conducted to the underlying substrates and cause deformations [62].

(d) *Large area sintering*: IPL sintering can sinter large areas quickly, which is highly advantageous and suitable for printed displays and solar cells applications.

6.3.3.2 *Weaknesses*

The key weaknesses of IPL sintering are:

(a) *Expensive equipment*: Commercial IPL sintering systems can be quite expensive and may not be cost-effective for low volume production.

(b) *A lot of process optimisations work required*: In general, each metallic nanoparticle ink formulation and substrate combination has its own set of optimal conditions for IPL sintering. Mismatch of sintering process parameters with materials properties of inks and substrates can cause defects in the sintered patterns. Therefore, a lot of time-consuming and tedious process optimisations work is required to ensure IPL sintering can achieve the optimal electrical properties with the least defects [23]. Lee *et al.* [56] also demonstrated that the surface morphologies and microstructures of the sintered silver nanoparticle ink were significantly affected by the film dimensions and the irradiation intensity during the IPL sintering process.

(c) *May cause undesirable defects in the sintered patterns*: Some of the common surface morphological defects on the printed patterns that can be observed during the IPL sintering process include swellings, film delamination, ruptures and high porosities. These morphological defects are unfavourable for the printed patterns to achieve good electrical properties, and sometimes causing open circuit faults.

Swellings are usually caused by the entrapment of the organic additive vapours when multiple unoptimised pulses are irradiated consecutively during the IPL sintering process [56]. The first pulse of irradiations usually sinters the metallic nanoparticles on the top surfaces of the printed patterns, transforming them into a thin metallic film that encapsulates the entire printed patterns. As pulses are consecutively irradiated onto the printed patterns again, the thin metallic film gets heated up and thermal energy is conducted to the unsintered nanoparticles beneath. The organic additives in the ink are thermally decomposed by the heat and get vaporised instantaneously. The organic additive vapours entrapped between the thin metallic film and substrate inevitably cause a pressure build-up that resulted in the film swelling. Rupturing of the printed patterns may occur if the thin metallic film is unable to tolerate the vapour pressure. Lee *et al.* [56] noted that the swellings of the printed patterns were less prominent when higher irradiation intensity was used. Higher irradiation intensity was able to induce more grain growths at the top surface layer. Thus, the mechanical properties of the top surface layer improved as the layer densified and grew thicker. Although subsequent pulses induced more vaporisation of organic additives and caused a build-up of the internal pressure trapped between the top surface layer and the substrate, the thicker and denser top layer was able to prevent film swelling. However, metallic nanoparticle inks that are sintered with very high-intensity irradiations tend to have highly porous and uneven microstructures. This is due to the explosive internal vaporisation of organic additives within the printed patterns during the IPL sintering process [56,60].

Many research studies have also explored how metallic nanoparticle inks can be sintered by IPL sintering to achieve better morphologies while reducing defect formations and preventing substrate damages. One feasible approach is using pulse management in IPL sintering, whereby a high-intensity pulse is split into several shorter and lower-intensity pulses [60]. Shorter and lower-intensity pulses can prevent organic additives from vaporising excessively and therefore, giving smoother morphologies and lesser defects. However, shorter and lower-intensity pulses are inadequate for inducing significant grain growth and densification within the printed patterns and resulting in lower achieved electrical conductivity [23].

Another approach [63] is to have a two-step IPL sintering process that comprises of pre-heating and main sintering steps. The pre-heating step consists of multiple short pulses of low-intensity irradiations. This step helps to remove the bulk of the organic additives present in the metallic nanoparticle inks and reduce cracks and swellings in the printed patterns by minimising abrupt vaporisations. The main sintering step, consisting of one high-intensity pulse, further sinters and densifies the metallic nanoparticles within the printed patterns and improve the final electrical conductivity. However, this two-step IPL sintering method is only able to achieve good electrical and morphological properties when appropriate process parameters are used for both pre-heating and main sintering steps. Experimental results had also demonstrated that delamination of the printed patterns might occur if higher irradiation energy was used for the pre-heating step.

(d) *May damage substrates*: Generally, a combination of high flashing frequency and high-intensity irradiations in IPL sintering contribute to overheating that results in substrate damage, deformation and degradation, especially temperature-sensitive polymers films like PET [23,64].

6.3.4 *Ultraviolet Sintering*

Ultraviolet (UV) sintering uses irradiations from the UV region (wavelengths from 100 nm–400 nm) for sintering [23]. However, UV sintering may not be applicable for selectively heating applications as most metallic nanoparticle inks' absorption maximum range fall within the visible region. Nevertheless, wavelength absorption regions of the metallic nanoparticle inks can be tuned since they are highly dependent on the particle size and materials. Hence, there are only a few reports [65,66] of successful sintering of metallic nanoparticle inks with independent UV sintering.

6.3.4.1 *Strengths*

The key strengths of UV sintering are:

(a) *Inexpensive*: Inexpensive LED-based UV light can also be used for UV sintering [66].

(b) *Fast and selective heating*: Saleh *et al.* [66] demonstrated the use of low power UV light source to sinter silver nanoparticle ink by matching the UV wavelength to the ink's absorption maximum for initiating photothermal sintering process. Their sintered silver tracks achieved an average electrical resistivity of 0.48 $\mu\Omega$m with the 30s of UV sintering. This shows that the UV sintering technique can achieve lower electrical resistivity in a much shorter time as compared to thermal sintering.

(c) *Ability to cure photopolymers*: UV sintering may be beneficial for the fabrication of embedded electronics, in which photopolymers can be cured alongside with the metallic nanoparticle inks with one UV sintering apparatus [66].

(d) *Assistive role*: The UV sintering process usually plays an assistive role in sintering metallic nanoparticle inks with other sintering techniques particularly if the inks have an absorption band in the UV region [23] (for example, thermal sintering, IPL sintering and reactive chemical sintering [67]). UV sintering can further help improve the electrical conductivity of the printed patterns [30], decrease sintering time, and as well as allowing thermal sintering at lower temperatures [23,67].

Hwang *et al.* [30] combined the IPL sintering process with deep-UV irradiations to sinter poly(N-vinylpyrrolidone) (PVP) functionalised copper nanoparticle ink. SEM images showed that greater coalescence and densification of copper nanoparticles took place in using UV-assisted IPL sintering. In comparison with their previous work, they demonstrated that the UV-assisted IPL sintering technique achieved significantly better electrical conductivity as compared to the IPL technique. This was because UV irradiation can effectively decompose the PVP coating on the copper nanoparticles as the coating had an absorption maximum in the UV spectrum at a wavelength of 198 nm [30]. Thus, the intense pulsed light is then able to sinter the exposed copper nanoparticles effectively by first reducing the copper oxides layer and then heating the individual nanoparticles through photothermal effects.

6.3.4.2 *Weaknesses*

The key weaknesses of UV sintering are:

(a) *May damage substrates*: Overexposure to UV irradiations can cause damages to the substrates, especially the degradation of polymer films [57].

(b) *Not applicable for sintering every type of metallic nanoparticle ink*: UV sintering may not be effective or efficient for sintering metallic nanoparticle ink if the absorption maximum of the inks does not fall within the UV region.

6.4 Microwave Sintering

Microwaves are electromagnetic irradiations with frequencies ranging from 300 MHz to 300 GHz [68]. Microwave sintering uses microwave energy for sintering metallic nanoparticle inks, in which metallic nanoparticles absorb and transform microwave energy into heat internally from within [31,69]. The generated heat then decomposes the organic materials present in the ink and coalesces the nanoparticles to form a continuous conductive pattern.

6.4.1 *Strengths*

The key strengths of microwave sintering are:

(a) *Uniform heating*: Microwave sintering can heat metallic nanoparticle inks uniformly through microwave irradiations. The heating uniformity is largely influenced by the microwave penetration depth and absorbed power [31].
(b) *Short sintering time*: Microwave sintering can sinter metallic nanoparticle inks in relatively short time with excellent material properties and electrical conductivity [16,68,69]. As compared to thermal sintering, microwave sintering can achieve more crystallisation and homogeneous nucleation in a shorter period [68]. Perelaer *et al.* [70] demonstrated the use of microwave irradiations for sintering silver nanoparticle ink and achieved 5% bulk conductivity of silver within 240 seconds.
(c) *More environmentally friendly*: Microwave sintering is more environmentally friendly than conventional thermal sintering [31,69], as the microwave sintering technique can target microwave energy directly onto the required areas without the need to heat the surrounding chamber areas. Thus, it is possible to significantly cut down energy consumption by eliminating unnecessary heating.

(d) *Selectively heating*: Microwave sintering can allow temperature sensitive thermoplastics films (for example, PET and polycarbonate (PC) [16]) to be used as low-cost substrates for 3D printed electronics. The microwave irradiations can pass through these substrates and prevent absorption, and thus avoid the substrates being heated up during the microwave sintering process [16,70].

(e) *Can be combined with other sintering techniques*: To date, there is literature proving that microwave sintering can be combined with other sintering techniques to achieve better sintering results while decreasing the sintering time significantly. For instance, Perelaer *et al.* [19] combined flash microwave sintering and plasma sintering processes for sintering silver nanoparticle ink, and had achieved 60% bulk conductivity of silver in less than 10 minutes. Perelaer *et al.* [71] had also combined flash microwave sintering with photonic sintering and had achieved 40% bulk conductivity of silver in less than 15 seconds.

6.4.2 *Weaknesses*

The key weaknesses of microwave sintering are:

(a) *Limited penetration depth*: Microwave energy has a limited penetration depth of a few microns on the printed patterns [7,31,70]. Due to the reflective nature of metallic nanoparticles, most incident microwaves are reflected away and only some can penetrate through the printed layers to heat the individual metallic nanoparticles. The penetration depth is also highly dependent on the magnetic and electrical properties of the metallic nanoparticles, in which materials with lower magnetic permeability and electrical conductivity can allow higher penetration depth [31].

Even though the penetration depth is low, microwave sintering is still feasible in 3D printed electronics as the layer thickness of the printed patterns is usually in the order of a few microns [70]. Also, metallic nanoparticles have excellent thermal conductivity and can help conduct heat from the topmost surfaces to the underlying layers. Hence, the sintering process is not significantly limited by the penetration depth of the microwave [7,31].

6.5 Electrical Sintering

As metallic nanoparticle inks are initially electrically non-conductive, the electrical sintering technique is unable to sinter the as-deposited inks directly and requires the help of another sintering technique to first introduce some electrical conductivity within the printed patterns. In the electrical sintering technique, a voltage is applied across the conductive pattern. The current passes through the conductive patterns and rapidly generates heat within the patterns through Joule heating effects. The generated heat decomposes any leftover organic additives within the printed patterns and further coalesces the exposed metallic nanoparticles [72], which leads to the neck formations and grain growths [73–75] within the patterns [23], and ultimately improving the electrical conductivity.

6.5.1 *Strengths*

The key strengths of electrical sintering are:

(a) *High-speed sintering*: The electrical sintering technique is one of the fastest sintering methods. Many research papers had reported that the electrical sintering technique can sinter metallic nanoparticles within seconds. Moon *et al.* [73] demonstrated that the silver nanoparticles ink printed patterns could be sintered within 10 milliseconds and achieved approximately 1.7 times of bulk silver resistivity. Kravchuk *et al.* [76] also reported that electrical sintering was able to achieve the desired electrical resistance for inkjet printed strain gauges within 5 seconds.

(b) *Selective sintering*: On top of its fast sintering speed, the electrical sintering technique is selective in which the required printed areas are only sintered. Thus, the unprinted areas are not affected by the sintering process and allow the sintering of conductive patterns on temperature-sensitive substrates without introducing unnecessary thermal stress on the substrates [77]. However, the generated heat can also be conducted to the underlying substrates and cause damages.

(c) *Better control*: Electrical sintering can also allow precise control of the final desired electrical resistances of the printed patterns [17,75]. Werner *et al.* [75] reported that a tolerance of 2% deviation from the desired electrical resistances for the printed patterns was achievable through electrical sintering. An electronic regulator module was utilised during the electrical sintering process to cut off power supply when the printed patterns obtained the desired electrical resistance. Hence, ceasing any further sintering process within the printed patterns. Therefore, this approach is beneficial for applications that require very tight tolerances for the electrical resistance, for instance, the manufacturing of strain gauges.

(d) *In-situ monitoring ability*: *In-situ* monitoring on the morphologies of the printed patterns is possible with electrical sintering. Thus, helping to better understand the morphological changes of the metallic nanoparticles under the influences of electrical sintering with various operating parameters [17,72,78].

6.5.2 *Weaknesses*

The key weaknesses of electrical sintering are:

(a) *Pre-sintering required*: Although electrical sintering can achieve precise control of the final desired electrical resistances of the printed patterns in a short time, alternative pre-sintering techniques are required to introduce some electrical conductivity within the printed patterns first by decomposing the organic additives and coalescing the metallic nanoparticles [77]. Some of the pre-sintering techniques can include thermal sintering, IPL sintering and IR sintering.

(b) *Needs physical contact*: Electrical sintering requires probes to be in physical contact with the printed patterns to conduct electrical currents, and thus, may inevitably cause damages to the printed patterns at the point of contact during the sintering process. However, Allen *et al.* [78] had demonstrated the possibility of contactless electrical sintering, by adopting microwave sintering heads to focus the electrical field onto the printed patterns. Contactless electrical sintering may have high potentials of improving efficiencies in the mass production of flexible printed electronics.

(c) *Not efficient for mass sintering production*: It is very cumbersome to sinter the printed patterns individually and it might not be very efficient in sintering large quantities of printed patterns for mass production.

Technically demanding: It is technically demanding for electrical sintering to work, as the printed patterns require certain operating conditions for successful sintering. Jang *et al.* [77] grouped the electrical sintering process into three main categories: blowout, no change and sintering. The printed patterns are highly likely to be blown out by input voltages that are greater than 40V, even though the initial electrical resistances are low. The high input voltages generate an enormous amount of heat within the printed patterns to cause rapid decomposition of the organic additives into vapours. The organic additives vapours are entrapped within the printed patterns, which are likely to be accounted for the blown out phenomenon [77]. They noted that if the initial electrical resistance of the printed patterns were greater than 100 kΩ, there were high probabilities of the electrical sintering process not taking place regardless of the magnitude of the applied input voltages. They also presented that the input voltages should be in the range between 20 V to 40 V, with the initial electrical resistance of the printed patterns under 100 kΩ, to achieve optimal electrical sintering without damaging both the printed patterns and substrates.

6.6 Plasma Sintering

Plasma sintering is one of the few sintering techniques that allow the sintering of metallic nanoparticle inks at low sintering temperatures. Thus, avoiding excessive degradations of substrates and allowing the use of cheap temperature-sensitive substrates for 3D printed electronics applications. In manufacturing, plasma treatments are commonly and conventionally used for surface treatments, cleaning of substrates [22] and surface energy manipulations [20].

The type of feed gas used for plasma generation determines the chemical nature of plasma produced: inert, reducing and oxidising. Nitrogen and argon gases are used to generate plasma with inert properties. Hydrogen gas is used to generate plasma with reducing properties, whereas air and oxygen are used to generate plasma with oxidising properties [23]. Oxygen is usually not preferred for plasma sintering of metallic nanoparticle inks due to its oxidative nature. Oxygen plasma can oxidise metallic nanoparticles into metal oxides and make them less conductive or even electrically non-conductive [22]. Oxygen plasma is also capable of degrading and etching the surfaces of polymer substrates. It is also interesting to note that undesirable functional groups are formed on the surfaces of polyethylene naphthalate (PEN) and polyimide (PI) if they interact with nitrogen and oxygen plasmas, and can affect their stability [22]. Argon plasma will not oxidise metallic nanoparticles since argon gas is inert. There is also a reduced etching effect on the substrates as compared to oxygen and nitrogen plasmas [20,22]. Thus, argon gas is the most suitable gas to be used in plasma sintering of metallic nanoparticles. The plasma sintering technique can be further classified into two different techniques; low-pressure plasma sintering and localised atmospheric pressure plasma sintering. Each plasma sintering technique is technically similar in various areas, but each technique has its advantages and disadvantages and they will be further discussed.

6.6.1 *Low-pressure Plasma Sintering*

Low-pressure plasma sintering is a plasma treatment process done in a vacuum chamber, in which the vacuum chamber is used for creating the low-pressure environment during the sintering process. There are various configurations of low-pressure plasma systems [79], such as the barrel plasma system and planar diode plasma system (see **Figure 6.5**).

A typical low-pressure planar diode plasma system includes a radio frequency (RF) generator, an electrode connected to the RF generator, a grounded electrode, a vacuum chamber, a feed gas supply, a flow valve to control the flow rate of the feed gas into the chamber, an inlet for feed gas and an outlet for the vacuum pump to evacuate the air from the vacuum chamber [18]. The air within the vacuum chamber is first evacuated, then

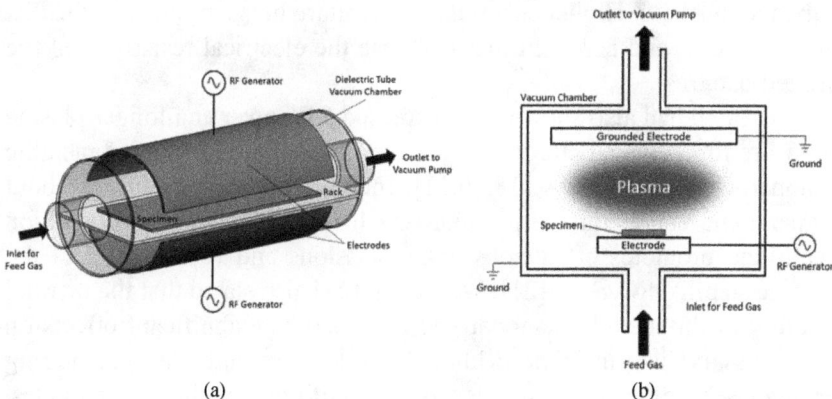

(a) (b)

Figure 6.5. Schematic diagrams of (a) barrel plasma system [79], (b) planar diode plasma system [18].

an appropriate flow of feed gas is released into the chamber precisely to create a low pressure environment through the flow valve.

The printed patterns that are required to be sintered are placed on the electrode that is connected to the RF generator, and plasma is ignited between it and the grounded electrode. As the grounded electrode has larger surface area than the electrode connected to the RF controller, highly energetic argon ions in the plasma tend to impact more towards the latter at a normal angle [18] which ultimately impact onto the printed patterns' surfaces. Energetic ions, UV radiations and radicals in the generated plasma expedite decompositions and chain scissions of the organic additives that encompass each metallic nanoparticle [18,20,22]. These decomposed organic additives are prone to easy removal in a low pressure environment. The exposed metallic nanoparticles in contact with the neighbouring particles after the decomposed organic additives are removed. Since the nanoparticles have high surface energies, the imparted energy from the plasma further promotes coalescence of the neighbouring nanoparticles. There is observable growth of grain boundaries and densification of metallic nanoparticles as the plasma sintering process proceeds, and hence forming a conductive percolating network [18,20,22].

Reinhold *et al.* [20] pointed out that different thermal dissipation of substrates could significantly affect the final resistivity of the printed patterns after low-pressure plasma sintering. For instance, a comparison between glass and polymer substrates, glass tends to have a lower thermal dissipation rate and hence retains a higher surface temperature than the polymer substrate. Higher substrate temperature helps to improve the diffusion process and hence further decrease the electrical resistivity of the printed patterns.

Research had also shown that increased RF power and longer plasma exposure time give a better electrical conductivity of the printed metallic nanoparticle inks patterns [18,20,21]. Increased RF power brings about higher ionic energy in the plasma, which increases the ion penetration depth and promotes efficacious chain scissions and vaporisation of the organic stabilising agents [21]. Wolf *et al.* [21] also stated that the thermal stability of the metallic nanoparticle inks also had significant effects on the necessary sintering time. Although the low-pressure plasma sintering technique can give significant densification of the metallic nanoparticles, the plasma species can only penetrate a limited depth. This effect is also known as the skin effect [23]. The skin effect impedes the bottommost layers of the printed patterns from sintering. Hence, the low-pressure plasma sintering technique is not very useful for sintering very thick printed patterns [20].

6.6.1.1 *Strengths*

The key strengths of low-pressure plasma sintering are:

(a) *Low sintering temperature*: Low-pressure plasma sintering is a low-temperature sintering process. However, it is still able to heat the substrates significantly depending on the input RF power [18,80]. Thus, special care has to be exercised to couple appropriate RF power to each corresponding substrate.

(b) *Can be combined with other sintering techniques*: Literature also demonstrated that electrical conductivity of the printed patterns can be further improved by combining plasma sintering with other sintering techniques. Perelaer *et al.* [19] demonstrated that by having plasma sintering as a pre-sintering step to sinter the printed patterns before the microwave flash sintering, the sintered patterns could achieve up to 60% bulk conductivity of silver. Dennelin *et al.* [14] also combined plasma sintering with IR sintering and observed significant improvements to the electrical conductivity of the printed patterns. They reported that it was only helpful to use plasma sintering as a pre-sintering step before IR sintering to improve the electrical conductivity of the printed patterns significantly. However, there were negligible improvements to the electrical conductivity of the printed patterns if plasma sintering was used as a post-sintering step after IR sintering. Wunscher *et al.* [80] also attempted to combine mild thermal sintering with plasma sintering. Their experiments concluded that with the help of the thermal sintering process, low-pressure plasma sintering could metallic nanoparticle inks in a shorter time frame while giving better electrical conductivity.

6.6.1.2 *Weaknesses*

The key weaknesses of low-pressure plasma sintering are:

(a) *Requires a mandatory vacuum chamber*: The low-pressure plasma sintering technique requires a mandatory vacuum chamber for the sintering process to work. The vacuum chamber helps to create a low pressure environment, but it is also associated with longer preparation time and higher equipment costs [22].

(b) *Time-consuming*: The low-pressure plasma sintering technique is very time-consuming. Wolf *et al.* [21] demonstrated 40.2% of silver's bulk conductivity was achievable after 60 minutes of low-pressure plasma sintering at 300W, and Reinhold *et al.* [20] also reported that they required 60 minutes to fully sinter the silver tracks.

(c) *Limited penetration depth*: The low-pressure plasma sintering technique is a top-to-bottom sintering approach and has limited penetration depth. Hence, it is unable to sinter thicker printed patterns [33, 34] effectively.

(d) *May etch away surfaces of polymer substrates*: The surfaces of polymer substrates may be etched away with extended exposures to the plasma species during the low-temperature plasma sintering process. Wunscher *et al.* [22] demonstrated the etching effects of plasma sintering on various polymer substrates (such as PEN, PET, PI, polypropylene (PP) and poly(methyl methacrylate) (PMMA)) by exposing them to plasma for 120 minutes. Their findings concluded significant etching effects were present in the plasma sintering process, especially for temperature-sensitive substrates like PMMA and PET, which could reach etching rates up to 0.32 μm h^{-1}. The etching effects of the plasma sintering are disadvantageous as it can significantly modify the optical properties of the substrates.

(e) *May damage multi-layers printed components*: Multi-layered printed electronics components may also be damaged when they undergo the low-temperature plasma sintering process [23]. For instance, dielectric layers in multi-layered components may be etched away and hence resulting in undesirable electrical properties at the end of the sintering process.

6.6.2 *Localised Atmospheric Pressure Plasma Sintering*

In addition to low-pressure plasma sintering, Wunscher *et al.* [22] also introduced another plasma sintering technique known as the localised atmospheric pressure plasma sintering technique for sintering silver nanoparticles. For this technique (see **Figure 6.6**), a plasma pencil is mounted onto a precision *xyz*-positioning translation stage for precise control. A focused jet of atmospheric pressure plasma is discharged from the

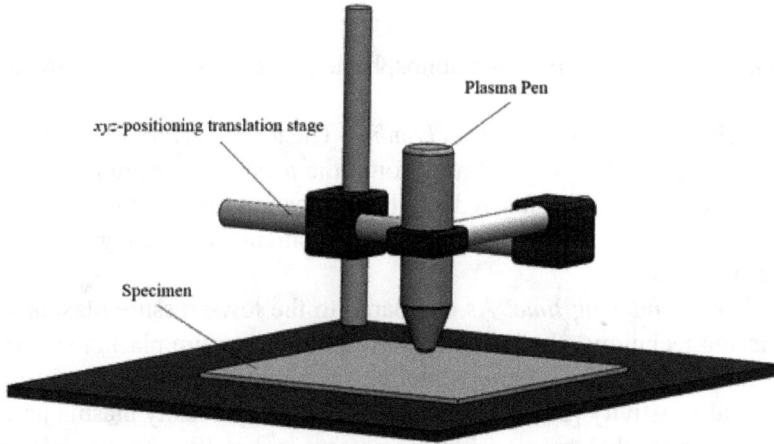

Figure 6.6. Schematic diagram of a localised atmospheric pressure plasma sintering system [22].

plasma pencil. The plasma pencil generates its plasma through a dielectric barrier discharge (DBD) configuration, in which the electrodes ignite the feed gas (argon) by applying a potential difference between them. The generated plasma is pushed out by the continuous flow of feed gas that is supplied into the plasma pencil and hence creating a constant jet of plasma [22].

Wunscher *et al.* [22] pointed out that the distance between the plasma pencil tip and printed patterns had significant effects on the final quality and electrical resistivity of the sintered patterns. They believed that if the plasma pencil tip was too far above the printed patterns, a significant amount of oxygen would be absorbed by the plasma jet and thus oxidising the silver nanoparticles into silver oxides. Silver oxides have higher electrical resistivity than silver and hence can cause the electrical resistivity of the entire sintered printed pattern to be of few orders of magnitude higher than those printed patterns that were sintered at a closer distance [22]. They also noted that the writing speeds and sintering cycles of the plasma jet played significant roles in influencing the morphologies and electrical resistivity of the final sintered patterns. The densification of the silver nanoparticles' structures increases with lower plasma jet writing speed as slower writing speed gives more exposure time to the printed patterns. Denser silver structures are also less porous and have lower electrical resistivity.

6.6.2.1 *Strengths*

The key strengths of localised atmospheric pressure plasma sintering are:

(a) *Omit the need for a vacuum chamber*: The localised atmospheric pressure plasma sintering technique omit the need for a vacuum chamber, and hence significantly reduce the fabrication time and costs associated with the technically complex equipment for creating a vacuum environment.

(b) *Shorter sintering time*: As compared to the low-pressure plasma sintering technique, the localised atmospheric pressure plasma sintering technique requires shorter sintering time and can achieve lower electrical resistivity [22]. The focused, high-energy density plasma jet can decompose the organic additives present in metallic nanoparticle inks at a faster rate and hence expedite the entire sintering process.

(c) *Selective sintering*: The localised atmospheric pressure plasma sintering technique can selectively sinter the printed patterns, where the plasma jet only traces over the areas that the metallic nanoparticle inks are deposited. Those areas that are not printed are not exposed to the plasma jet and thus minimises the degradation of the substrates.

(d) *Ideal for temperature-sensitive substrates*: The localised atmospheric pressure plasma sintering technique is ideal for sintering metallic nanoparticle inks on temperature-sensitive substrates as the temperature of the plasma jet from the plasma pencil does not exceed 60°C [22].

(e) *Can be integrated with ink deposition systems*: It is technically possible to integrate the localised atmospheric pressure plasma sintering system with various ink deposition systems (such as inkjet printer and aerosol jet printer) to create an *in situ* printing and sintering system, in which the deposited inks can be sintered immediately [22].

6.6.2.2 *Weaknesses*

The key weaknesses of localised atmospheric pressure plasma sintering are:

(a) *Limited penetration depth*: The localised atmospheric pressure plasma sintering technique is also a top-to-bottom sintering approach. A metallic layer is formed as the top surfaces are sintered and thus preventing the plasma species to interact with the underlying unsintered ink [22].

(b) *A lot of process optimisations work required*: A lot of process optimisations work required to find out the optimal operating parameters (for instance, the distance between the plasma pencil tip and printed patterns, sintering cycles and writing speed) for every unique ink formulation [22].

(c) *Many sintering passes required*: The localised atmospheric pressure plasma sintering technique requires many sintering passes, ranging from 20 to 100 passes [22], over the printed patterns to achieve the desired electrical conductivity.

Wunscher *et al.* [80] further looked into this problem and proposed two solutions. One of the solutions incorporated the localised atmospheric pressure plasma sintering technique with mild thermal treatment of the substrates. The substrates were first heated up to 110°C and the printed patterns were able to achieve improved electrical resistivity within a single pass of the localised atmospheric pressure plasma sintering. This was because thermal treatment of the printed patterns aided the sintering process of the metallic nanoparticles that were located on the bottom of the printed layer, by providing a bottom-to-top sintering approach. The other solution was to replace the cold plasma with warm plasma in the localised atmospheric pressure plasma sintering system for sintering, and nitrogen gas was used as feed gas instead. Although the warm plasma can heat the substrates to 200°C, degradations of the substrates can be minimised if the warm plasma combined with a fast writing speed of the localised atmospheric pressure plasma sintering system. They also demonstrated it was possible to sinter the printed patterns within a single pass, without the need of having additional thermal sintering process [80].

6.7 Chemical Sintering

Chemical sintering allows sintering of metallic nanoparticle inks at room temperature without the need to introduce external energy [4,23]. Most chemical sintering methods require the addition of sintering agents to the printed patterns to destabilise the encompassing organic additives and stabilising agents around the metallic nanoparticles. These organic compounds get dissolved and detached from the metallic nanoparticles' surfaces and the exposed metallic nanoparticles are then able to form contact points with adjacent nanoparticles. The diffusive Ostwald ripening phenomenon and the need to reduce total interfacial energy enable these contacting nanoparticles to coalesce with each other by forming sinter necks. Hence, the printed patterns can now conduct electricity as the nanoparticles within the printed patterns have integrated into a single continuous entity to form a continuous path [4,27,28].

6.7.1 *Strengths*

The key strengths of chemical sintering are:

(a) *Low temperature sintering*: Chemical sintering allows coalescence of metallic nanoparticles at room temperature [23].

(b) *Allow usage of temperature-sensitive substrates*: Chemical sintering can allow low-cost temperature-sensitive substrates to be used for many different 3D printed electronics applications and promote additional cost savings. Temperature-sensitive substrates will not be damaged, deformed or degraded during the chemical sintering process due to low working temperatures.

6.7.2 *Weaknesses*

The key weaknesses of chemical sintering are:

(a) *Requires tailor-made sintering agents*: Chemical sintering does not apply to all commercially available metallic nanoparticles inks and requires tailor-made sintering agents to decompose organic compounds present in a particular metallic nanoparticle ink formulation. For instance, Magdassi *et al.* [81] used poly(diallyldimethylammonium chloride) as a sintering agent to chemically sinter silver nanoparticle ink which contained poly(acrylic acid) as a stabilising agent. In addition, the properties of organic compounds present in commercially available metallic nanoparticle inks are usually classified and undisclosed. Therefore, chemical sintering is not a one-size-fits-all solution for sintering all kinds of metallic nanoparticle inks that are available in the commercial market.

(b) *Poor electrical conductivity*: Metallic nanoparticle inks that are sintered through chemical sintering mostly tend to have poorer electrical conductivity as compared to other sintering techniques. For instance, Wakuda *et al.* [82,83] chemically sintered dodecylamine-stabilised silver nanoparticle ink by dipping the printed patterns into several types of alcohol. The alcohols dissolved the dodecylamine present in the silver nanoparticle ink and helped the exposed silver nanoparticles to form contact points with each other. The achieved electrical conductivity this chemically sintered silver film was between 0.7–2.2% of bulk silver only. Grouchko *et al.* [84] also reported that their chemically sintered silver film only achieved 5% of bulk silver. However, chemical sintering may be combined with other sintering techniques to improve the electrical conductivity of the sintered films further.

References

[1] Tan, H. W., An, J., Chua, C. K. and Tran, T. (2019). Metallic nanoparticle inks for 3D printing of electronics, *Adv. Electron. Mater.*, 5, p. 1800831.

[2] Tan, H. W., Saengchairat, N., Goh, G. L., An, J., Chua, C. K. and Tran, T. (2020). Induction sintering of silver nanoparticle inks on polyimide substrates, *Adv. Mater. Technol.*, 5, p. 1900897.

[3] Tan, H. W., Tran, T. and Chua, C. K. (2016). A review of printed passive electronic components through fully additive manufacturing methods, *Virtual Phys. Prototyp.*, 11, pp. 271–288.

[4] Perelaer, J. and Schubert, U. S. (2013). Novel approaches for low temperature sintering of inkjet-printed inorganic nanoparticles for roll-to-roll (R2R) applications, *J. Mater. Res.*, 28, pp. 564–573.

[5] Ian M., H. and Graham D., M. (2012). *Inkjet Technology for Digital Fabrication* (John Wiley & Sons Ltd, United Kingdom).

[6] Shimoni, A., Azoubel, S. and Magdassi, S. (2014). Inkjet printing of flexible high-performance carbon nanotube transparent conductive films by "coffee ring effect", *Nanoscale*, 6, pp. 11084–11089.

[7] Kamyshny, A. and Magdassi, S. (2014). Conductive nanomaterials for printed electronics, *Small*, 10, pp. 3515–3535.

[8] German, R. M. (2010). *Sintering of Advanced Materials*, eds. Zhigang Zak Fang, Fundamentals of Sintering (Woodhead Publishing, United Kingdom).

[9] Halonen, E., Viiru, T., Ostman, K., Cabezas, A. L. and Mantysalo, M. (2013). Oven sintering process optimization for inkjet-printed Ag

nanoparticle ink, *IEEE Trans. Compon., Packag., Manuf. Technol.*, 3, pp. 350–356.

[10] Myong-Ki, K., Jun Young, H., Heuiseok, K., Kyungtae, K., Sang-Ho, L. and Seung-Jae, M. (2009). Laser sintering of the printed silver ink, *presented at the IEEE International Symposium on Assembly and Manufacturing (ISAM)*, Seoul, South Korea.

[11] Niizeki, T., Maekawa, K., Mita, M., Yamasaki, K., Matsuba, Y., Terada, N. and Saito, H. (2008). Laser sintering of Ag nanopaste film and its application to bond-pad formation, *presented at the 58th Electronic Components and Technology Conference (ECTC)*, Lake Buena Vista, Florida, USA.

[12] Kumpulainen, T., Pekkanen, J., Valkama, J., Laakso, J., Tuokko, R. and Mäntysalo, M. (2011). Low temperature nanoparticle sintering with continuous wave and pulse lasers, *Opt. Laser Technol.*, 43, pp. 570–576.

[13] Michael, Z., Oleg, E., Amir, S. and Zvi, K. (2014). Laser sintering of copper nanoparticles, *J. Phys. D: Appl. Phys.*, 47, p. 025501.

[14] Denneulin, A., Blayo, A., Neuman, C. and Bras, J. (2011). Infra-red assisted sintering of inkjet printed silver tracks on paper substrates, *J. Nanopart. Res.*, 13, pp. 3815–3823.

[15] Kim, H. S., Dhage, S. R., Shim, D.-E. and Hahn, H. T. (2009). Intense pulsed light sintering of copper nanoink for printed electronics, *Appl. Phys. A*, 97, pp. 791–798.

[16] Perelaer, J., Klokkenburg, M., Hendriks, C. E. and Schubert, U. S. (2009). Microwave flash sintering of inkjet-printed silver tracks on polymer substrates, *Adv. Mater.*, 21, pp. 4830–4834.

[17] Allen, M. L., Aronniemi, M., Mattila, T., Alastalo, A., Ojanperä, K., Suhonen, M. and Seppä, H. (2008). Electrical sintering of nanoparticle structures, *Nanotechnology*, 19, p. 175201.

[18] Ma, S., Bromberg, V., Liu, L., Egitto, F. D., Chiarot, P. R. and Singler, T. J. (2014). Low temperature plasma sintering of silver nanoparticles, *Appl. Surf. Sci.*, 293, pp. 207–215.

[19] Perelaer, J., Jani, R., Grouchko, M., Kamyshny, A., Magdassi, S. and Schubert, U. S. (2012). Plasma and microwave flash sintering of a tailored silver nanoparticle ink, yielding 60% bulk conductivity on cost-effective polymer foils, *Adv. Mater.*, 24, pp. 3993–3998.

[20] Reinhold, I., Hendriks, C. E., Eckardt, R., Kranenburg, J. M., Perelaer, J., Baumann, R. R. and Schubert, U. S. (2009). Argon plasma sintering of inkjet printed silver tracks on polymer substrates, *J. Mater. Chem.*, 19, pp. 3384–3388.

[21] Wolf, F. M., Perelaer, J., Stumpf, S., Bollen, D., Kriebel, F. and Schubert, U. S. (2013). Rapid low-pressure plasma sintering of inkjet-printed silver nanoparticles for RFID antennas, *J. Mater. Res.*, 28, pp. 1254–1261.

[22] Wunscher, S., Stumpf, S., Teichler, A., Pabst, O., Perelaer, J., Beckert, E. and Schubert, U. S. (2012). Localized atmospheric plasma sintering of inkjet printed silver nanoparticles, *J. Mater. Chem.*, 22, pp. 24569–24576.

[23] Wunscher, S., Abbel, R., Perelaer, J. and Schubert, U. S. (2014). Progress of alternative sintering approaches of inkjet-printed metal inks and their application for manufacturing of flexible electronic devices, *J. Mater. Chem. C*, 2, pp. 10232–10261.

[24] Chua, C. K. and Leong, K. F. (2017). *3D Printing and Additive Manufacturing — Principles and Applications*, 5th edn. (World Scientific Publishing, Singapore).

[25] Galagan, Y., Coenen, E. W. C., Abbel, R., van Lammeren, T. J., Sabik, S., Barink, M., Meinders, E. R., Andriessen, R. and Blom, P. W. M. (2013). Photonic sintering of inkjet printed current collecting grids for organic solar cell applications, *Org. Electron.*, 14, pp. 38–46.

[26] Ingham, B., Lim, T. H., Dotzler, C. J., Henning, A., Toney, M. F. and Tilley, R. D. (2011). How nanoparticles coalesce: An *in situ* study of Au nanoparticle aggregation and grain growth, *Chem. Mater.*, 23, pp. 3312–3317.

[27] Voorhees, P. W. (1985). The theory of Ostwald ripening, *J. Stat. Phys.*, 38, pp. 231–252.

[28] Kang, S.-J. L. (2005). *Sintering: Densification, Grain Growth and Microstructure*, 1st edn, (Butterworth-Heinemann).

[29] Perelaer, B. J., de Laat, A. W. M., Hendriks, C. E. and Schubert, U. S. (2008). Inkjet-printed silver tracks: Low temperature curing and thermal stability investigation, *J. Mater. Chem.*, 18, pp. 3209–3215.

[30] Hwang, H. J., Oh, K.-H. and Kim, H.-S. (2016). All-photonic drying and sintering process via flash white light combined with deep-UV and near-infrared irradiation for highly conductive copper nano-ink, *Sci. Rep.*, 6, p. 19696.

[31] Oghbaei, M. and Mirzaee, O. (2010). Microwave versus conventional sintering: A review of fundamentals, advantages and applications, *J. Alloys Compd.*, 494, pp. 175–189.

[32] Hsien-Hsueh, L., Kan-Sen, C. and Kuo-Cheng, H. (2005). Inkjet printing of nanosized silver colloids, *Nanotechnology*, 16, p. 2436.

[33] Gaspar, C., Passoja, S., Olkkonen, J. and Smolander, M. (2016). IR-sintering efficiency on inkjet-printed conductive structures on paper substrates, *Microelectron. Eng.*, 149, pp. 135–140.

[34] Kim, D. and Moon, J. (2005). Highly conductive ink jet printed films of nanosilver particles for printable electronics, *Electrochem. Solid-State Lett.*, 8, pp. J30–J33.

[35] Greer, J. R. and Street, R. A. (2007). Thermal cure effects on electrical performance of nanoparticle silver inks, *Acta Mater.*, 55, pp. 6345–6349.

[36] Öhlund, T., Ortegren, J., Andersson, H. and Nilsson, H.-E. (2009). *Sintering Methods for Metal Nanoparticle Inks on Flexible Substrates, presented at the 25th NIP & Digital Fabrication Conference*, Louisville, Kentucky, USA.

[37] Niittynen, J. and Mantysalo, M. (2014). Characterization of laser sintering of copper nanoparticle ink by FEM and experimental testing, *IEEE Trans. Compon., Packag., Manuf. Technol.*, 4, pp. 2018–2025.

[38] Jain, P. K., El-Sayed, I. H. and El-Sayed, M. A. (2007). Au nanoparticles target cancer, *Nano Today*, 2, pp. 18–29.

[39] Cheng, Y.-T., Uang, R.-H., Wang, Y.-M., Chiou, K.-C. and Lee, T.-M. (2009). Laser annealing of gold nanoparticles thin film using photothermal effect, *Microelectron. Eng.*, 86, pp. 865–867.

[40] Lee, K.-C., Lin, S.-J., Lin, C.-H., Tsai, C.-S. and Lu, Y.-J. (2008). Size effect of Ag nanoparticles on surface plasmon resonance, *Surf. Coat. Technol.*, 202, pp. 5339–5342.

[41] Zhang, J. Z. and Noguez, C. (2008). Plasmonic optical properties and applications of metal nanostructures, *Plasmonics*, 3, pp. 127–150.

[42] Liz-Marzán, L. M. (2006). Tailoring surface plasmons through the morphology and assembly of metal nanoparticles, *Langmuir*, 22, pp. 32–41.

[43] Koga, K., Ikeshoji, T. and Sugawara, K.-I. (2004). Size- and temperature-dependent structural transitions in gold nanoparticles, *Phys. Rev. Lett.*, 92, p. 115507.

[44] Anker, J. N., Hall, W. P., Lyandres, O., Shah, N. C., Zhao, J. and Van Duyne, R. P. (2008). Biosensing with plasmonic nanosensors, *Nat. Mater.*, 7, pp. 442–453.

[45] Link, S. and El-Sayed, M. A. (1999). Size and temperature dependence of the plasmon absorption of colloidal gold nanoparticles, *J. Phys. Chem. B*, 103, pp. 4212–4217.

[46] Halonen, E. and Heinonen, E. (2013). The effect of laser sintering process parameters on Cu nanoparticle ink in room conditions, *Opt. Photonics J.*, 3, pp. 40–44.

[47] Tobjörk, D., Aarnio, H., Pulkkinen, P., Bollström, R., Määttänen, A., Ihalainen, P., Mäkelä, T., Peltonen, J., Toivakka, M., Tenhu, H. and Österbacka, R. (2012). IR-sintering of ink-jet printed metal-nanoparticles on paper, *Thin Solid Films*, 520, pp. 2949–2955.

[48] Ko, S. H., Pan, H., Grigoropoulos, C. P., Luscombe, C. K., Fréchet, J. M. J. and Poulikakos, D. (2007). All-inkjet-printed flexible electronics fabrication on a polymer substrate by low-temperature high-resolution selective laser sintering of metal nanoparticles, *Nanotechnology*, 18, p. 345202.

[49] Laakso, P., Ruotsalainen, S., Halonen, E., Mäntysalo, M. and Kemppainen, A. (2009). Sintering of printed nanoparticle structures using laser

treatment, *presented at the 28th International Congress on Applications of Lasers & Electro-Optics (ICALEO)*, Orlando, Florida, USA.

[50] Yang, W., Yi, L., Torah, R. and Tudor, J. (2015). Laser curing of screen and inkjet printed conductors on flexible substrates, *presented at the Symposium on Design, Test, Integration and Packaging of MEMS/MOEMS (DTIP)*, Montpellier, France.

[51] Garnett, E. C., Cai, W., Cha, J. J., Mahmood, F., Connor, S. T., Greyson Christoforo, M., Cui, Y., McGehee, M. D. and Brongersma, M. L. (2012). Self-limited plasmonic welding of silver nanowire junctions, *Nat. Mater.*, 11, pp. 241–249.

[52] Weibing, G. and Zheng, C. (2016). Photonic sintering of nano-silver conductive ink for printed electronics, *presented at the 6th Electronic System-Integration Technology Conference (ESTC)*, Grenoble, France.

[53] Sowade, E., Kang, H., Mitra, K. Y., Wei, Weber, J. and Baumann, R. R. (2015). Roll-to-roll infrared (IR) drying and sintering of an inkjet-printed silver nanoparticle ink within 1 second, *J. Mater. Chem. C*, 3, pp. 11815–11826.

[54] Vena, A., Perret, E., Tedjini, S., Tourtollet, G. E. P., Delattre, A., Garet, F. and Boutant, Y. (2013). Design of chipless RFID tags printed on paper by flexography, *IEEE Trans. Antennas Propag.*, 61, pp. 5868–5877.

[55] Cherrington, M., Claypole, T. C., Deganello, D., Mabbett, I., Watson, T. and Worsley, D. (2011). Ultrafast near-infrared sintering of a slot-die coated nano-silver conducting ink, *J. Mater. Chem.*, 21, pp. 7562–7564.

[56] Dong Jun, L., Sung Hyeon, P., Shin, J., Hak Sung, K., Je Hoon, O. and Yong Won, S. (2011). Pulsed light sintering characteristics of inkjet-printed nanosilver films on a polymer substrate, *J. Micromech. Microeng.*, 21, p. 125023.

[57] Kang, J. S., Ryu, J., Kim, H. S. and Hahn, H. T. (2011). Sintering of inkjet-printed silver nanoparticles at room temperature using intense pulsed light, *J. Electron. Mater.*, 40, pp. 2268–2277.

[58] Dharmadasa, R., Jha, M., Amos, D. A. and Druffel, T. (2013). Room temperature synthesis of a copper ink for the intense pulsed light sintering of conductive copper films, *ACS Appl. Mater. Interfaces*, 5, pp. 13227–13234.

[59] Colorado, H. A., Dhage, S. R., Yang, J. M. and Hahn, H. T. (2012). Intense pulsed light sintering technique for nanomaterials, *presented at the 141st Meeting the Minerals, Metals, and Materials Society*, Warrendale, PA, USA.

[60] Won-Suk, H., Jae-Min, H., Hak-Sung, K. and Yong-Won, S. (2011). Multi-pulsed white light sintering of printed Cu nanoinks, *Nanotechnology*, 22, p. 395705.

[61] Jang, Y.-R., Joo, S.-J., Chu, J.-H., Uhm, H.-J., Park, J.-W., Ryu, C.-H., Yu, M.-H. and Kim, H.-S. (2020). A review on intense pulsed light sintering

technologies for conductive electrodes in printed electronics, *Int. J. Preci. Eng. Manuf.-Green Technol.,* doi: 10.1007/s40684-020-00193-8.

[62] Abbel, R., van Lammeren, T., Hendriks, R., Ploegmakers, J., Rubingh, E. J., Meinders, E. R. and Groen, W. A. (2012). Photonic flash sintering of silver nanoparticle inks: A fast and convenient method for the preparation of highly conductive structures on foil, *MRS Commun.*, 2, pp. 145–150.

[63] Sung-Hyeon, P., Shin, J., Dong-Jun, L., Jehoon, O. and Hak-Sung, K. (2013). Two-step flash light sintering process for crack-free inkjet-printed Ag films, *J. Micromech. Microeng.*, 23, p. 015013.

[64] Chung, W.-H., Hwang, H. J., Lee, S. H. and Kim, H. S. (2013). *In situ* monitoring of a flash light sintering process using silver nano-ink for producing flexible electronics, *Nanotechnology*, 24, p. 035202.

[65] Polzinger, B., Schoen, F., Matic, V., Keck, J., Willeck, H., Eberhardt, W. and Kueck, H. (2011). UV-sintering of inkjet-printed conductive silver tracks, *presented at the 11th IEEE Conference on Nanotechnology (IEEE-NANO)*, Portland, Oregon, USA.

[66] Ehab, S., Fan, Z., Yinfeng, H., Jayasheelan, V., Ledesma, F. J., Ricky, W., Ian, A., Richard, H., Phill, D. and Christopher, T. (2017). 3D inkjet printing of electronics using UV conversion, *Adv. Mater. Technol.*, 2, p. 1700134.

[67] Oh, Y., Lee, S.-N., Kim, H.-K. and Kim, J. (2012). UV-assisted chemical sintering of inkjet-printed TiO_2 photoelectrodes for low-temperature flexible dye-sensitized solar cells, *J. Electrochem. Soc.*, 159, pp. H777–H781.

[68] Penchal Reddy, M., Madhuri, W., Sadhana, K., Kim, I. G., Hui, K. N., Hui, K. S., Siva Kumar, K. V. and Ramakrishna Reddy, R. (2014). Microwave sintering of nickel ferrite nanoparticles processed via sol-gel method, *J. Sol-Gel Sci. Technol.*, 70, pp. 400–404.

[69] Anklekar, R. M., Agrawal, D. K. and Roy, R. (2001). Microwave sintering and mechanical properties of PM copper steel, *Powder Metall.*, 44, pp. 355–362.

[70] Perelaer, J., deGans, B. J. and Schubert, U. S. (2006). Ink-jet printing and microwave sintering of conductive silver tracks, *Adv. Mater.*, 18, pp. 2101–2104.

[71] Perelaer, J., Abbel, R., Wünscher, S., Jani, R., van Lammeren, T. and Schubert, U. S. (2012). Roll-to-roll compatible sintering of inkjet printed features by photonic and microwave exposure: From non-conductive ink to 40% bulk silver conductivity in less than 15 seconds, *Adv. Mater.*, 24, pp. 2620–2625.

[72] Hummelgård, M., Zhang, R., Nilsson, H.-E. and Olin, H. (2011). Electrical sintering of silver nanoparticle ink studied by TEM probing, *PLoS ONE*, 6, p. 17209.

[73] Moon, Y. J., Lee, S. H., Kang, H., Kang, K., Kim, K. Y., Hwang, J. Y. and Cho, Y. J. (2011). Electrical sintering of inkjet-printed silver electrode for c-Si solar cells, *presented at the 37th IEEE Photovoltaic Specialists Conference (PVSC)*, Seattle, Washington, USA.

[74] Alastalo, A. T., Mattila, T., Allen, M. L., Aronniemi, M. J., Leppäniemi, J. H., Ojanperä, K. A., Suhonen, M. P. and Seppä, H. (2011). Rapid electrical sintering of nanoparticle structures, *MRS Proc.*, 1113, pp. 2–7.

[75] Werner, C., Godlinski, D., Zöllmer, V. and Busse, M. (2013). Morphological influences on the electrical sintering process of aerosol jet and nk jet printed silver microstructures, *J. Mater. Sci.: Mater. Electron.*, 24, pp. 4367–4377.

[76] Kravchuk, O., Bobitski, Y. and Reichenberger, M. (2016). Electrical sintering of inkjet printed sensor structures on polyimide substrate, *presented at the 36th International Conference on Electronics and Nanotechnology (ELNANO)*, Kiev, Ukraine.

[77] Jang, S., Lee, D. J., Lee, D. and Oh, J. H. (2013). Electrical sintering characteristics of inkjet-printed conductive Ag lines on a paper substrate, *Thin Solid Films*, 546, pp. 157–161.

[78] Allen, M., Alastalo, A., Suhonen, M., Mattila, T., Leppaniemi, J. and Seppa, H. (2011). Contactless electrical sintering of silver nanoparticles on flexible substrates, *IEEE Trans. Microwave Theory Tech.*, 59, pp. 1419–1429.

[79] Shul, R. J. and Pearton, S. J. (2000). *Handbook of Advanced Plasma Processing Techniques*, 1st edn. (Springer Nature, New York).

[80] Wunscher, S., Stumpf, S., Perelaer, J. and Schubert, U. S. (2014). Towards single-pass plasma sintering: temperature influence of atmospheric pressure plasma sintering of silver nanoparticle ink, *J. Mater. Chem. C*, 2, pp. 1642–1649.

[81] Magdassi, S., Grouchko, M., Berezin, O. and Kamyshny, A. (2010). Triggering the sintering of silver nanoparticles at room temperature, *ACS Nano*, 4, pp. 1943–1948.

[82] Wakuda, D., Kim, K. S. and Suganuma, K. (2009). Room-temperature sintering process of Ag nanoparticle paste, *IEEE Trans. Compon. Packag. Technol.*, 32, pp. 627–632.

[83] Wakuda, D., Kim, K.-S. and Suganuma, K. (2008). Room temperature sintering of Ag nanoparticles by drying solvent, *Scr. Mater.*, 59, pp. 649–652.

[84] Grouchko, M., Kamyshny, A., Mihailescu, C. F., Anghel, D. F. and Magdassi, S. (2011). Conductive inks with a "built-in" mechanism that enables sintering at room temperature, *ACS Nano*, 5, pp. 3354–3359.

Problems

1. Using a sketch to illustrate your answer, describe the sintering process of metallic nanoparticle inks.
2. List the advantages and disadvantages of the thermal sintering process.

3. What are some of the critical factors that influence laser sintering?
4. Discuss the advantages and disadvantages of the IPL sintering process.
5. Describe the differences between IR sintering and IPL sintering.
6. Using a sketch to illustrate your answer, describe the low-pressure plasma sintering process in a low-pressure planar diode plasma system.
7. Which sintering technique do you think is most suitable for sintering silver nanoparticle ink on PET substrates and why?
8. Do you think the electrical sintering technique is a suitable technique for sintering large quantities of printed patterns and what are the advantages that come with it? Justify your answers.

Chapter 7

Designs and Simulations for 3D Printed Electronics

Additive manufacturing (AM) technology has revolutionized the conventional way of fabricating and designing electronics, by providing many opportunities in fabricating 3D printed electronics with interesting and complex designs. With the emergence of new printing materials and processes, the technology for fabricating 3D printed electronics has improved immensely over the years [1–3]. However, the complexity of the interdisciplinary interactions taking place during 3D electronics printing processes also has increased considerably. Therefore, there is a need for computational modelling and simulations to help investigate the underlying mechanisms and optimise the printing processes more effectively and efficiently. The direction for 3D printed electronics has also gradually shifted to the fabrication of components with multi-material, multi-functionality, and multi-scale, to meet the needs and demands of more advanced applications [4].

The potentials of modelling and simulations can be applied at different stages throughout the 3D printed electronics process flow, including process modelling of various printing techniques, functional modelling of 3D printed electronics components, and finally, the guided design of 3D printed electronics. This chapter highlights the latest advances in these areas with an emphasis on modelling and simulations for 3D printed electronics.

7.1 Modelling and Simulation of 3D Electronics Printing Processes

There are many complex interactions taking place between different process parameters during a 3D electronics printing process. Furthermore, it is a labour-intensive and time-consuming task to optimise the printing processes empirically [5]. Therefore, there is a need to have modelling and simulations tools to assist us in guiding the design of process and inks, optimising the printing processes, reducing physical experiments, ensuring quality and enabling *in-situ* monitoring and process capability. This section discusses the general process of modelling, limitations of modelling, and analytical modelling and numerical simulations for 3D electronics printing processes.

7.1.1 *Process of Modelling*

It is difficult to perform physical experiments without inadvertently disrupting the system when making the necessary observations and manipulations. Furthermore, some systems can be very complex, and the observed effects may not be translated explicitly [6]. Therefore, effective mathematical tools are often required to help analyse and interpret data from physical experiments. After computational modelling, physical experiments can be repeated to allow validation of the computer model. Some of the noteworthy benefits of computational modelling include increasing the number of experiments done in a shorter time, conducting any arbitrary number of experiments with the same setup configurations, allowing observations and manipulations without disrupting or degrading the system, direct exporting of data and results to compatible graphical display or analysis software, easy modifications of state variables which are usually not easy to do so in physical experiments (for instance, using very small user-defined magnitudes or changing different material properties). The schematic workflow of the modelling process is shown in **Figure 7.1**.

In the initial step of computational modelling, a conceptual model has to be formulated first. The conceptual model should comprise the fundamental operations of the system of interest, or a list of interacting components that are involved and how they interact with each other [6]. Proper assumptions are required to improve the model's simplicity and clarity. However, overgeneralising assumptions could contribute to inaccuracies

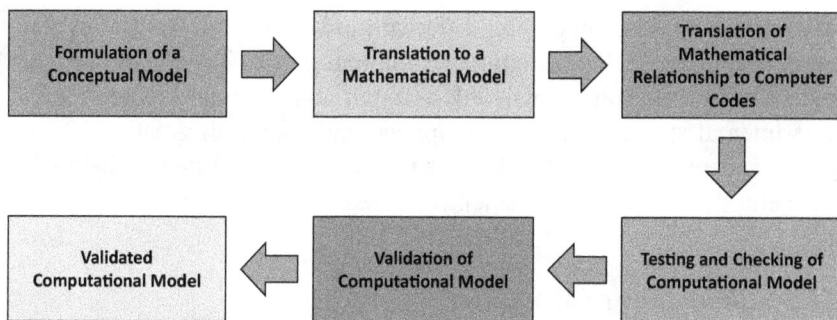

Figure 7.1.   Schematic workflow of the modelling process.

of the model as it is unable to mimic the actual physical system as much as possible. To make reasonable and appropriate assumptions, a good understanding and knowledge of the physical system are needed.

The next step is to translate the conceptual model into a mathematical model that explicitly lists all state variables. Any interaction that takes place within the concept model is required to be represented as a mathematical relationship. The mathematical relationships are then translated into computer codes as the final stage of model construction. The users need to choose the most suitable computational framework, algorithms and computer language for their computational models. Poor selection of these choices could lead to unnecessary waste of time and frustrations, and even poor result outputs [6].

When a computational model is ready, thorough testing and checks are necessary for ensuring the computer code correctly reflects the mathematical relationships over a certain range of parameters. This verification process is critical for ensuring good reliability and stability of the computational model. The computational model is then required to simulate simple cases whose behaviour is already known, so as to compare the experimental findings and model results. The computational model is said to be validated when there are good agreements between experimental findings and model results over a range of cases. Hence, the assumptions and simplifications made at the beginning for the computational model are deemed to be appropriate. Continued verification and validation are necessary to ensure the computational model is used confidently [6].

In general, computational models can also be used for testing hypotheses, refining experiments, interpreting experimental results, carrying sensitivity analyses, integrating different kinds of knowledge and exploring new approaches [6]. Computer modelling may offer valuable insights into an issue that is not obtainable through conventional theoretical and experimental studies. Besides, computer modelling will greatly improve the advancement of research and it has already become one of the essential methods for mainstream scientific research.

7.1.2 Limitations of Modelling

Despite its advantages, computational models do have their shortcomings and limitations too. Good computational models can assist in inspiring new experiments and increasing the number of experiments done in a shorter time [6]. Ultimately, they are only complementary tools that are not intended to be complete substitutes for actual experiments and testing. Furthermore, they are also unable to identify the actual underlying mechanisms that cause the observed phenomenon. However, computational models are particularly useful to prove if the proposed mechanisms are adequate to describe an observed result and vice versa.

7.1.3 Analytical Modelling and Numerical Simulations for 3D Electronics Printing Processes

Analytical modelling and numerical simulations are common modelling approaches used in modelling 3D electronics printing processes. This subsection briefly discusses each modelling approach and some case studies are provided.

7.1.3.1 Analytical modelling of 3D printed electronics processes

An analytical model is a set of mathematical equations that are used to represent a system and it is inherently computational or quantitative [7]. The mathematical equations identify the critical parametric relationships in the system, typically as a function of space, time and other variables. The defined mathematical equations need to have a sufficiently precise representation of the system for more effective and accurate modelling. Analytical modelling typically models the underlying mechanisms of a system to predict the system's performances and other characteristics. A closed-form solution can be used to solve an analytical model and deriving at an exact solution. The analytical results are often expressed in functions comprising time and other parameters and are often presented in tables, plots and charts. Analytical models may be categorised as static or dynamic, in which the former represents system properties that are time-independent and the latter represents the time-varying state of the system [8]. This sub-section showcases some case studies which use analytical modelling to model 3D printing electronics processes such as aerosol jet printing and inkjet printing processes.

Case Study 1:
Goh [9] proposed an analytical modelling approach to investigate the effects of sheath flow on the ink deposition rate in the aerosol jet printing process. In his analytical model, he considered the aerosol flow rate, the effect of print speed and the wetting behaviour of the substrate as critical parameters. Complete aerosols coalescence into a single drop, stable printed lines with cylindrical caps, and printed liquid beads do not bleed into the trenches due to gas flow are some of the assumptions made for his analytical model. The analytical model was then validated with the actual printed lines using carbon nanotube (CNT) ink and it has shown good data fitting for the low print speed regime. His model was able to predict the ink deposition rate and explain how sheath flows can affect the ink deposition rate. To further validate his analytic model, Goh compared the printed lines using silver nanoparticle ink with his analytical model again. It has been observed that the ink deposition rate decreases with rising sheath flow at all atomiser flow rates as the sheath flow affects the evaporation rates of the aerosols before the aerosols impacting the substrate. From this study, it can be suggested that analytical modelling can be used to obtain valuable insights and a better understanding of the aerosol jet printing process to achieve good print quality.

Case Study 2:
In another study by Salmerón *et al.* [10], they developed an analytical model based on the ink composition, substrates and printing conditions to predict the thickness of inkjet printed patterns. The predicted thickness of the inkjet printed patterns from their analytical model were well-validated by the experimental results of two different types of silver nanoparticle inks (see **Figure 7.2(a) and (b)**). The thickness of the inkjet printed patterns are highly dependent on the ink composition, substrate used, and the number of layers printed. They utilised the analytical model to find out the

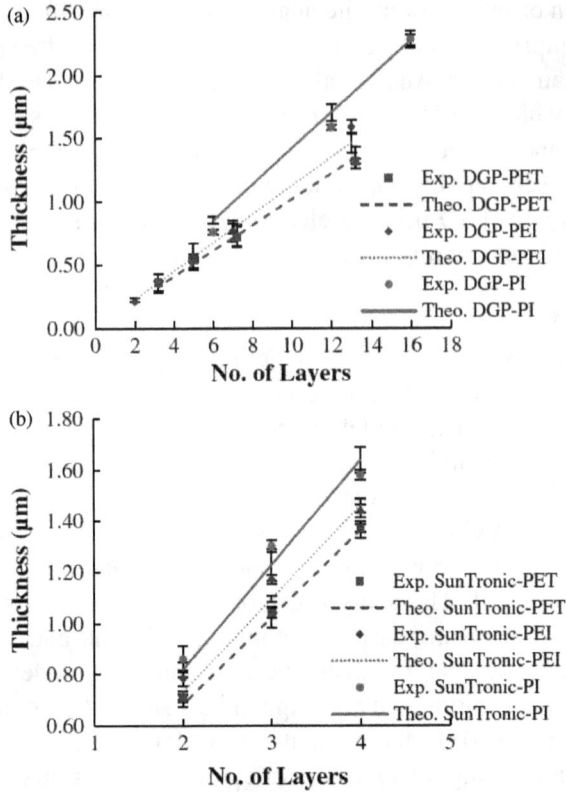

Figure 7.2.   Modelled and experimental data of the thickness of the inkjet printed patterns with a) DGP silver nanoparticle ink and b) SunTronic silver nanoparticle ink, with respect to different layers and substrates. Reprinted with permission from Ref. [10]. Copyright (2014) Springer Nature.

optimal conditions need to achieve the required thickness for fabricating the printed coil of a radio-frequency identification (RFID) tag. This study demonstrated the applicability of using analytical modelling to determine the optimal printing parameters to achieve the required geometric dimensions for 3D printed electronics applications.

Case Study 3:

Analytical models for the stability of inkjet printed lines were developed by Stringer and Derby [11] in one study. Their models were able to predict the lower and upper bound conditions for drop width with stable line

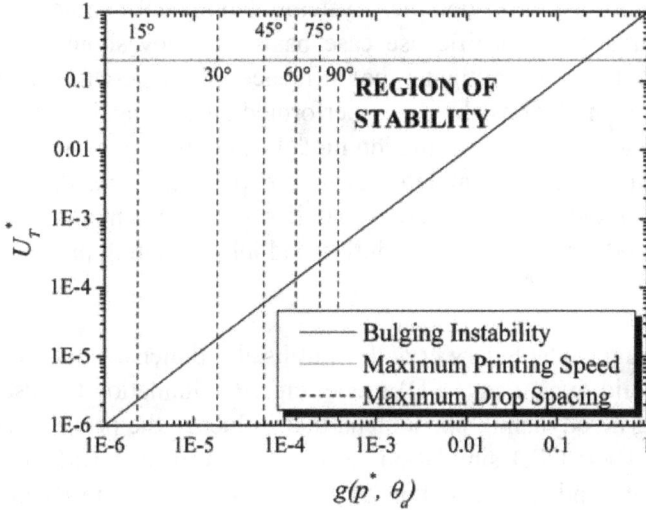

Figure 7.3. Graphical presentation of the analytical model for predicting the region of stability during inkjet printing. Reprinted with permission from Ref. [11]. Copyright (2010) with American Chemical Society.

formation. The lower bound is dependent on the interactions between the contact line pinning and the equilibrium surface tension, whereas the upper bound is affected by drop coalescence and kinetic instability during deposition. These bounds were defined with a common dimensionless formulation and a graphical presentation was used to predict the region of stability (see **Figure 7.3**). Generally, the model gave good agreement with the experimental results for a wide range of inks, except for ink with high vapour pressure solvent. The graphical presentation of the analytical models allows users to determine the region of stability for inkjet printed lines easily.

7.1.3.2 *Numerical simulation of 3D printed electronics processes*

A numerical simulation is a calculation that relies on computational power to solve mathematical models to determine an approximate solution for a physical problem [12]. Numerical simulations are often used when analytical models are too complex to solve (for instance, when there can be no exact solution or when the model scale is too big to solve). Hence, numerical simulations are typically used to simulate and study a wide variety of problems in complex systems. Numerical simulation models yield results on a specific use case basis and they should be iterated numerous times to mitigate the influence of numerical calculations. Numerical simulations should be performed all over again for a separate functioning use case. A simulation model can only be approved when the simulation results are validated against experimental results [13]. This sub-section showcases some case studies which use numerical simulations to model the aerosol jet printing and inkjet printing processes.

Case Study 1:
In a separate study, Salary *et al.* [14] utilised 2-dimensional (2D) computational fluid dynamics (CFD) modelling and simulation to describe the governing aerodynamic phenomena that influence the printed lines morphology. Their CFD simulation results were validated with the experimental data and indicated that the sheath gas flow rate (ShGFR) and carrier gas flow rate (CGFR) can affect the printed lines morphology. In the future, there is a huge potential to use CFD model-derived predictions to prescribe appropriate remedial actions for closed-loop process control in aerosol jet printing when drifting in the printed lines morphology is observed from a charge-coupled device (CCD) camera.

Case Study 2:
Chen *et al.* [15] also developed a 3D CFD model to investigate how CGFR and ShGFR affect the overspray in aerosol jet printed traces. The CFD model was used to describe the fundamental fluid mechanics mechanisms that affect overspray in aerosol jet printing as a function of ShGFR and droplet size distribution. **Figure 7.4** shows the CFD simulated trajectories of different diameter ink particles at various ShGFR, while the CGFR was held constant at 30 ccm. They validated experimental observations with the CFD simulation results and demonstrated that inks with smaller particles tend to have overspray when the ShGFR is low. They expected their study to provide significant insights into the optimization of operating parameters for reducing overspray in aerosol jet printing.

Case Study 3:
Feng [16] employed the Galerkin finite-element approach with boundary-fitted quadrilateral mesh to determine sessile drop deformations on a flat solid surface under an impinging aerosol jet. The simulations were used to help better understand the basic behaviour of free surface flows and

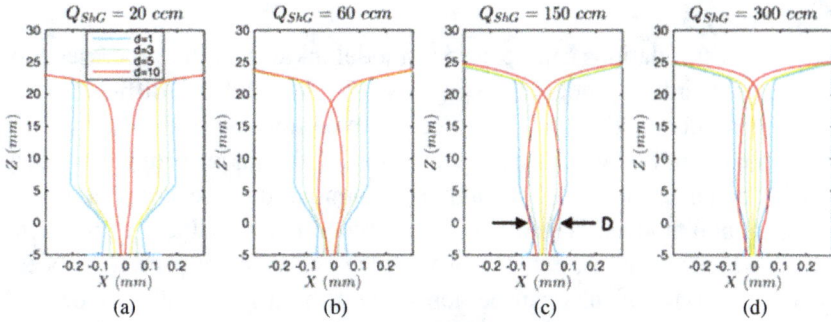

Figure 7.4. CFD simulated trajectories of different diameter ink particles at various ShGFR: (a) 20 ccm, (b) 60 ccm, (c) 150 ccm and (d) 300 ccm, while the CGFR was held constant at 30 ccm. Reprinted with permission from Ref. [15]. Copyright (2018) John Wiley and Sons.

develop theoretical fundamentals design process for the aerosol jet printing process. An empirical formula involving critical capillary number for sessile drops of 45° contact angle was derived for estimating aerosol jet's maximum mass deposition rate under certain ink properties and process parameters This study has helped to shed some light on the fundamental aspects of the aerosol jet's material deposition behaviour.

Case Study 4:
Secor [17] studied the annular drying effects induced by the sheath gas and impaction during an aerosol jet printing process. He designed focused experiments particularly to study the drying effects induced by the sheath gas. He then utilised numerical modelling to explain the fundamental mechanics of the aerosol jet printing process, replicating experimental results and generalising the simulated results for arbitrary ink systems. His numerical modelling and experimental studies have also shown that any small variations in the atomisation yield and ink composition can be easily amplified by the annular drying effects and leading to undesired process drift and sensitivity. Thus, the morphology, resolution and deposition rate of the aerosol jet printing process were significantly affected too. This research demonstrates the huge potentials of getting a better understanding of process variables, designing of ink formulations, and optimising print reliability and quality of aerosol jet printing process based on both experimental studies and numerical modelling.

Case Study 5:
Numerical simulations can be used to model inkjet printing processes too. The inkjet printing process is extremely complex, and it is difficult to get a deeper understanding solely based on experimental studies. However, the fundamental mechanisms at each step of the inkjet printing process can be better explored when numerical simulations are combined with experimental studies. Wijshoff [18] combined numerical and experimental techniques to investigate the drop dynamics in inkjet printing process so that any type of inks can be deposited at high rates with any desired dimension, shape and speed. His numerical simulations with lattice Boltzmann, lubrication theory and volume of fluid method studies have demonstrated a strong correlation with experimental data and revealed critical insights regarding the breakup mechanisms, velocity and mass distributions.

7.2 Data-Driven Approaches in 3D Electronics Printing Processes

Analytical modelling and numerical simulation approaches are generally restrictive for optimising 3D electronics printing processes, as they can only assist in determining the general trend and optimising the printing processes qualitatively. Furthermore, they also rely heavily on individual printing systems and their realistic implementations are significantly hampered by the limited computational resources available [19]. Therefore, there is a need to look into other approaches to better optimise the printing processes to achieve the most optimal printing quality.

Data-driven-based process modelling and optimisation are already widely received in many industries due to their high-cost efficiency and high performances. Hence, there are many opportunities to integrate data-driven approaches into 3D electronics printing systems to optimise the printing process and achieve the optimal printing quality [19]. In general, the data-driven approaches can model the quantitative relationship between the printed features and the influencing printing parameters. A case study is discussed below to better illustrate how the data-driven approaches are used for printing process modelling.

Zhang *et al.* [20] proposed a hybrid machine learning technique to determine the optimal operating process window for the aerosol jet printing process within a 2D design space in achieving good line morphology.

Classical machine learning methods such as experimental sampling, classification, data clustering and knowledge transfer are used in their study. The aerosol jet printed line morphology is primarily influenced by the ink's material properties, the substrate's surface energy and the printing process parameters. The printing process parameters such as print speed, CGFR and ShGFR, can be easily controlled with the aerosol jet printing system.

A 2D Latin hypercube sampling (LHS) experimental design was used to investigate and maximize the uniformity of a design space at a specific printing speed. A K-means clustering algorithm was then used to analyse and group the printed lines that were under different influences of CGFR and SHGFR into four different clusters: discontinuous lines, high roughness and high overspray, high roughness and low overspray and normal line (see **Figure 7.5(a)–(d)**). A support vector machine (SVM), a type of supervised learning model with associated learning algorithms, helped to identify the optimal operating process window that allowed the aerosol jet printer to achieve good line morphology at that specified print speed. An inductive transfer learning approach is used to leverage relatedness between multiple operating process windows. This approach can help to identify new operating process windows at multiple print speeds more efficiently by merging the prior data sets with new sample results (see **Figure 7.5(e)**). Hence, this study showcased the potentials of using data-driven approaches for printing process modelling.

Separately, there are also several studies on using data-driven approaches to model other printing processes [19,21,22]. For instance, Wu *et al.* [22] have proposed a new predictive model for predicting the inkjet droplet volume and velocity by using an ensemble learning algorithm. Their predicted results have shown sufficient accuracy when compared with experimental data too.

Data-driven based process modelling and optimisation for 3D electronics printing processes is one exciting area to look out for in the near future. Data-driven approaches have immense promise and potentials to outperform analytical modelling and numerical simulation, especially when there is a great abundance of data readily available for extraction [23]. Note that large datasets with good quality input data and good classification accuracy are essential prerequisites to achieve high prediction accuracies. With more advanced technology and computational power in the future, real-time *in-situ* monitoring and closed-loop feedback control may also be used to extract data and detecting anomaly during the printing

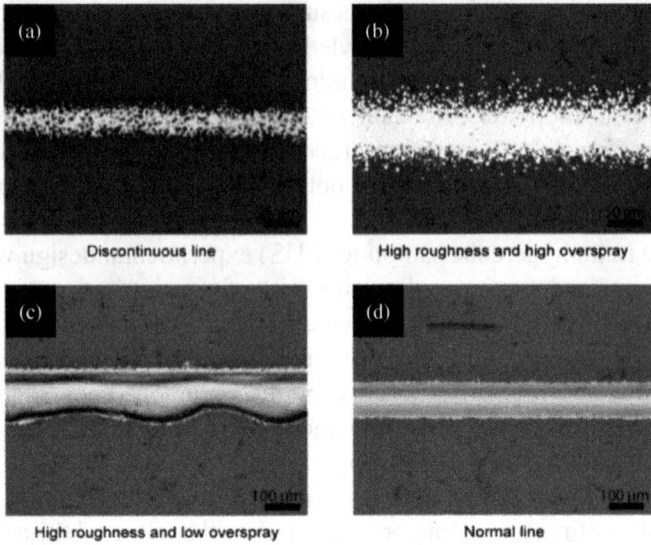

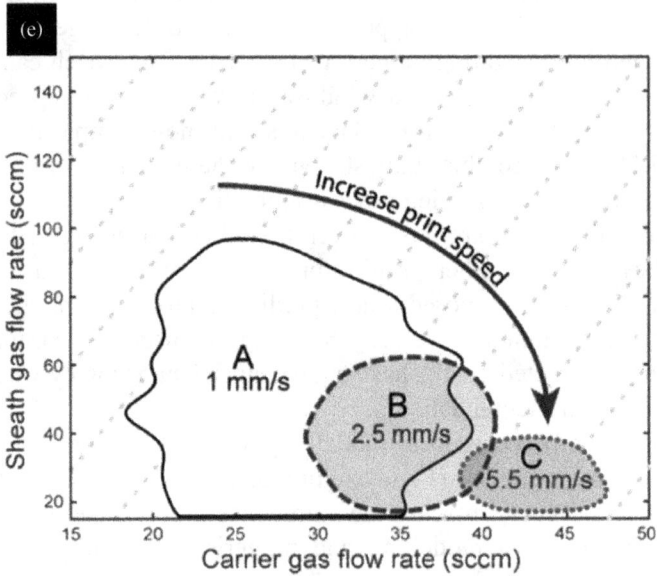

Figure 7.5.   Various print line morphologies: (a) discontinuous lines, (b) high roughness and high overspray, (c) high roughness and low overspray and (d) normal line; and (e) multiple operating process windows at various printing speeds. Reprinted with permission from Ref. [20]. Copyright (2019) American Chemical Society.

process and making the necessary changes to printing parameters to maintain good printing quality.

7.3 Functional Simulations of Electrical Components in 3D Electronics Printing

It is very challenging to design 3D printed electronics first time right that has good manufacturability and performances. To solve this issue, simulation can be introduced into the design phase, in which an iterative approach can be used in the design methodology to optimize the predict the physical results of changes. Finite element method (FEM), or finite element analysis, is one of the commonly used multiphysics simulation tools for estimating the behaviour of an object [4] and can also help to verify various design properties for design optimization [24]. These simulation tools can be particularly helpful for 3D printed electronics applications, as they can address the coupled effects among mechanics, electromagnetics, fluid mechanics, thermodynamics and other properties.

FEM is a computational technique for obtaining numerical solutions to time- and space-dependent partial differential equations (PDEs) of physical problems, of which analytical solutions cannot be easily obtained [25–27]. Geometric representation, material representation and boundary condition are the three main aspects of FEM [12]. Rather than solving the problem for the entire body within a single operation, the FEM discretises a large body into smaller interconnected elements, known as "finite elements", by creating a mesh of the object. The elements should be sufficiently small to achieve functional results, but not too small for unnecessary excessive computational efforts. Each finite element has a displacement function (typically linear, quadratic and cubic polynomials functions) and it is related to each other through common interfaces such as boundary lines, surfaces and nodes. Calculations are performed for each element one at a time to find the element properties, and ultimately calculating for all the elements within the mesh. In order to determine the global equation system for the entire body, all local elements must be assembled, and all boundary conditions need to be well-defined. Direct and iterative methods are usually used for solving the finite element global equation system. Finally, the results are interpreted and analysed for use in the design or analysis process [25–27].

Case Study 1:
In one research, Agarwala *et al.* [28] used simulation computational studies to optimize the physical dimensions of the aerosol jet printed strain gauge (such as the number of grids, gauge length grid line width and end loop length), while evaluating how different design parameters can influence its performances (gauge factor and sensitivity) (see **Figure 7.6**). The optimized strain gauge design with the best-simulated performances was then fabricated with aerosol jet printing. Simulation studies remove the need for physically manufacturing and testing different designs, and thereby allowing more cost and time-saving.

Case Study 2:
In another study by Hou *et al.* [29], they utilized FEM simulations and experiments to verify their mathematical models in calculating the self- and mutual- inductance of the 3D printed coils for wireless power transfer (WPT). Their study showed that the absolute mean errors between

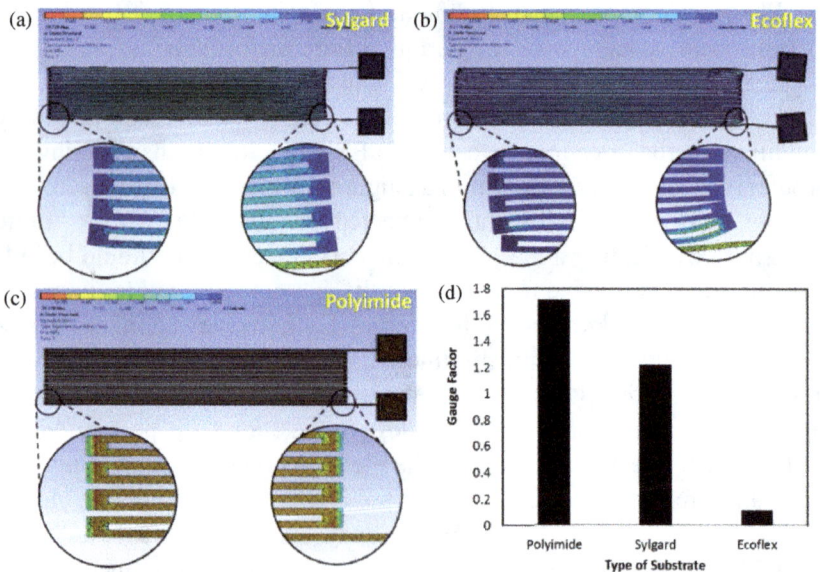

Figure 7.6.   (a)–(c) Stress analysis for strain sensors that were fabricated on different substrates; and (d) Gauge factors for strain sensors that were fabricated on the different type of substrates. Reprinted with permission from Ref. [28]. Copyright (2018) IEEE.

the measurements and simulation results were below 5%. This also demonstrated that there are huge potentials in using multiphysics simulation tools in optimizing the design parameters for 3D printed electronics first before fabrication. This is particularly advantageous for 3D printed electronic devices with very complex designs and structures, in which the fabrication process is not simple and time-consuming. Note that multiphysics simulations are usually computationally expensive and often require precise inputs, such as material properties and boundary conditions, to generate accurate results [4].

Case Study 3:

Zhou *et al.* [30] have also demonstrated the use of FEM simulations to guide the design and fabrication of a four-fingered soft gripper. The soft gripper (see **Figure 7.7(a)**) was fabricated by 3D printing flexible structures directly onto a pre-stretched dielectric elastomer membrane, based on the fused deposition modelling (FDM) technology. Each finger

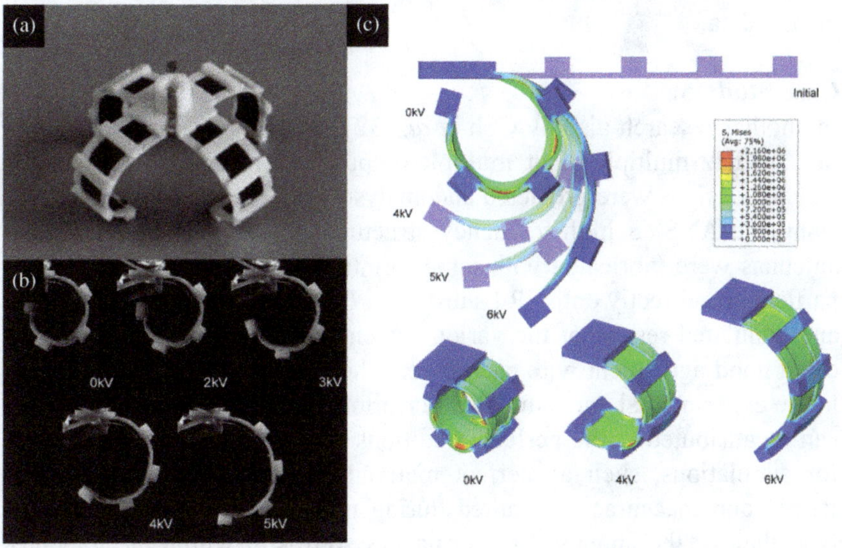

Figure 7.7. (a) Image of a four-fingered soft gripper; (b) bending states of a finger actuator when subjected to different applied voltage; and (c) FEM simulated results of a finger actuator when subjected to a different applied voltage. Reprinted with permission from Ref. [30]. Copyright (2019) Elsevier.

actuator is initially in an equilibrium bending state when no voltage is applied. The pre-tension and bending energy in the finger actuator are released as a voltage is applied, and the bending deformation increases when the applied voltage increases (see **Figure 7.7(b)**). FEM simulations were used to investigate the actuating behaviour of a finger actuator with different applied voltage and results showed a good fit with the experimental data too. This case study, therefore, shows the huge potentials of using FEM simulations to guide the design and fabrication of electrical devices, as they can predict the general behaviour and offer insights under various input conditions quickly.

Case Study 4:
Mohammed *et al.* [31] also have 3D printed stretchable strain sensors for wind sensing applications and employed FEM to analyse the sensors' behaviours under various airflow conditions. They fixed both ends of the wind sensor for their FEM simulation and subjected a wind load perpendicularly to it. The simulation helped to predict the sensor response at different wind velocities and the results showed a good fit with the experimental data.

Case Study 5:
In another research also by Goh *et al.* [32], the theoretical performances of 2.4 GHz multiple-input multiple-output (MIMO) and 6 GHz patch antenna designs were predicted and analysed with numerical simulations, using the ANSYS high-frequency structural simulator software. These antennas were fabricated with inkjet printing by depositing silver nanoparticle inks directly onto FR4 substrates. **Figure 7.8** shows the measured and simulated results of the various patch antennas and they have indicated good agreement with each other. There were only slight variations in the experimental and simulated performances. These small differences can be attributed to the perfect conditions that software usually takes on for simulations, such as perfect material properties and no electrical losses, and inaccuracies resulted during measurements and fabrication. Nonetheless, the antennas' performances were still within the tolerance range and experimental results' operating bandwidth also aligns with the simulated test results. This work thus demonstrated the possibilities of using numerical simulations for simulating antennas and can allow optimisations of the antenna design to achieve the desired properties before fabrication.

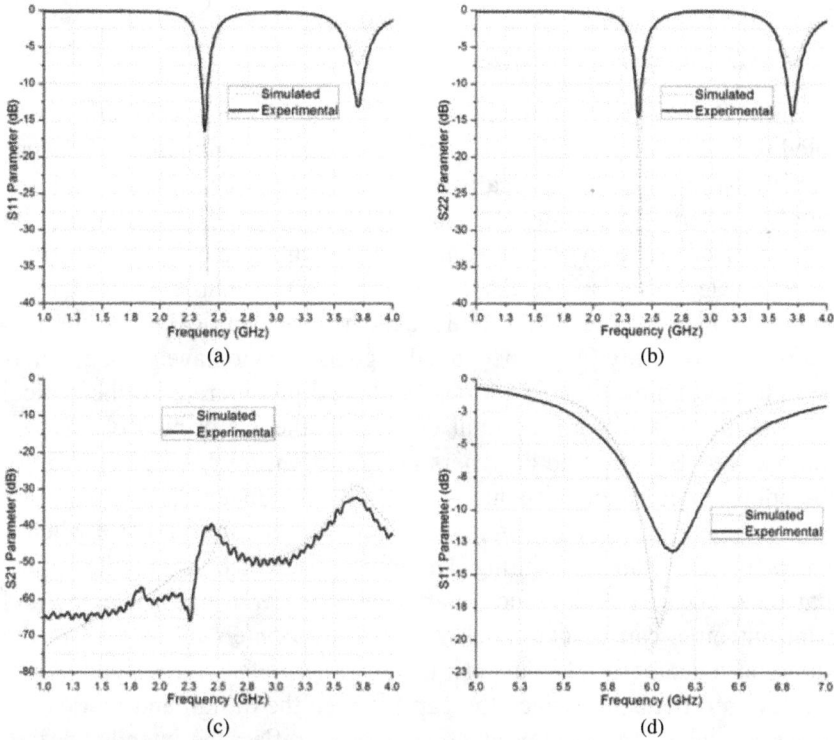

Figure 7.8. Plotted (a) S11 parameter, (b) S22 parameter, (c) S21 parameter for 2.4 GHz MIMO patch antenna against frequency and (d) S11 for 6 GHz antenna against frequency. Reprinted with permission from Ref. [32], Copyright (2016) Taylor & Francis Ltd.

7.4 Designing 3D Printed Electronics

As more design freedom and flexibility can be offered with the AM technology, electronics can be either integrated into complex structures or fabricated directly onto 3D substrate geometries. Thus, enabling designers to stay concentrated in optimizing the 3D printed electronics' functionality, shape and aesthetical looks. However, at the same time, 3D printed electronics are moving forward for more advanced applications and it becomes increasingly challenging for users to incorporate multi-material, multi-scale and multi-functionality features in device design with the current computer-aided design (CAD) state-of-the-art software [4]. This section discusses some of the challenges and opportunities for designing 3D printed electronics.

7.4.1 *Multi-material, Single-functionality 3D Printed Electronics*

3D printed electronics components on rigid substrates are usually fabricated with different materials (e.g. conductive, semiconducting, dielectric and insulating inks) and typically have single functionality. Although the state-of-the-art CAD software can define the discrete multi-material regions of the 3D printed electronics components and assign the various materials accordingly, the CAD software is still unable to design the amount of materials needed to achieve the precise electrical properties and desired functionality [4]. Hence, the designers need to have a good understanding and knowledge in various fields and incorporate all the critical information in achieving optimal designs of the 3D printed electronics components. Besides, the designers also must ensure good compatibility and adhesion between adjacent layers of different materials.

The process parameters may also be more difficult to control when it comes to multi-material printing. Therefore, it will be helpful if the important data, such as AM process parameters, material specifications and print outcome, can be extracted from the 3D printers and stored in the library as a database. The data can then be interpreted and integrated into the design process, to reduce the gap between the design and fabrication processes, where the final product can closely reflect the intended design with the expected properties, functionality and repeatability.

7.4.2 *Multi-material, Multi-functionality 3D Printed Electronics*

Research has also begun to focus more on producing complex multi-functional 3D printed electronics devices for more advanced applications. Multi-material additive manufacturing extends the spatial variety of materials, and thereby improves the freedom of design and allow multi-functionality [4]. Multi-functional 3D printed electronics devices usually require embedding active electrical components within structures to give extra functional capability (such as electrical, optical, electromagnetic, thermal, electro-mechanical and chemical properties) [33]. For instance, Valentine *et al.* [34] fabricated a wearable strain monitoring device on a soft substrate with direct ink writing and surface mount electrical components were pick-and-place automatically. This wearable device is comprised of a large-area soft sensor array and a strain sensor. It is to be worn posterior to the elbow joint and can read the strain data corresponding to the various elbow joint angles. Hence, this application has demonstrated the ability to produce multi-functional 3D printed electronics with state-of-the-art technology, by integrating functional inks, electrical components and soft substrates.

Geometric models of 3D printed electronics usually comprise both electronic parts, and as well as, mechanical parts of the design. Furthermore, the design processes are not straightforward and may also result in a complex design representation. More than one CAD software is usually required for designing multi-functionality features (e.g. mechanical, electrical, thermal and optical features) in 3D printed electronics, especially for structural and embedded electronics [4]. For instance, a printed circuit board (PCB) design software typically can only design 2D circuits. Some 3D mechanical software platforms may not support the importing of electronic CAD files and hence causing incompatibility issues between different CAD software. Researchers are currently using different approaches to address these shortcomings. MacDonald *et al.* [35] projected 2D circuits generated from electrical CAD software directly onto the surfaces of the intended 3D models. However, this approach only allows electronics to be integrated onto the surfaces of the mechanical part and did not take full advantage of 3D designs [24]. Furthermore, it also limits the ability to have interconnections between different layers. Some 3D mechanical software platforms also have additional customized plug-ins to generate compatible printing instructions from the design data directly to the AM systems.

Panesar *et al.* [33] also had proposed a strategy for optimizing multi-functional 3D printed structures, in which both the structural and system design aspects are coupled together (see **Figure 7.9**). The structural design is based on mechanical behaviour, whereas the system design is based on functional features and performances. The two main aspects that allow this strategy to work are intelligent components placement and generations of connections between them. The electrical components are placed based on the locations which fulfil the geometry and performance criteria. A thinning algorithm was used to extract the skeletal information of a part's topology and the information was used for the evaluation of the electrical components' orientation. The approximate routing (Dijkstra's algorithm for generating shortest path) or accurate routing (Fast Marching (FM) method) computation approaches were then used to focus on generating connections between electrical components to form a circuit and

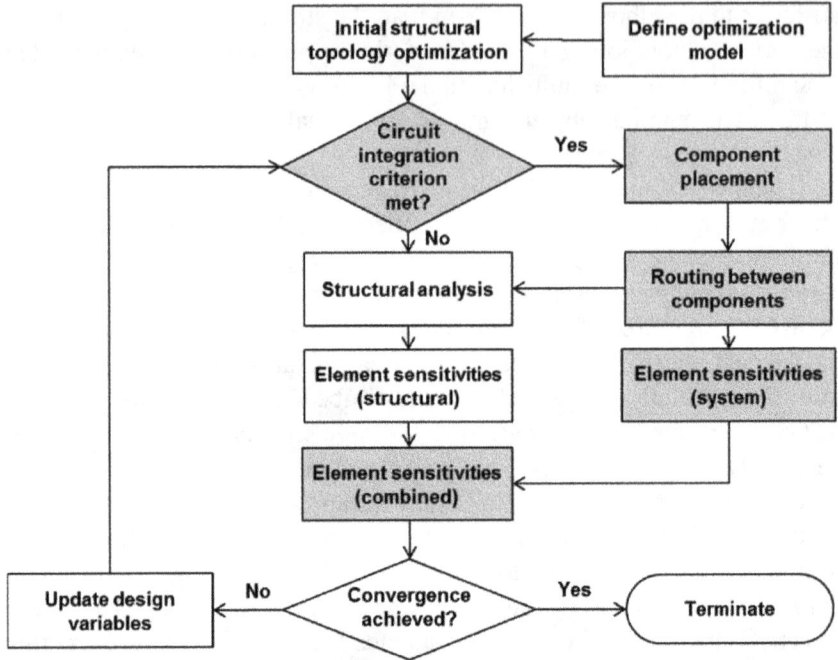

Figure 7.9.   Design framework depicting the coupled optimization procedure. Reprinted with permission from Ref. [33]. Copyright (2017).

improving circuit efficiency. The system design is conducted simultaneously with structural optimization and takes into account the system effects on the structural response for every iteration within a modified bidirectional evolutionary structural optimization.

References

[1] Tan, H. W., An, J., Chua, C. K. and Tran, T. (2019). Metallic nanoparticle inks for 3D printing of electronics, *Adv. Electron. Mater.*, 5, p. 1800831.

[2] Tan, H. W., Saengchairat, N., Goh, G. L., An, J., Chua, C. K. and Tran, T. (2020). Induction sintering of silver nanoparticle inks on polyimide substrates, *Adv. Mater. Technol.*, 5, p. 1900897.

[3] Tan, H. W., Tran, T. and Chua, C. K. (2016). A review of printed passive electronic components through fully additive manufacturing methods, *Virtual Phys. Prototyp.*, 11, pp. 271–288.

[4] Leung, Y.-S., Kwok, T.-H., Li, X., Yang, Y., Wang, C. C. L. and Chen, Y. (2019). Challenges and status on design and computation for emerging additive manufacturing technologies, *J. Comput. Inf. Sci. Eng.*, 19(2), p. 021013.

[5] Tan, J. H. K., Sing, S. L. and Yeong, W. Y. (2019). Microstructure modelling for metallic additive manufacturing: a review, *Virtual Phys. Prototyp.*, 15, pp. 87–105.

[6] Brodland, G. W. (2015). How computational models can help unlock biological systems, *Semin. Cell Dev. Biol.*, 47–48, pp. 62–73.

[7] Friedenthal, S., Moore, A. and Steiner, R. (2015). *A Practical Guide to SysML* 3rd edn, eds. Sanford Friedenthal, Alan Moore and Rick Steiner, Chapter 18: Integrating SysML into a systems development environment (Morgan Kaufmann, Boston) pp. 507–541.

[8] Friedenthal, S., Moore, A. and Steiner, R. (2012). *A Practical Guide to SysML*, 2nd edn, eds. Sanford Friedenthal, Alan Moore and Rick Steiner, Chapter 18: Integrating SysML into a systems development environment (Morgan Kaufmann, Boston) pp. 523–556.

[9] Goh, G. L. (2019). *Aligning Carbon Nanotubes via Aerosol Jet Printing for Flexible Electronics* (Doctoral Dissertation, Nanyang Technological University).

[10] Salmerón, J. F., Molina-Lopez, F., Briand, D., Ruan, J. J., Rivadeneyra, A., Carvajal, M. A., Capitán-Vallvey, L. F., de Rooij, N. F. and Palma, A. J. (2014). Properties and rintability of inkjet and screen-printed silver patterns for RFID antennas, *J. Electron. Mater.*, 43, pp. 604–617.

[11] Stringer, J. and Derby, B. (2010). Formation and stability of lines produced by inkjet printing, *Langmuir*, 26, pp. 10365–10372.

[12] Zafarparandeh, I. and Lazoglu, I. (2012). *The Design and Manufacture of Medical Devices*, eds. J. Paulo Davim, Chapter 4: Application of the finite element method in spinal implant design and manufacture (Woodhead Publishing, United Kingdom) pp. 153–183.

[13] Estabragh, A. R., Pereshkafti, M. R. S. and Javadi, A. A. (2013). Comparison between analytical and numerical methods in evaluating the pollution transport in porous media, *Geotech. Geol. Eng.*, 31, pp. 93–101.

[14] Salary, R., Lombardi, J. P., Samie Tootooni, M., Donovan, R., Rao, P. K., Borgesen, P. and Poliks, M. D. (2016). Computational fluid dynamics modelling and online monitoring of aerosol jet printing process, ASME. *J. Manuf. Sci. Eng.*, 139(2), p. 021015.

[15] Chen, G., Gu, Y., Tsang, H., Hines, D. R. and Das, S. (2018). The effect of droplet sizes on overspray in aerosol-jet printing, *Adv. Eng. Mater.*, 20, p. 1701084.

[16] Feng, J. Q. (2015). Sessile drop deformations under an impinging jet, *Theor. Comput. Fluid Dyn.*, 29, pp. 277–290.

[17] Secor, E. B. (2018). Guided ink and process design for aerosol jet printing based on annular drying effects, *Flexible Printed Electron.*, 3, p. 035007.

[18] Wijshoff, H. (2018). Drop dynamics in the inkjet printing process, *Curr. Opin. Colloid Interface Sci.*, 36, pp. 20–27.

[19] Zhang, H. (2020). *A Printing Quality Optimization Framework for Non-Contact Ink Writing Techniques* (Doctoral Dissertation, Nanyang Technological University).

[20] Zhang, H., Moon, S. K. and Ngo, T. H. (2019). Hybrid machine learning method to determine the optimal operating process window in aerosol jet 3D printing, *ACS Appl. Mater. Interfaces*, 11, pp. 17994–18003.

[21] Wang, T., Kwok, T.-H., Zhou, C. and Vader, S. (2018). *In-situ* droplet inspection and closed-loop control system using machine learning for liquid metal jet printing, *J. Manuf. Syst.*, 47, pp. 83–92.

[22] Wu, D. and Xu, C. (2018). Predictive Modeling of droplet formation processes in inkjet-based bioprinting, ASME. *J. Manuf. Sci. Eng.*, 140(10), p. 101007.

[23] Goh, G. D., Sing, S. L. and Yeong, W. Y. (2020). A review on machine learning in 3D printing: Applications, potential, and challenges, *Artif. Intell. Rev.*, doi: 10.1007/s10462-020-09876-9.

[24] Song, Y., Boekraad, R. A., Roussos, L., Kooijman, A., Wang, C. C. L. and Geraedts, J. M. P. (2017). 3D printed electronics: Opportunities and challenges from case studies, *presented at the ASME 2017 International Design Engineering Technical Conferences and Computers and Information in Engineering Conference*, Cleveland, Ohio, USA.

[25] Li, G. (2020). *Introduction to the Finite Element Method and Implementation with MATLAB®* (Cambridge University Press, United Kingdom).

[26] Nikishkov, G. (2004). Introduction to the finite element method, *University of Aizu*, pp. 1–70.

[27] Logan, D. L. (2011). *A First Course in the Finite Element Method* (Cengage Learning, USA).

[28] Agarwala, S., Goh, G. L. and Yeong, W. Y. (2018). Aerosol jet printed strain sensor: Simulation studies analyzing the effect of dimension and design on performance (September 2018), *IEEE Access*, 6, pp. 63080–63086.

[29] Hou, T., Xu, J., Elkhuizen, W. S., Wang, C. C. L., Jiang, J., Geraedts, J. M. P. and Song, Y. (2019). Design of 3D wireless power transfer system based on 3D printed electronics, *IEEE Access*, 7, pp. 94793–94805.

[30] Zhou, F., Zhang, M., Cao, X., Zhang, Z., Chen, X., Xiao, Y., Liang, Y., Wong, T.-W., Li, T. and Xu, Z. (2019). Fabrication and modeling of dielectric elastomer soft actuator with 3D printed thermoplastic frame, *Sens. Actuators, A*, 292, pp. 112–120.

[31] Al-Rubaiai, M., Tsuruta, R., Gandhi, U., Wang, C. and Tan, X. (2019). A 3D-printed stretchable strain sensor for wind sensing, *Smart Mater. Struct.*, 28, p. 084001.

[32] Goh, G. L., Ma, J., Chua, K. L. F., Shweta, A., Yeong, W. Y. and Zhang, Y. P. (2016). Inkjet-printed patch antenna emitter for wireless communication application, *Virtual Phys. Prototyp.*, 11, pp. 289–294.

[33] Panesar, A., Ashcroft, I., Brackett, D., Wildman, R. and Hague, R. (2017). Design framework for multifunctional additive manufacturing: Coupled optimization strategy for structures with embedded functional systems, *Addit. Manuf.*, 16, pp. 98–106.

[34] Valentine, A. D., Busbee, T. A., Boley, J. W., Raney, J. R., Chortos, A., Kotikian, A., Berrigan, J. D., Durstock, M. F. and Lewis, J. A. (2017). Hybrid 3D printing of soft electronics, *Adv. Mater.*, 29, p. 1703817.

[35] Macdonald, E., Salas, R., Espalin, D., Perez, M., Aguilera, E., Muse, D. and Wicker, R. B. (2014). 3D printing for the rapid prototyping of structural electronics, *IEEE Access*, 2, pp. 234–242.

Problems

1. What is the process of modelling?
2. What are the limitations of modelling?
3. Briefly describe what are the main differences between analytical modelling and numerical simulations.
4. How are data-driven approaches more advantageous than the analytical modelling and numerical simulation approaches in optimising the 3D printing processes?
5. Briefly describe what are some of the current challenges in designing multi-material, multi-functional 3D printed electronics?
6. Discuss how are the potential solutions aim to mitigate the challenges faced in designing 3D printed electronics?
7. Briefly describe how simulations can be applied to 3D printed electronics.
8. How are simulations advantageous for designing 3D printed electronic devices?

Chapter 8

Applications of 3D Printed Electronics and Future Outlook

In recent years, great progress has been made in 3D printing of electronics. Significant efforts have been put into developing advanced printing systems and functional inks, refining printing processes, optimising sintering processes, using computer simulations for optimising designs and even integrating machine learning algorithms for printing. 3D printed electronics has been considered to be one of the next upcoming new frontiers of additive manufacturing and it is set to revolutionise the electronics industry in the near future [1–4]. With the advanced software, machines and functional inks, our imagination is the only limit to what we can fabricate for 3D printed electronics applications.

In literature, there are myriads of 3D printing electronics applications and this chapter is unable to cover them exhaustively. Hence, this chapter only highlights some of the electrical components and devices that are commonly used for 3D printed electronics, including passive components, active components and sensors. Their basic architecture designs, operating mechanisms and functionalities are discussed greater in details and showcase the potentials of using additive manufacturing technologies for motivating further advances in 3D printed electronics. This chapter lastly also discusses the future outlook of 3D printed electronics.

8.1 Passive Electrical Components

Passive electrical components are the most basic and simplest electrical circuit components which do not produce power or gain in their operations [4,5]. Passive electrical components, including resistors, capacitors and inductors, primarily function as filters, tuners, converters and protection for active circuitries, and constitute the bulk of the overall electrical components in any standard printed circuit board (PCB). Apart from the conventional passive electrical components, passive electrical components can also be fabricated by the additive manufacturing technology. The 3D printed passive electrical components eliminate the need for solder joints and can be fabricated onto different substrates (e.g. flexible, rigid, opaque or transparent substrates) directly [6]. This section discusses some of the commonly fabricated passive electrical components using additive manufacturing technologies.

8.1.1 *Resistors*

Resistors are basic passive electrical components primarily used for regulating current flows in electrical circuits [4,7]. Each resistor is characterised by electrical resistance (R), which is defined as the ratio of voltage across the resistor (ΔV) to the current going through the resistor (I):

$$R = \frac{\Delta V}{I}. \tag{8.1}$$

Fixed resistors can be easily fabricated with additive manufacturing technologies. **Figure 8.1** shows the schematic diagram of a 3D printed fixed resistor. Electrically conductive traces, with a gap separation in the middle, are first fabricated by depositing metallic nanoparticle inks on the substrate (see **Figure 8.1(a)**). Electrically resistive material is then deposited in the gap to connect the discontinuous conductive traces (see **Figure 8.1(b)**). Structural conductive polymers inks (e.g. PEDOT:PSS) or composite conductive polymers inks (e.g. carbon inks) are usually used as the resistive material, as they have significantly higher electrical resistivity as compared to the electrically conductive traces. Both conductive traces and resistive layers are then sintered together simultaneously [4,8].

(a)

(b)

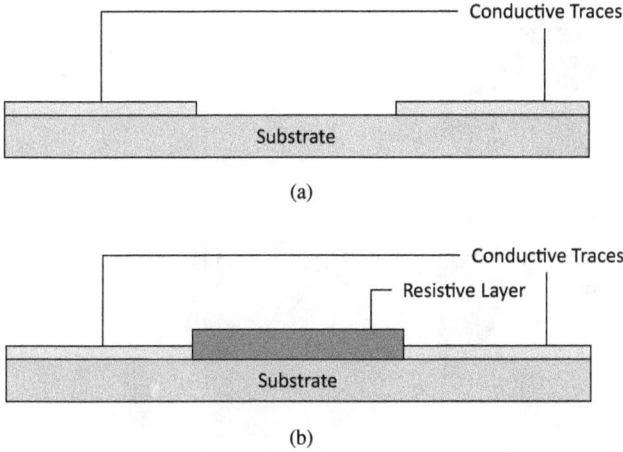

Figure 8.1. Schematic diagram of the printing process of resistor: (a) conductive traces are first fabricated on the substrate, and (b) electrically resistive material deposited in the gap to connect the discontinuous conductive traces [4,8].

The electrical resistance of the 3D printed resistor can be expressed as in terms of the electrical resistivity of the resistive material (ρ), and the length (l) and cross-sectional area of the resistive layer (A):

$$R = \rho \frac{l}{A}. \tag{8.2}$$

By varying the geometric dimensions of the resistive layer, the 3D printed resistors can be designed with the desired electrical resistance. The 3D printed resistors can be measured accurately by the 4-point probe method, as this method eliminates the contact resistance at the contacting points [9–11].

Flowers *et al.* [12] demonstrated the fabrication of 3D printed resistors with various electrically conductive filaments (see **Figure 8.2(a)**). They obtained 3D printed resistors with varying resistance values ranging from 10 Ω to 10 KΩ by altering the resistors' geometric dimensions and materials. The authors measured the electrical resistance of each 3D printed resistor and plotted them as a function of the inversed cross-sectional area (see **Figure 8.2(b)**). The data showed a good linear relationship for each corresponding electrically conductive filament, and this linear relationship can allow one to design and achieve reproducible and predictable resistance values for any desired application.

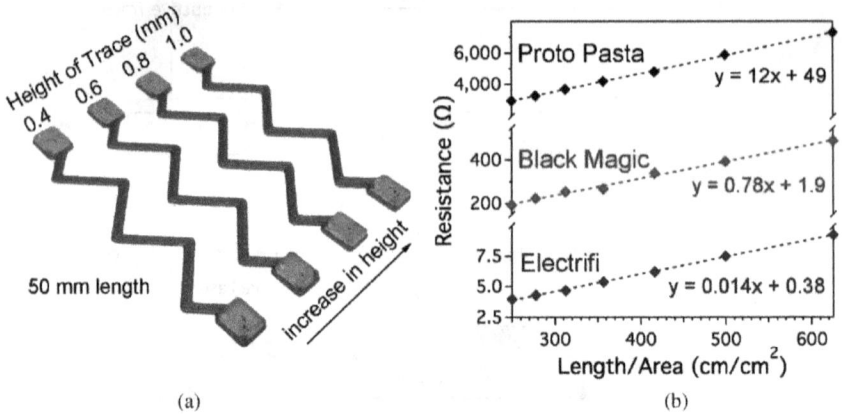

Figure 8.2.   (a) 3D printed resistors with varying thickness; and (b) electrical resistance of the 3D printed resistors, that were fabricated with various materials, plotted as a function of the inversed cross-sectional area. Reprinted with permission from Ref. [12]. Copyright (2017) from Elsevier.

8.1.2 *Capacitors*

Capacitors are primarily used for temporary storing of electrical charges or signal filtering in electrical circuits [4,8,13–14]. Each capacitor is characterised by the capacitance (C), which is defined as the ratio of the charge stored in the capacitor (Q) to the potential difference across the capacitor (ΔV):

$$C = \frac{Q}{\Delta V}. \qquad (8.3)$$

A basic 3D printed capacitor comprises a dielectric layer that is sandwiched between two conductive layers (see **Figure 8.3**). The top and bottom conductive layers are usually printed with metallic nanoparticle inks (e.g. silver nanoparticle inks) so that they have good electrical conductivity. The bottom conductive layer is printed onto the substrate directly and it is sintered first before the next layer of dielectric material is deposited on top of it. The sintering process converts the metallic nanoparticle inks into a thin metallic film, and it helps to prevent the dielectric material from absorbing into the bottom conductive layer and thereby preventing cracks. The dielectric layer and top conductive layer are then subsequently printed on top of the bottom conductive layer. The dielectric

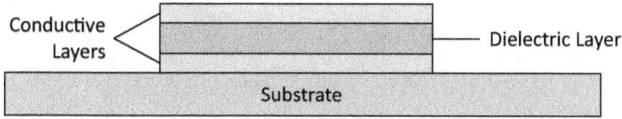

Figure 8.3.   Schematic diagram of a 3D printed capacitor [4].

and top conductive layers need to be sintered again after the printing is completed [4,8,15].

The capacitance of the 3D printed capacitor is heavily dependent on the material properties and geometric parameters of the dielectric layer, and can be expressed as in terms of the dielectric constant of the dielectric layer (κ), permittivity of free space (ε_0), the overlapping area between the two conductive layers (A) and the thickness of the dielectric layer (d) [14]:

$$C = \kappa \frac{\varepsilon_0 A}{d}. \tag{8.4}$$

8.1.3 *Inductors*

The inductor is one of the passive electrical components that is frequently used in electrical circuits. The inductor accumulates electromagnetic energy in a magnetic field as the electric current flows through it and it is characterised by its inductance (L). The self-induced electromotive force (emf), ε_L is proportional to the time rate of change of current ($\frac{di}{dt}$) flowing through the coil, in which the inductance (L) is a proportionality constant [16]:

$$\varepsilon_L = -L \frac{di}{dt}. \tag{8.5}$$

Re-arranging the above equation, the inductance can be expressed as:

$$L = -\frac{\varepsilon_L}{di / dt}. \tag{8.6}$$

According to Faraday's law, the self-induced emf can also be expressed as:

$$\varepsilon_L = -N \frac{d\Phi_B}{dt}, \tag{8.7}$$

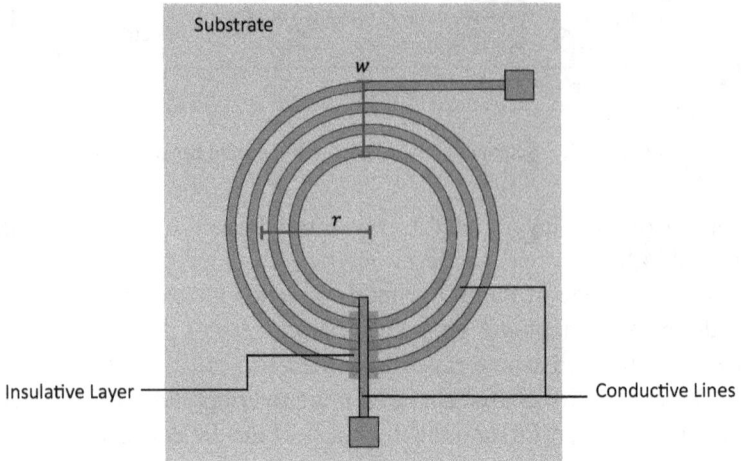

Figure 8.4.   Schematic diagram of a 3D printed inductor.

where N is the number of loops and $\frac{d\Phi_B}{dt}$ is the time rate of change of magnetic flux through one loop.

A basic printed flat spiral inductor typically comprises of few turns of spiral conductive lines (see **Figure 8.4**) which help to enhance the electromagnetic field [4,8,15]. The spiral conductive lines are first printed and sintered on the substrate. An insulating layer is subsequently printed on part of the spiral conductive lines to prevent short-circuiting. Another conductive line is then printed on top of this insulating layer, in which it extends the centre of the spiral to the outer perimeter of the printed inductor. To further improve the inductance, a layer of ferrite film may also be printed over the printed inductor to increase the coil's magnetic permeability.

The inductance of the printed flat spiral inductor (in μH) can be predicted empirically by the Wheeler's approximation [17] for single-layer helical coil:

$$L = \frac{r^2 n^2}{8r + 11w}, \tag{8.8}$$

where n is the number of turns, r is the mean radius of the coil (mean radius of the outer and inner edges of the spiral) and w is the width of the coil (see **Figure 8.4**) [12]. Note that the dimensions are measured in inches for this empirical formula.

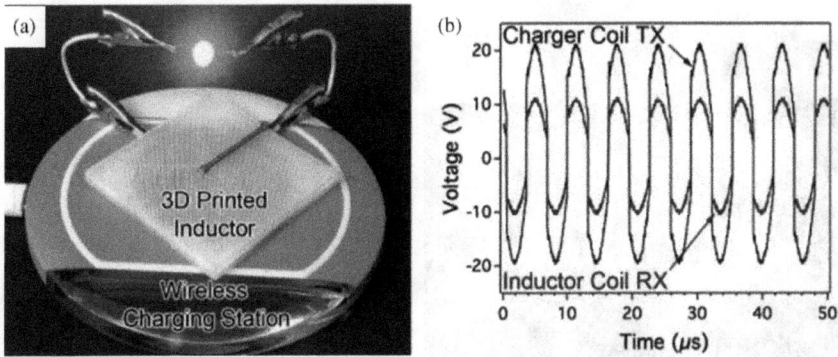

Figure 8.5.   (a) 3D printed inductor receiving wireless power from a wireless charging station to light up a LED, and (b) oscilloscope measurements of the transmitted waveforms of the charger coil and received waveforms of the induction coil during the wireless power transfer process. Reprinted from Ref. [12]. Copyright (2017) from Elsevier.

Flowers *et al.* [12] also demonstrated the fabrication of 3D printed air-core spiral inductors, in which Electrifi (Multi3D LLC) and polylactic acid (PLA) were used as the electrically conductive and insulative dielectric materials respectively. The 3D printed inductor was able to receive wireless power from a commercial wireless phone charger and lit a light-emitting diode (LED) that was connected to it (see **Figure 8.5(a)**). The transmitted waveforms of the charger coil and received waveforms of the induction coil during the wireless power transfer process are showed in **Figure 8.5(b)**.

8.1.4 *Passive Filters*

High-pass filters and low-pass filters can also be designed with 3D printed passive electrical components such as resistors, capacitors and inductors. A high-pass filter only allows signals with frequencies higher than the cut-off frequency to pass through while blocking off lower frequency signals. Similarly, a low-pass filter only allows signals with frequencies lower than the cut-off frequency to pass through. Flowers *et al.* [12] demonstrated a fully 3D printed a high-pass filter circuit that comprises of 3D printed conductive lines, inductor and capacitor (see **Figure 8.6(a)**). They input a sinusoidal waveform to the 3D printed high-pass filter circuit and measured the output waveform. From **Figure 8.6(b)**, it can be observed that the cut-off frequency was at around 7MHz and successfully filtered out the low frequencies that may cause signal interferences.

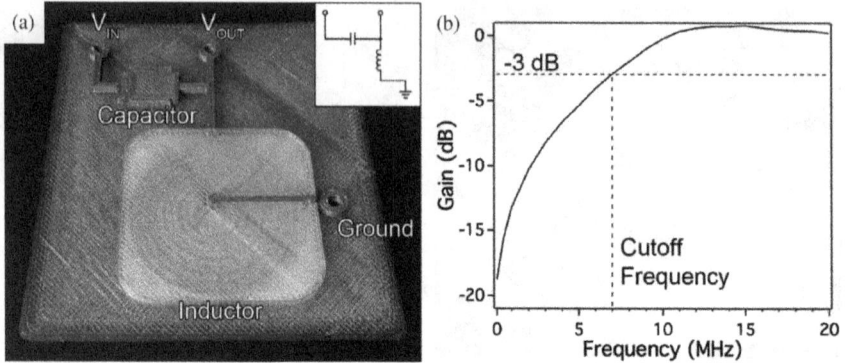

Figure 8.6.   (a) A 3D printed high-pass filter circuit, and (b) the measured filtering characteristics of the 3D printed high-pass filter circuit. Reprinted from Ref. [12], Copyright (2017), with permission from Elsevier.

8.2 Active Electrical Components

Active electrical components are electrical circuits components that need power to work or can supply power in operations [18]. In general, active electrical components have multi-material structures and they are more complex than passive electric components. Similarly, active electrical components can also be fabricated by additive manufacturing technology. This section discusses some of the commonly fabricated active electrical components using additive manufacturing technologies.

8.2.1 *Transistors*

Transistors are one of the most critical active electrical components in electronics devices. They are mainly used for amplifying or switching electrical power and signals [19]. Due to the far less complicated processes than the traditional silicon technology, there has been an increasing interest in recent years to fabricate organic thin-film transistors (OTFTs) with additive manufacturing technology. Besides that, organic materials generally have better mechanical flexibility, and inherently making OTFTs to be more compatible with flexible substrates [20]. OTFT is a general term for representing most types of organic transistors (such as organic field-effect transistors (OFETs), organic electrochemical transistors (OECTs), electrolyte-gated organic field-effect transistors

(EGOFETs), etc.) [21]. The mechanisms used to achieve current modulation vary in the various types of organic transistors.

OTFT has three terminals just like any conventional transistors and its design architecture is similar to the metal-oxide-semiconductor field-effect transistor (MOSFET). An OTFT comprises thin layers of dielectric, organic semiconducting and conductive materials (namely source, drain and gate electrodes). Highly conjugated small molecules or polymers are usually used as the organic semiconducting materials for OTFTs. Note that the active semiconducting layer is in contact with both source and drain electrodes, while the dielectric layer separates the gate electrode from the active semiconducting layer [22]. The top contact and bottom contact type are the two key device configurations for OTFTs (see **Figure 8.7**). Top contact OTFTs have source and drain electrodes

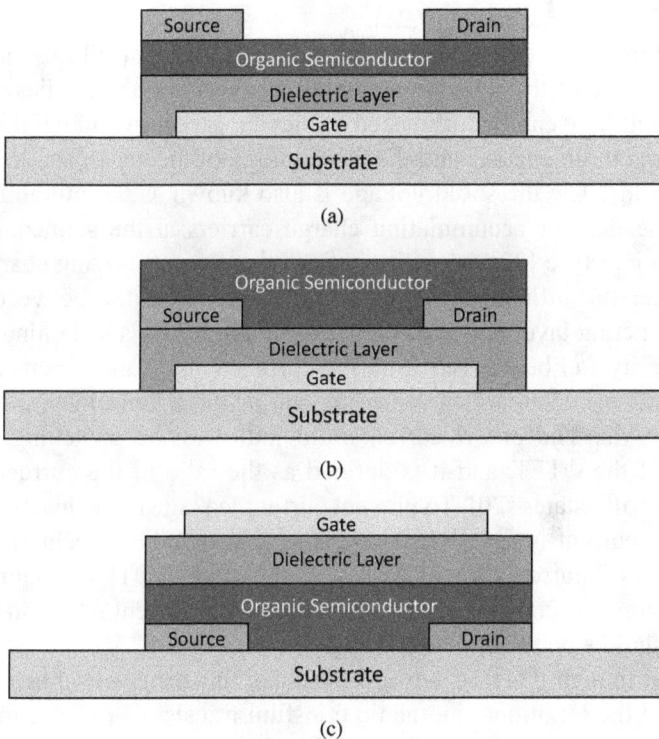

(a)

(b)

(c)

Figure 8.7. Schematic diagram of the various OTFTs design architectures: (a) top-contact, bottom-gate; (b) bottom-contact, bottom-gate and (c) bottom-contact, top-gate [19].

fabricated on top of the organic semiconducting layer (see **Figure 8.7(a)**), whereas the bottom contact OTFTs have the organic semiconducting layer is fabricated over the source and drain electrodes (see **Figure 8.7(b)**).

The electrical current flows between the source and drain electrodes under an imposed bias, and this current is modulated by the gate electrode by having a voltage applied to it [20]. There is an accumulation of mobile charges near the semiconductor-dielectric interface when a bias is applied between the gate and source, and thereby allowing current flow through the active semiconducting layer on applying a suitable drain to source potential [22]. For instance, the OTFT device is in an "off" state when no voltage is applied between the source and gate electrodes, and minimum current is measured between the source and drain electrodes. The OTFT device is in an "on" state when voltage is applied between the source and gate electrodes, and increasing current is measured between the source and drain electrodes due to the induced holes or electrons at the semiconductor-dielectric interface [19,23].

The threshold voltage, mobility, on/off current ratio and sub-threshold slope are some of the critical parameters determining the applicability of OTFTs, and they can be influenced by device geometry, material properties structural dimensions and the morphology of the active semiconducting materials. The threshold voltage is also known as the minimum gate voltage needed for accumulating charge carriers at the semiconductor-dielectric interface [22]. Mobility defines the ease of moving charge carriers under the influence of an electric field within the active organic semiconducting layer and it directly affects the OTFT's switching speed. The mobility can be derived from the current-voltage measurements and the organic semiconducting materials can achieve mobilities as high as 1–10 cm^2/V.s. The on/off current ratio indicates the switching performances of the OTFT, and it is defined as the ratio of the current in the "on" and "off" states [20]. To prevent current leakage in the inactive state, low "off" current is usually desired. The current OTFTs technology can achieve on/off current ratio as high as 10^{-6}. An ideal OTFT should have a low threshold voltage, high mobility, large on/off current ratio and steeper sub-threshold slope [22].

Even though OTFTs have drawbacks in their switching speeds relative to the traditional inorganic thin-film transistors (TFTs), there are still immerse potentials in using OTFTs for many applications, such as sensors, smart tags, electronic papers and flexible displays [20,24,25], due to their mechanical flexibility, low cost and lightweight. Ha *et al.* [26]

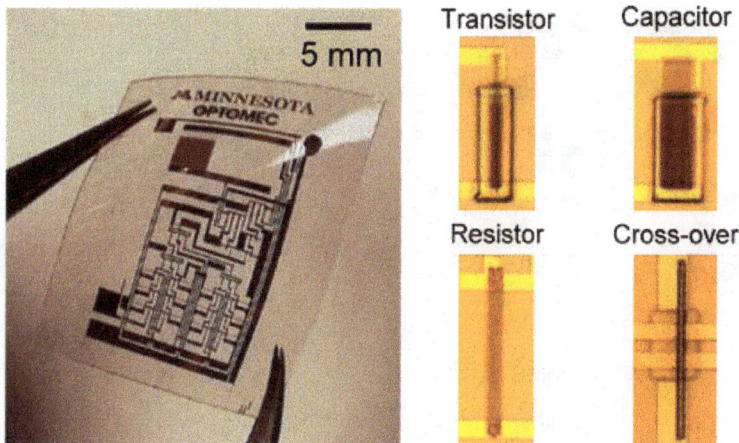

Figure 8.8. Aerosol printed complex drive circuitry comprising of resistors, capacitors and electrolyte-gated transistors (EGTs). Reprinted with permission from Ref. [26]. Copyright (2013) from American Chemical Society.

demonstrated the use of aerosol printing to fabricate key components of a complex drive circuitry (resistors, capacitors and electrolyte-gated transistors (EGTs)) on a flexible polyethene terephthalate (PET) (see **Figure 8.8**). This low-voltage circuit can drive a 4 mm^2 polymer electrochromic (EC) pixel. The semiconductor channels and gate insulator layers were fabricated with poly(3-hexylthiophene) and printable ion gels respectively.

8.2.2 *Polymer Organic Light-Emitting Diodes (P-OLEDs)*

Organic light-emitting diodes (OLEDs) are electroluminescent devices that emit light in response to a current flow, and they are commonly used in solid-state lightings and digital displays. The emissive materials in OLEDs are made of organic materials, and they can be further categorised into either small molecule or polymer type [27]. Small-molecule OLEDs have complex multilayer device architecture, with different layers of small molecules materials stacked on top of each other (see **Figure 8.9(a)**). Small molecules OLED materials are usually insoluble in solvents and thereby restricting their printability for printing processes. Hence, the fabrication process of small molecule OLEDs usually requires complex vacuum evaporation process for material deposition. Furthermore, chemical

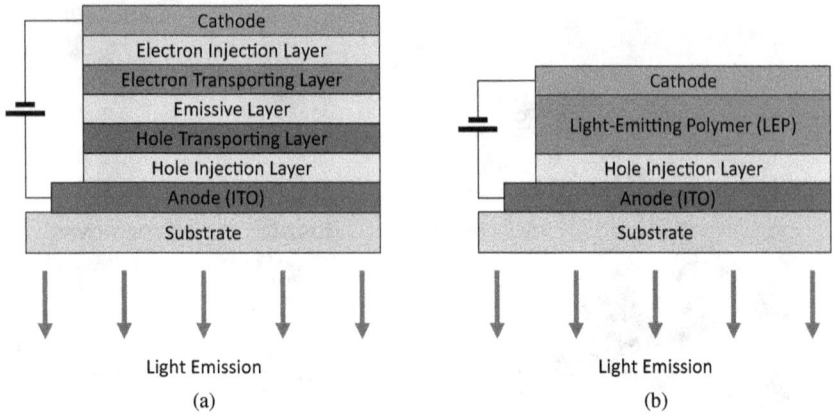

Figure 8.9.   Schematic diagram of (a) small molecule OLEDs; and (b) polymer OLEDs (P-OLEDs) [27].

modifications are also needed to ensure good compatibility and adhesions between different layers of small molecules OLED materials [27].

Polymer OLEDs (P-OLEDs) have a layer of conductive electroluminescent polymer that emits light in response to current flow. They usually require low power consumption for light emission and have a high contrast ratio and high-speed image switching [27]. Conversely, P-OLEDs have relatively simple device architectures that only comprises of an anode, a hole injection layer, a light-emitting polymer layer and a cathode (see **Figure 8.9(b)**). The anode is usually made of transparent indium tin oxide (ITO), whereas the cathode is made of reflective metal. Note that the substrate needs to be transparent for the emitted light to pass through. Both conjugated polymers and non-conjugated polymers can be used as P-OLEDs materials for the light-emitting polymer layer. These P-OLEDs materials are usually soluble in solvents and they also can easily be deposited by various 3D electronics printing processes such as aerosol jet and inkjet printing. This unique feature can allow the P-OLEDs materials to be deposited directly onto various flexible substrates such as polymer films and paper, and at the same time, enjoying higher cost-effectiveness. The light-emitting polymer layer's thickness usually ranges from 100–150 nm, and this layer incorporates all the charge, emitter and host transport functions within it [27].

When a voltage is applied between the anode and cathode, charges are injected into the light-emitting polymer layer from each of the two electrodes (electrons from the cathode and holes from the anode) [28]. The electrons and holes recombine in the recombination zone within the light-emitting polymer layer to form excitons. An exciton is defined as a charge-less electron-hole pair that can carry energy [29]. The light-emitting polymer is now in a high energy state, which is also known as the "excited state". The light-emitting polymer then releases the excess energy as light when it radiatively relaxes back to the stable state, also known as the "ground state" [29]. Apart from red, green and blue (RGB) colour emissions, P-OLEDs can also exhibit white colour emission as well [27]. The emitted light colour is determined by the light-emitting polymer's energy gap and the colour can be changed by modifying the polymer's chemical structure [29]. The emission efficiency of P-OLED can be calculated as follow:

$$\varphi = \gamma \cdot \eta_{e-h} \cdot \varphi_{ph} \cdot (1 - Q), \qquad (8.9)$$

where φ is the electroluminescence quantum efficiency, γ is carrier balance of electrons and holes, η_{e-h} is the recombination rate, φ_{ph} is the photoluminescence efficiency and Q is the quenching factor by the cathode [27]. It can be deduced from the equation that the overall emission efficiency of the P-OLED can be improved by increasing the recombination rate and photoluminescence efficiency while suppressing the cathode quenching and having good control over the injection of electrons and holes.

8.2.3 *Organic Photovoltaic (OPV)*

Photovoltaic (PV) devices, also commonly known as solar cells, are electrical devices which can convert light energy into electrical energy. Conventional silicon solar cells typically have complex fabrication processes that usually require high processing temperatures. Also, silicon solar cells have limited mechanical flexibility and higher price tags [30]. In recent years, organic photovoltaic (OPV) devices (also known as organic solar cells (OSCs)) are gaining significant attention as a potentially cheaper alternative to silicon solar cells. Besides their thin, flexible, and lightweight features, OPVs can be easily fabricated directly on flexible polymer substrates by 3D electronics printing techniques to fabricate flexible large-area OPVs and do not require high processing temperatures [27].

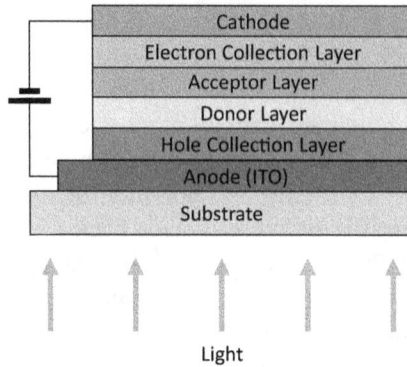

Figure 8.10.   Schematic diagram of a bilayer OPV device [33].

Depending on their device architecture, OPV devices can be further classified into single layer OPV, bilayer OPV, bulk heterojunction OPV, tandem OPV and many other more [30]. A basic bilayer OPV device has a device architecture that comprises of an anode, a hole collection layer, a donor layer, an acceptor layer, an electron collection layer and a cathode (see **Figure 8.10**). The donor and acceptor layers are made of two dissimilar photoactive materials. Conjugated polymers are used as donor material due to their high ionisation potential, whereas fullerene derivatives are used as acceptor material due to their high electron affinity properties [31].

The working mechanism of OPV devices is somewhat reverse to that of OLEDs. In general, photoactive materials produce excitons when they absorb the incident. The concentration gradient allows the excitons to diffuse within the active layer until they reach the donor-acceptor interface. The excitons dissociation takes place at the interfaces, in which the excitons are separated into electrons (negative charge carriers) and free holes (positive charge carriers). Electrons are produced in the lowest unoccupied molecular orbital (LUMO) of the acceptor layer, while holes are created in the highest occupied molecular orbital (HOMO) of the donor layer. The electrons and holes then move towards the cathode and anode respectively under internal electric fields and generating a photocurrent when the various charges are collected by the respective anode and cathode [30,32–34].

The power conversion efficiency, η of OPVs can be expressed as:

$$\eta = \frac{V_{oc} \times J_{sc} \times FF}{P_{in}}, \qquad (8.10)$$

where V_{oc} is the open-circuit voltage, J_{sc} is the short circuit current density, FF is the fill factor and P_{in} is the incident light power [27]. The power conversion efficiency of OPVs can be improved by increasing V_{oc}, J_{sc} and FF. Open circuit voltage, V_{oc} is defined as the voltage across the OPV device under illumination with a zero current, at which the dark current cancels out the photocurrent [30]. V_{oc} is associated with the energy gap between HUMO of the donor material and LUMO of an acceptor material (see **Figure 8.11**). Thus, it is desirable to have donor material with a deep HUMO level and acceptor material with a shallow LUMO level for increasing V_{oc} [27,30]. Factors such as the rate of the photogenerated charge separation process, the suppression of the recombination of hole and electron and the absorbance of the photoactive materials can be enhanced for improving J_{sc}. FF is strongly related to the resistance of the OPV. Increasing the FF value can be accomplished by the use of high mobility carrier materials in the active layer and reducing the resistance between the layer's interface [27]. Moreover, the donor and acceptor materials should generally have high stabilities, good extinction coefficients and good film morphologies for good performances.

Eggenhuisen *et al.* [35] inkjet-printed electrodes and the photoactive layers directly onto glass substrates for the fabrication of OPVs devices. Their fabricated OPVs were ITO-free, in which the front and back

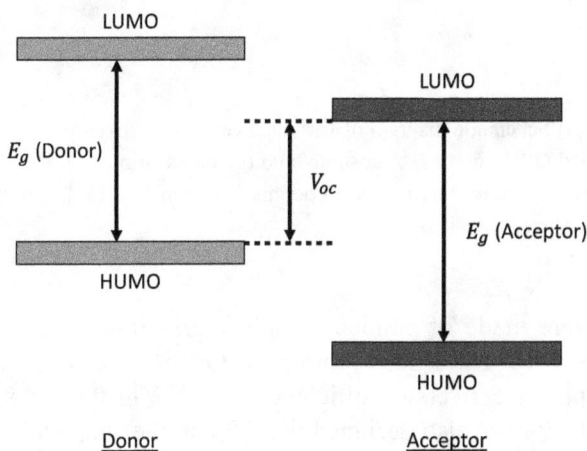

Figure 8.11.   Schematic diagram of the energy level diagram of the donor and acceptor system [30].

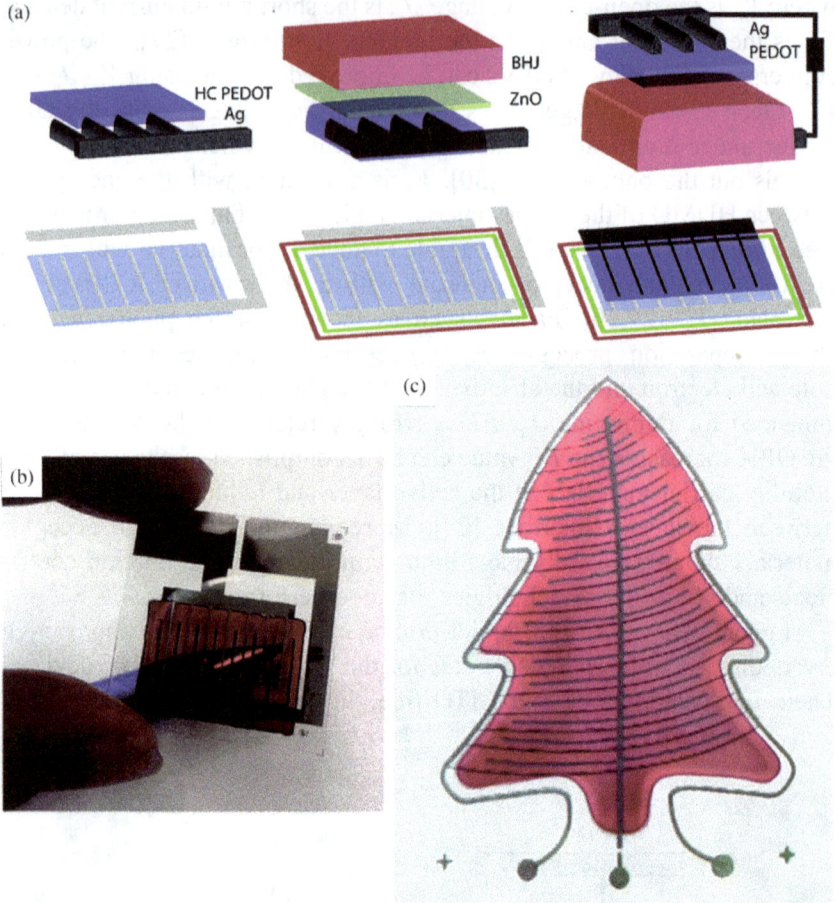

Figure 8.12.   (a) Schematic diagram of various layers of the inkjet printed OPV device, (b) Inkjet printed OPV on a glass substrate, and (c) Inkjet printed OPV device with the shape of a Christmas tree. Reprinted with permission from Ref. [35]. Copyright (2015), from the Royal Society of Chemistry.

electrodes were made of conductive silver grid lines and PEDOT:PSS (see **Figure 8.12(a)**). Their fabricated OPVs (see **Figure 8.12(b)**) achieved a power conversion efficiency of 4.1% in the air atmosphere. Separately, they have also designed the OPV in the shape of a Christmas tree (see **Figure 8.12(c)**) and thereby demonstrating the potentials of having the freedom of shape and designs by inkjet printing.

8.2.4 *Batteries*

The battery is also another type of interesting active device that can be fabricated by 3D electronics printing techniques. Batteries are electrical devices that can store energy and supply the circuit with power [19]. During the battery discharging process, the stored chemical energy in the batteries is converted to electrical energy, and vice versa for the charging process. The sandwich-type (see **Figure 8.13(a)**) and the interdigitated-type (see **Figure 8.13(b)**) are the two common design architectures for 3D printed batteries.

The sandwich-type printed battery has a design architecture that is similar to conventional batteries, in which each layer (current collector for cathode, cathode, separator with electrolyte, anode and current collector for the anode) is stacked up on top of one another on a flexible substrate [36]. This kind of design architecture is relatively simple and allows for more cost-effective mass production. However, a limited amount of energy can only be stored within a sandwich-type printed battery with a small footprint [37]. The interdigitated-type printed battery is another variation of a sandwich-type printed battery, in which two individual interdigitated electrodes are fabricated on the same substrate plane. The distance between the interdigitated electrodes can be more accurately controlled, and thus allowing the electrodes to place closer to each other and only require thin layers of electrolyte/separator material to be deposited in between them. As a result, there is a significant reduction in the electrical resistance across the battery due to shorter paths for ionic transport during the charge and discharge processes, which is advantageous for

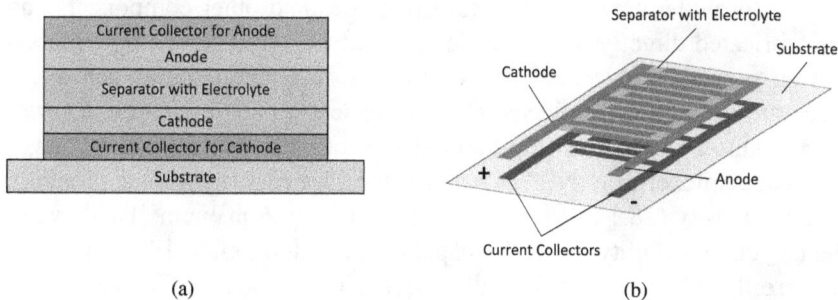

(a) (b)

Figure 8.13. Schematic diagram of (a) sandwich-type printed battery, and (b) interdigitated-type printed battery [36].

increasing the battery power. The use of interdigitated electrodes also helps to increase the surface areas of the electrodes [36,38].

The anode and cathode are the most critical components of a printed battery, in which their electrochemical properties and microscopic/macroscopic structures can influence the battery capacity and performance [36,39]. Lithium-based, sodium-based, metal-based (metals, metal oxides or metal sulfides) and carbon-based materials (carbon nanotubes (CNTs), graphene and activated carbon) are some of the commonly-used materials for fabricating the various electrodes [39]. Typically, the materials used for fabricating the anode should be a good reducing agent that can readily release electrons, whereas the materials used for fabricating the cathode should accept electrons readily [40]. The electrodes can also have unique microstructures and surface structures for improving the batteries' electrochemical performances. The current collectors are highly conductive films that serve as supports for the electrodes and they help to collect the accumulated electrical energy from the electrodes [41]. The separator with electrolyte is a permeable membrane that separates the anode from the cathode, and at the same time allowing ionic transport of charge carriers while preventing short-circuiting [19]. The substrates used for the printed battery should ideally be chemically stable and flexible too. For instance, during the charging process for a rechargeable lithium-ion battery, lithium ions are released from the cathode when a suitable voltage is applied. These lithium ions then diffuse through the electrolyte to the anode and remain there, and thus allowing the battery to store the energy. The reverse process occurs for a discharging process and releases electric energy [38].

The use of additive manufacturing technologies for fabricating batteries can provide many advantages. Electrodes and other components can be fabricated directly onto flexible substrates without any complex fabrication processes. Also, additive manufacturing technologies can allow accurate layer-by-layer deposition for fabricating complex architectures. For instance, Kong *et al.* [42] fabricated a compressible 3D printed quasi-solid-state nickel-iron (Ni−Fe) battery (see **Figure 8.14**) with an ultrahigh energy density (28.1 mWh cm^{-3} at a power of 10.6 mW cm^{-3}) and excellent cycling stability (~91.3% capacity retentions after 10000 cycles). **Figure 8.14(b–c)** shows the full recovery of the quasi-solid-state Ni-Fe battery without structural deformation when stress was removed.

Only less than 10% of the maximum capacity was lost when the battery experienced a 60% compressive strain ratio. To fabricate the

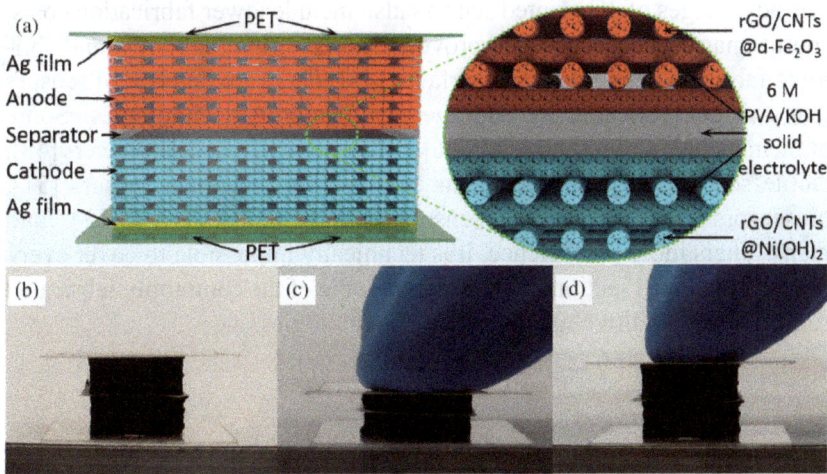

Figure 8.14. (a) Schematic diagram of the compressible 3D printed quasi-solid-state nickel-iron (Ni−Fe) battery; and (b–d) compression and recovery of the compressible 3D printed quasi-solid-state nickel-iron (Ni−Fe) battery as seen in real-time images. Reprinted with permission from Ref. [42]. Copyright (2020) from American Chemical Society.

electrodes for this battery, they first 3D printed a scaffold, composing graphene oxide/carbon nanotubes (GO/CNT) hybrid aerogel, though direct ink writing. 3D printing can allow easy fabrication of the scaffold, in which its size and morphology can be precisely controlled. The scaffold was then annealed under argon atmosphere to convert GO to reduced GO (rGO). Holey α-Fe_2O_3 nanorod arrays and ultrathin $Ni(OH)_2$ nanosheet were then grown in the interior and on the surfaces of the scaffold through a simple solvothermal method, forming the battery's anode and cathode. This compressible 3D printed battery would be very suitable for wearable electronics applications where high compression tolerance, good electrochemical stability and high mechanical squeezability are required.

8.3 Sensors

Sensors are electrical devices which can detect changes in the environment and send corresponding real-world information to a computer for computation and analysis [43]. Recently, there are increasing interests to use additive manufacturing technologies to fabricate sensors either directly onto the part surfaces or embed them within printed structures.

The advantages of 3D printed sensors also include lower fabrication costs, shorter manufacturing time, improved manufacturing scalability and ability to fabricate onto conformal surfaces [44]. There are myriads of sensors that can be fabricated by additive manufacturing technologies. Some of them include strain sensors [45,46], pH sensors [47], accelerometers [48], tactile sensors [49], displacements sensors [50], chemical sensors [51], biosensors [52], temperature sensors [53], humidity sensors [54] and many others more [43]. Hence, it is technically impossible to cover every one of them. This section only discusses some of the commonly fabricated sensors using additive manufacturing technologies.

8.3.1 *Strain Gauges*

A strain gauge, or strain sensor, is a sensor mainly used for measuring the strain experienced by a test specimen, whereby strain is a measure of the deformation caused by an external force [55–57]. As the strain gauge is mounted onto the surface of the test specimen, any strain experienced by the test specimen is transferred directly to the strain gauge's resistive wires and causes a change in the electrical resistance. Thus, the strain experienced by the specimen can be calculated by measuring the change in the electrical resistance of the strain gauge. The gauge factor, *GF* is a dimensionless parameter for measuring the sensitivity of a metal strain gauge and can be expressed as:

$$GF = \frac{\Delta R / R_0}{\Delta L / L_0} = \frac{(R - R_0)/R_0}{(L - L_0)/L_0} = \frac{(R - R_0)/R}{\varepsilon}, \tag{8.11}$$

where R_0 is the initial electrical resistance of the strain gauge, $\Delta R/R$ is the relative change of electrical resistivity of the strain gauge, L_0 is the initial gauge length and ε is the strain experienced by the strain gauge which is also equivalent to the relative change in the length of the strain gauge ($\Delta L/L$) [45,55,56].

Metal foil strain gauges are the most commonly used strain gauges among various types of strain gauges (such as capacitive strain gauges, piezoelectric strain gauges, photoelastic strain gauges and semiconductor strain gauges) due to their ease of fabrication and high flexibility [55]. *GF* values for metal foil strain gauges are typically around 2 [56]. **Figure 8.15**

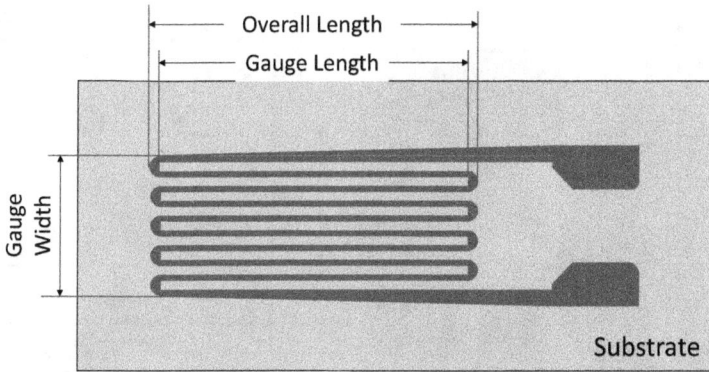

Figure 8.15.   Schematic diagram of a metal foil strain gauge.

shows a schematic diagram of a metal foil strain gauge, in which geometric parameters such as gauge length and gauge width can critically affect the strain gauges' properties. The chosen metal chosen for fabricating metal foil strain gauge should ideally exhibit a linear relationship between the strain experienced and the relative change in its electrical resistance [55], to allow more accurate calculations and readings.

Conventionally, a series of tedious etching and trimming steps are required to fabricate metal foil strain gauges with the desired designs and electrical resistances. Furthermore, it is also very labour intensive and time-consuming to mount conventional strain gauges for large scale installations [58]. The difficulty for mounting strain gauges increases further for conformal surfaces as well. Hence, many researchers are looking into using additive manufacturing technologies to fabricate customisable strain gauges directly onto the test specimens on-demand.

Agarwala *et al.* [59] fabricated a strain sensor on a thin flexible self-adhesive polyurethane (PU) membrane bandage using aerosol jet printing for low-cost home healthcare monitoring wearables application (see **Figure 8.16(a)**). Laser sintering was used as a low-temperature sintering technique for sintering the deposited silver nanoparticle inks on the bandage. **Figure 8.16(b)** shows the optical image of the laser-sintered silver nanoparticle ink and no visible cracks was observed. **Figure 8.16(c–e)** shows that the fabricated strain sensor can be rolled over a curved surface, adhered conformably to the wrist and stretched longitudinally while maintaining functionality. **Figure 8.16(f)** shows the change in the electrical resistance of the strain sensor in response to different wrist movements. The electrical resistance increased when the wrist was bent forward, and

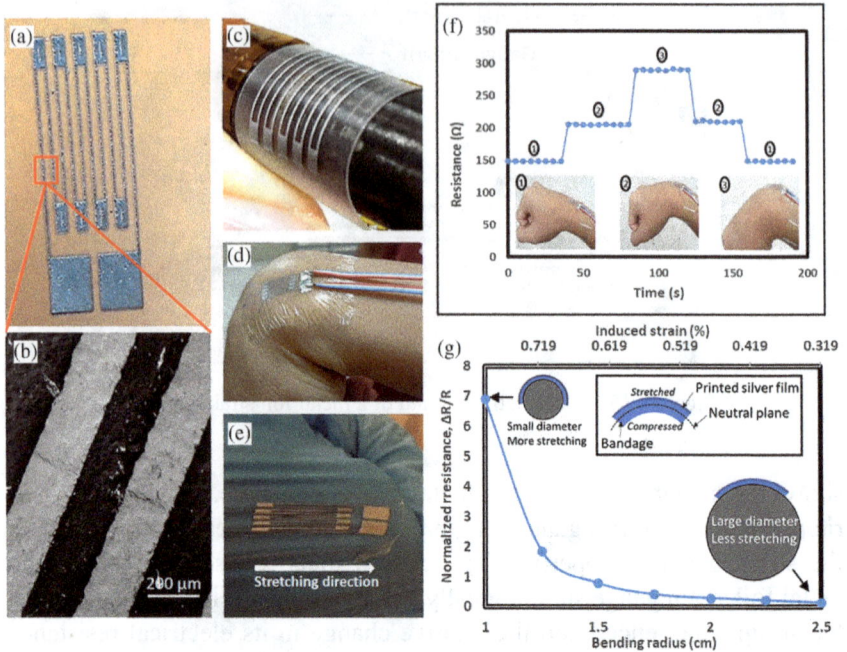

Figure 8.16.   (a) Aerosol jet printed strain sensor fabricated on a flexible bandage, (b) the optical image of the laser-sintered silver nanoparticle ink and no visible cracks was observed; strain sensor (c) rolled over a curved surface, (d) adhered conformally to the wrist and e) stretched longitudinally, (f) the change in the electrical resistance of the strain sensor in response to different wrist movements and (g) the normalised electrical resistance of the strain sensor as a function of bending radius. Reprinted with permission from Ref. [59]. Copyright (2019) from American Chemical Society.

the electrical resistance returned to the initial values when the wrist was relaxed. **Figure 8.16(g)** shows the normalised electrical resistance of the strain sensor as a function of the bending radius. It can be observed from the graph that a smaller bending radius can induce higher strain and resulting in higher electrical resistance, and vice versa.

Borghetti *et al.* [60] also demonstrated the fabrication of a strain gauge on polyvinyl chloride (PVC) conduit using the aerosol jet technology (see **Figure 8.17**). This application showcased the ability of 3D electronics printing techniques to fabricate strain gauges directly onto conformal surfaces without the need for labour-intensive mounting.

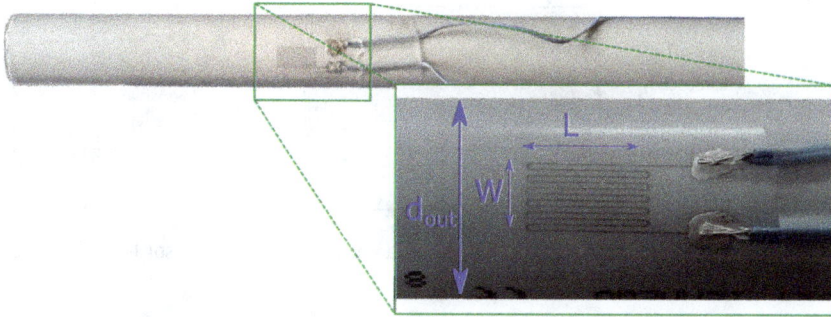

Figure 8.17. Aerosol jet printed strain gauge fabricated directly onto a PVC conduit. Reprinted with permission from Ref. [60]. Copyright (2019) from Multidisciplinary Digital Publishing Institute (MDPI).

8.3.2 *pH Sensors*

pH monitoring and control are essential for many important chemicals, biological and environmental processes which can directly influence the quality of human life [61]. pH sensors are electrical devices capable of measuring the pH values of aqueous solutions. The pH value is a measure of the molar concentration of hydronium ions (H_3O^+) in an aqueous solution and it is defined as [62]:

$$pH = -\log[H_3O^+]. \qquad (8.12)$$

In other words, the pH value is a numerical measure of acidity/alkalinity of an aqueous solution in which it ranges from 0–14. The pH values of acidic solutions are lower than 7, whereas pH values of alkaline solutions are higher than 7. pH neutral solutions have a pH value of 7.

The various types of pH sensors available include chemiresistive, capacitive, potentiometric, optical, luminescence and shape/mass pH sensors. Chemiresistive pH sensors are the most commonly used pH sensors. The design architecture of the chemiresistive pH sensor is relatively simple, which it only comprises two electrically conductive electrodes with a pH sensitive material deposited in between them (see **Figure 8.18**). Besides, it also does not require a reference electrode [62]. When subjected to the aqueous solutions of different pH values, the intrinsic electrical resistance of the pH sensitive material varies accordingly to

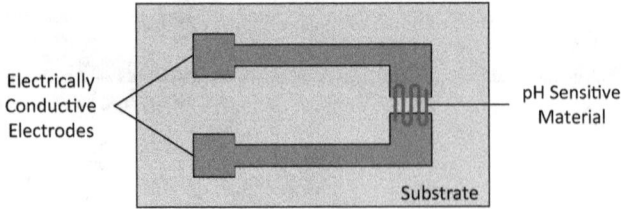

Figure 8.18.   Schematic diagram of a chemiresistive pH sensor [47].

the change in molar concentration of hydronium ions. Hence, the chemire-sistive pH sensor monitors the pH levels of an aqueous solution by meas-uring the change of electrical resistance of the pH sensitive material directly [61,62].

Goh *et al.* [47] fabricated a low-cost flexible chemiresistive pH sensor using the aerosol jet technology, with a miniaturised CNT-based serpen-tine sensing element printed in between two silver electrodes. With fine resolution printing of the aerosol jet technology, the effective sensing area of the CNT sensing element can be as small as 500 μm × 100 μm and the overall sensor dimensions can be better controlled as compared to drop casting methods. Their chemiresistive sensor achieved high sensitivity (up to 59 kΩ/pH) and repeatability (coefficient of variance <1.15%), as well as good biocompatibility for live cells applications. Their sensor also maintained its full functionality with a minimum variation of electrical resistivity after flexing for 1000 cycles at a bending radius of 5 mm. In their work, single-walled CNTs (SWCNTs) were chosen as the pH sensi-tive material for fabricating the sensing element of the chemiresistive sensor because they have tuneable electrical properties, ease of chemical functionalisation, high surface-to-volume ratio and good mechanical reliability. The electrical conductivity of SWCNTs is directly influenced by the molar concentration of hydroxide ions (OH$^-$) and hydronium ions (H$_3$O$^+$) in the aqueous solution, as they dope the CNT walls by behaving as electron donors and acceptors respectively [47,62,63].

8.4 Challenges of 3D Printed Electronics

The state-of-the-art technology for 3D printed electronics is still relatively new, so it is likely to still face challenges from circuits design to fabrica-tion and post-processing of the 3D printed electronics, and even at their end-of-life phase. The wider adoption and full potentials of 3D printed electronics may be impeded by these challenges. This section aims to discuss some of the existing challenges faced by 3D printed electronics.

8.4.1 *Printing Resolution*

The demand for miniaturisation of electronic systems, especially in medical and mobile devices, is ever-growing. Therefore, there is a critical need to shrink the size of electrical components for more efficient packing within a limited space. In comparison, the printing resolution of 3D electronics printing techniques is still much poorer than conventional electronics fabrication methods such as photolithography etching and laser ablation. The resolution of electrical components directly influences their electrical performances, in which higher resolution electrical components tend to have improved electrical performances [4]. Hence, 3D electronics printing techniques may not be appropriate for the fabrication of high radio frequency (RF) and high-speed communication applications [64,65]. There is still a significant technological gap for advanced electronics packaging applications due to the lower printing resolution and integration density.

8.4.2 *Manufacturability Concerns*

Indeed, 3D printed electronics' morphology and geometric characteristics play a critical role in influencing their electrical properties and performances. To achieve the most optimal electrical properties, a 3D printed electrical component should ideally adhere to strict geometrical and morphology specifications, such as precise dimensions and thickness, well-defined boundaries, smooth surfaces and no missing spots. Currently, there are still many challenges to achieve good geometrical and morphology control for 3D printed electronics.

The morphology and geometric characteristics of 3D printed electronics are mainly affected by the printing parameters, inks properties and sintering processes. During the printing processes, it is critical to ensure a precise amount of materials are deposited to have tight control of the dimensions and thickness of the printed components. Ink properties, such as dispersion stability, surface tension, wettability, viscosity and solid loading (for suspension-based inks), have substantial effects on the morphology of the 3D printed components. The intrinsic material properties of the inks also directly affect the mechanical and electrical properties of the 3D printed components. The sintering processes can also affect the electrical properties and microstructures of the sintered inks. Optimisations of sintering processes are usually required to prevent damaging temperature-sensitive substrates while having the most optimised electrical properties. The key challenge is to achieve the most optimised parameters for printing and sintering processes while achieving good geometrical and morphology for 3D printed electronics. It involves the accumulation of experience and expertise, as well as the physical understandings between different materials and processes.

8.4.3 *Standards for 3D Printed Electronics*

Standards are structured frameworks to ensure the manufactured products comply with the guidelines on product safety, product functionality, product compatibility, health safety and environmental sustainability. In addition, they also contribute in reducing incompatibility issues, giving better quality assurance, improving time-to-market and promoting interoperability between different devices.

In the field of 3D printed electronics, the technologies and processes are still relatively new and therefore there is still a lack of industrial standards and guidelines for fabricating reproducible and reliable electrical circuitries and components. Furthermore, it will be even more challenging to develop new standards for 3D printed electronics due to continual evolutions, developments and advancements of new materials and processes. Thus, the lack of standards presents one of the most challenging barriers for adopting 3D printed electronics as a mainstream technology for consumer electronics products.

8.4.4 *Environmental Impacts*

Although 3D printed electronics are more environmentally friendly than conventional electronics, they still face many challenges in terms of environmental impacts. A single 3D printed electronic component can comprise many different materials and separating each material for recycling can be difficult [66], especially for embedded electronics. Most 3D printed electronics are intended to be low cost and disposable, so recycling them may not be the most cost-effective. At their end-of-life phase, these 3D printed electronics are usually disposed through landfill or incineration. However, negative impacts on the environment may occur during these disposing processes, such as soil contamination due to landfill leaching and release of toxic gases during incineration [4,66].

8.5 Future Outlook of 3D Printed Electronics

3D printed electronics is a disruptive technology that holds tremendous promise for revolutionising the electronics industry in the near future. In contrast to traditional methods of electronics fabrication, this emerging technology seeks to reduce wastage, fabrication costs and time bottlenecks while improving efficiencies. It also offers other incentives such as on-demand fabrications of customisable electronics, shorter prototyping time and enabling highly innovative applications [2].

Increasing growth and adoptions of the 3D printed electronics sector within the additive manufacturing industry must be supported by continuous research and development of new advanced functional inks [2]. Different inks with unique electrical, mechanical and material properties are required to suit different applications. There also is a need to call for greater collaborative efforts among machines and inks manufacturers to share their knowledge, to better formulate and optimise the inks to suit a specific printer on a given substrate for better printability and print quality. Concurrent research is needed for optimising the synergy between the sintering processes and the inks' sinterability to further advance the 3D printed electronics technologies. In general, an ideal functional ink for 3D printed electronics applications should require low sintering temperatures while providing high cost-effectiveness, good material and electrical properties and good printability for the printers. The ideal printer must also have fast printing speed, good printing resolution, on-demand non-contact printing, and allow easy changing and scalability of designs. Besides, the sintering process should have fast sintering speed, deliver great sintering properties and minimising damages on temperature-sensitive substrates and printed patterns.

In literature, there are countless of examples demonstrating the use of additive manufacturing technologies for the fabrication of flexible 3D printed electronics. The progress of flexible 3D printed electronics is expected to improve tremendously and potentially lead to large-scale

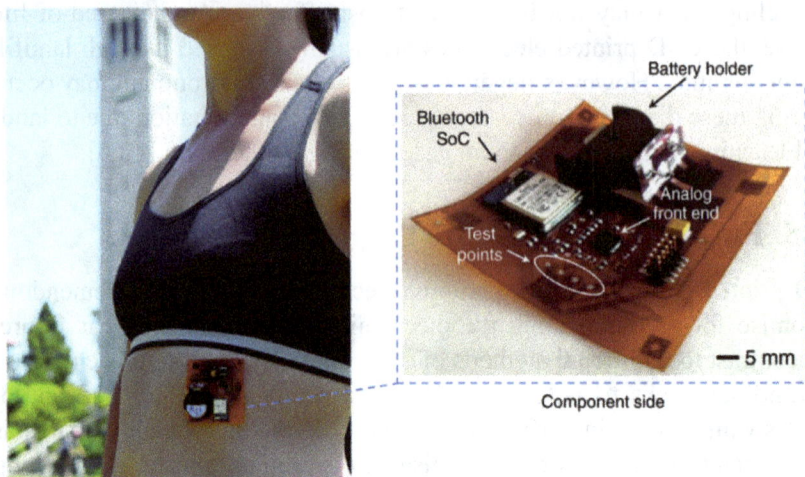

Figure 8.19.   Image of a flexible wearable sensor patch mounted on a volunteer's lower left rib cage. Reprinted with permission from Ref. [69]. Copyright (2016) from John Wiley and Sons.

fabrication of highly effective electrical devices at low cost [67]. There is enormous potential in using flexible 3D printed electronics over conventional rigid electronics for many industries, including healthcare and energy [68]. For instance, **Figure 8.19** shows a flexible wearable sensor patch capable of monitoring skin temperature and electrocardiography (ECG) signals. The ECG electrodes were inkjet printed directly onto a polyimide substrate with gold nanoparticle ink, and several conventional electrical components were integrated into the flexible wearable sensor too [69]. This kind of electronics is also known as a flexible hybrid electronics, whereby there is direct interfacing of soft and hard electronics. **Figure 8.20** shows a flexible, transparent PET-based electroluminescence device, in which the conductive traces were inkjet printed with silver nanoparticle ink [70].

The current research trend is also gradually shifting towards stretchable 3D printed electronics for wearable healthcare monitoring applications and soft robotics. The key technical challenge of fabricating stretchable 3D printed electronics usually lies in the mismatch of mechanical properties of the inks and stretchable soft substrates [2]. Hence, there is also a need to develop novel soft and stretchable functional and

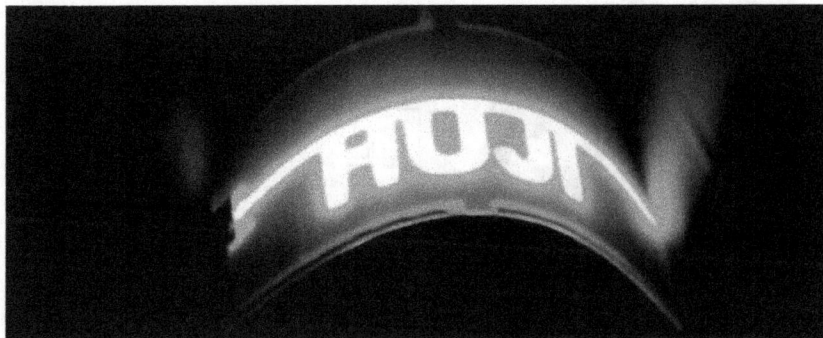

Figure 8.20. Image of a flexible, transparent PET-based electroluminescence device. Reprinted with permission from Ref. [70]. Copyright (2010) from American Chemical Society.

electrically conductive inks that are specially targeted for stretchable 3D printed electronics applications. Typically, conductive fillers are added to intrinsically stretchable materials for the formulation of stretchable inks. The latest developments in stretchable 3D printed electronics are seeing new technologies and innovation emerging, and significant research and improvements are needed to ensure these devices can perform well under stretching conditions [71]. **Figure 8.21(a–d)** show some examples of stretchable 3D printed embedded electronics [72].

Some 3D electronics printing techniques allow the deposition of functional inks directly onto conformal surfaces, enabling more creative designs and applications to be explored [4] while enjoying additional benefits such as space utilisation and weight reduction [73]. 3D conformal printing is usually used for fabricating 3D printed antennas, in which uniquely shaped antennas can be fabricated with the shape factor as a design consideration [74]. **Figure 8.22** shows electrically small antennas printed directly onto conformal surfaces of a hemispherical glass substrate and demonstrated good mechanical robustness [75]. The state-of-the-art technology for 3D printing of electronics can also allow multi-material, multi-functional and multi-scale printing. Hence, multi-functional embedded or structural electrical devices (see **Figure 8.23**) can be easily fabricated within a single print job. Conductive traces, passive components and active components can be integrated and embedded within complex

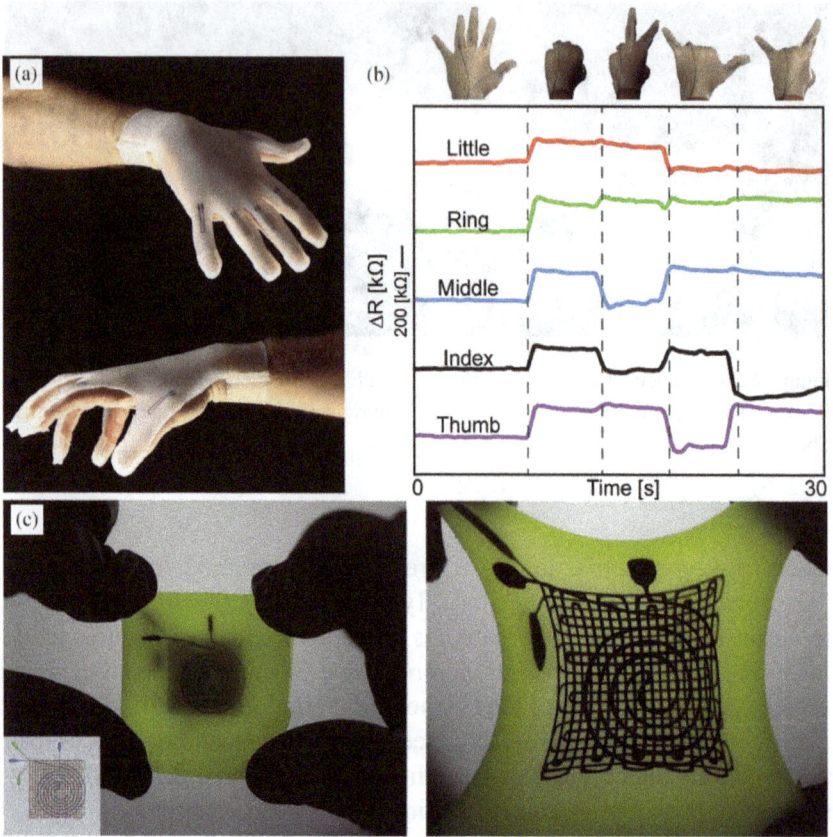

Figure 8.21.   (a) Image of a stretchable glove with embedded strain sensors; (b) electrical resistance of the strain sensors within the glove as a function of time at various hand positions; (c) image of an unstretched three-layer strain and pressure sensor and (d) image of an unstretched three-layer strain and pressure sensor. Reprinted with permission from Ref. [72]. Copyright (2014) John Wiley and Sons.

non-planar part geometries to reduce weight, save space and get protected from the external environment.

8.6 Summary

3D printed electronics is a disruptive technology which has already shown great potentials to revolutionize the electronics market in near future.

Figure 8.22.   Image of electrically small antennas printed directly onto conformal surfaces of a hemispherical glass substrate. Reprinted with permission from Ref. [75]. Copyright (2011) from John Wiley and Sons.

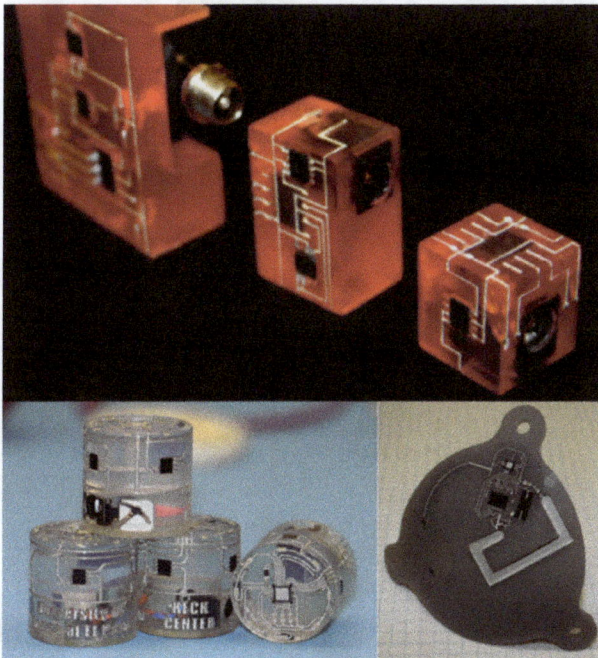

Figure 8.23.   Examples of 3D printed embedded and structural electronics. Reprinted with permission from Ref. [76]. Copyright (2012) from Emerald Publishing Limited.

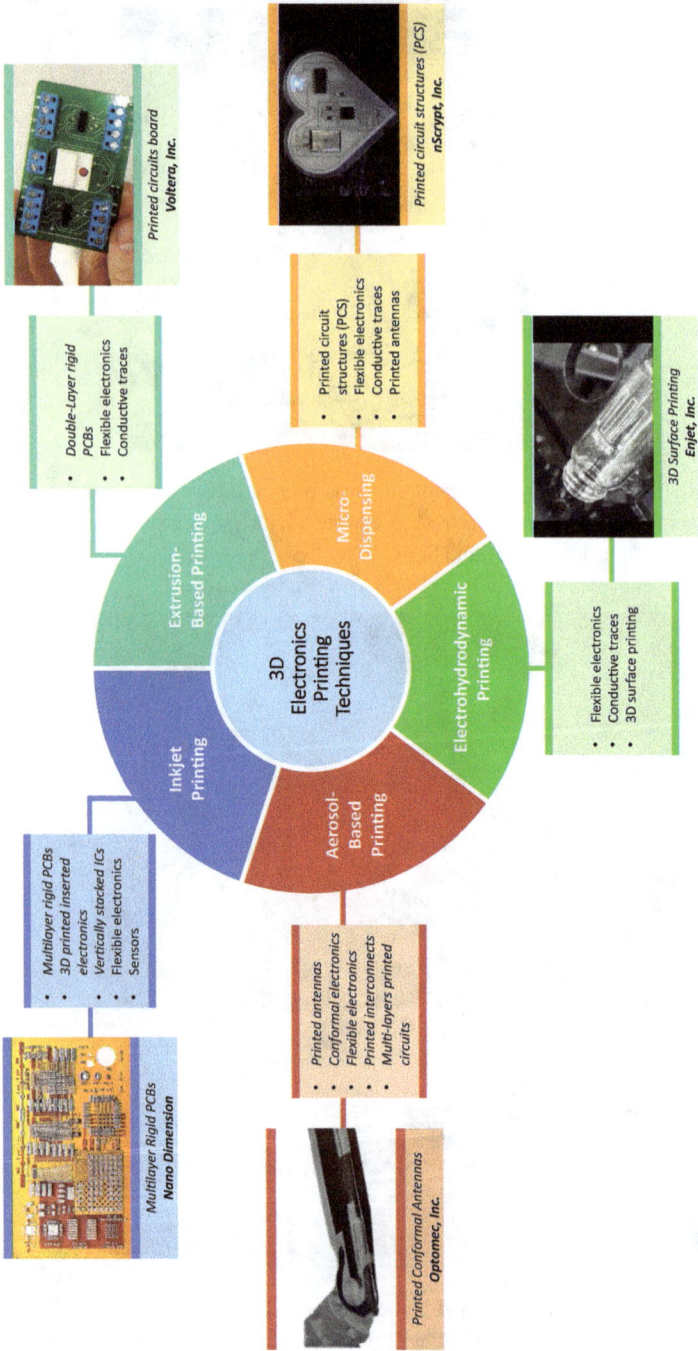

Figure 8.24.   Schematic diagrams of various 3D electronics printing techniques and their notable applications. Courtesy of NANO DIMENSION — ELECTRIFYING ADDITIVE MANUFACTURING, Optomec Inc., Voltera, Inc., nScrypt Inc. and Enjet Inc.

This book has provided a comprehensive coverage on the state-of-the-art printing technologies and their working principles, printable functional materials and application areas in 3D electronics printing.

The state-of-the-art 3D electronics printing techniques are mainly categorised into inkjet printing, aerosol-based printing, extrusion-based printing, electrohydrodynamic (EHD) printing and micro-dispensing technologies. Based on the discussed technical specifications, strengths and weaknesses of each printing technique in this book, users can now make a more informed decision in choosing the most appropriate printing technique, materials and substrates for their applications. The 3D electronics printing technology is typically used for fabricating conductive traces, printed circuited boards (PCBs), flexible electronics, conformal electronics and printed antennas. Some of the notable applications are presented in **Figure 8.24**. Various modelling and simulation methods can also be used to further enhance the print stability and optimise the electrical performance of the fabricate electrical devices. Nevertheless, continuous research and development in software, printing capabilities and printable functional materials, while concurrently overcoming existing challenges, will soon inspire new industries, markets and applications for wider adoption of this emerging technology.

References

[1] Lu, B., Lan, H. and Liu, H. (2018). Additive manufacturing frontier: 3D printing electronics, *Opto-Electron. Adv.*, 1, p. 170004.

[2] Tan, H. W., An, J., Chua, C. K. and Tran, T. (2019). Metallic nanoparticle inks for 3D printing of electronics, *Adv. Electron. Mater.*, 5, p. 1800831.

[3] Tan, H. W., Saengchairat, N., Goh, G. L., An, J., Chua, C. K. and Tran, T. (2020). Induction sintering of silver nanoparticle inks on polyimide substrates, *Adv. Mater. Technol.*, 5, p. 1900897.

[4] Tan, H. W., Tran, T. and Chua, C. K. (2016). A review of printed passive electronic components through fully additive manufacturing methods, *Virtual Phys. Prototyp.*, 11, pp. 271–288.

[5] Gerke, R. D. (2005). Embedded passives technology, *Resistor*, 146, p. 635.

[6] Bonadiman, R. and Salazar, M. M. P. (2010). Reliability of Ag ink jet printed traces on polyimide substrate, *presented at the 3rd Electronics System Integration Technology Conference ESTC*, Berlin, Germany.

[7] Serway, R. A. and John W. Jewett, J. (2014). *Physics for Scientists and Engineers with Modern Physics*, 9th edn., Chapter 27: Current and resistance, pp. 808–832.

[8] Kang, B. J., Lee, C. K. and Oh, J. H. (2012). All-inkjet-printed electrical components and circuit fabrication on a plastic substrate, *Microelectron. Eng.*, 97, pp. 251–254.

[9] Hesse, E. (1982). A four-point probe method with increased accuracy for the local determination of the thickness of thin, electrically conducting layers, *IEEE Trans. Instrum. Meas.*, IM-31, pp. 166–175.

[10] Lahti, M., Lantto, V. and Leppavuori, S. (2000). Planar inductors on an LTCC substrate realized by the gravure-offset-printing technique, *IEEE Trans. Compon. Packag. Technol.*, 23, pp. 606–610.

[11] Vaziri, S. (2011). *Fabrication and Characterization of Graphene Field Effect Transistors* (Master Thesis, Royal Institute of Technology (KTH)).

[12] Flowers, P. F., Reyes, C., Ye, S., Kim, M. J. and Wiley, B. J. (2017). 3D printing electronic components and circuits with conductive thermoplastic filament, *Addit. Manuf.*, 18, pp. 156–163.

[13] Serway, R. A. and Jewett, J. J. W. (2014). *Physics for Scientists and Engineers with Modern Physics*, 9th edn., Chapter 26: Capacitance and dielectric, pp. 777–807.

[14] Bishop, O. (2011). *Electronics: A First Course (Third Edition)*, eds. Owen Bishop, Chapter 16: Capacitors (Newnes, Oxford) pp. 57–58.

[15] Redinger, D., Molesa, S., Shong, Y., Farschi, R. and Subramanian, V. (2004). An ink-jet-deposited passive component process for RFID, *IEEE Trans. Electron Devices*, 51, pp. 1978–1983.

[16] Serway, R. A. and John W. Jewett, J. (2014). *Physics for Scientists and Engineers with Modern Physics*, 9th edn., Chapter 32: Inductance, pp. 970–997.

[17] Wheeler, H. A. (1928). Simple inductance formulas for radio coils, *Proceed. Inst. Radio Engin.*, 16, pp. 1398–1400.

[18] Wilkinson, N. J., Smith, M. A. A., Kay, R. W. and Harris, R. A. (2019). A review of aerosol jet printing — a non-traditional hybrid process for micro-manufacturing, *Int. J. Adv. Manuf. Technol.*, 105, pp. 4599–4619.

[19] Saengchairat, N., Tran, T. and Chua, C.-K. (2017). A review: Additive manufacturing for active electronic components, *Virtual Phys. Prototyp.*, 12, pp. 31–46.

[20] Reese, C., Roberts, M., Ling, M. -m. and Bao, Z. (2004). Organic thin film transistors, *Mater. Today*, 7, pp. 20–27.

[21] Elkington, D., Cooling, N., Belcher, W., Dastoor, P. C. and Zhou, X. (2014). Organic thin-film transistor (OTFT)-based sensors, *Electronics*, 3, pp. 234–254.

[22] Kumar, B., Kaushik, B. K. and Negi, Y. S. (2014). Organic thin film transistors: Structures, models, materials, fabrication, and applications: A review, *Polym. Rev.*, 54, pp. 33–111.

[23] Facchetti, A. (2007). Semiconductors for organic transistors, *Mater. Today*, 10, pp. 28–37.

[24] Sheraw, C. D., Zhou, L., Huang, J. R., Gundlach, D. J., Jackson, T. N., Kane, M. G., Hill, I. G., Hammond, M. S., Campi, J., Greening, B. K., Francl, J. and West, J. (2002). Organic thin-film transistor-driven polymer-dispersed liquid crystal displays on flexible polymeric substrates, *Appl. Phys. Lett.*, 80, pp. 1088–1090.

[25] Gelinck, G. H., Huitema, H. E. A., van Veenendaal, E., Cantatore, E., Schrijnemakers, L., van der Putten, J. B. P. H., Geuns, T. C. T., Beenhakkers, M., Giesbers, J. B., Huisman, B.-H., Meijer, E. J., Benito, E. M., Touwslager, F. J., Marsman, A. W., van Rens, B. J. E. and de Leeuw, D. M. (2004). Flexible active-matrix displays and shift registers based on solution-processed organic transistors, *Nat. Mater.*, 3, pp. 106–110.

[26] Ha, M., Zhang, W., Braga, D., Renn, M. J., Kim, C. H. and Frisbie, C. D. (2013). Aerosol-jet-printed, 1 volt H-bridge drive circuit on plastic with integrated electrochromic pixel, *ACS Appl. Mater. Interfaces*, 5, pp. 13198–13206.

[27] Sekine, C., Tsubata, Y., Yamada, T., Kitano, M. and Doi, S. (2014). Recent progress of high performance polymer OLED and OPV materials for organic printed electronics, *Sci. Technol. Adv. Mater.*, 15, p. 034203.

[28] Geffroy, B., le Roy, P. and Prat, C. (2006). Organic light-emitting diode (OLED) technology: Materials, devices and display technologies, *Polym. Int.*, 55, pp. 572–582.

[29] Ma, R. (2012). Organic Light Emitting Diodes (OLEDS), *Handbook of Visual Display Technology*, 2nd edn., pp. 1209–1221.

[30] Xu, T. and Qiao, Q. (2012). *Encyclopedia of Nanotechnology*, eds. Bharat Bhushan, "Organic photovoltaics: Basic concepts and device physics" (Springer Netherlands, Dordrecht) pp. 2022–2031.

[31] Arbouch, I., Karzazi, Y. and Hammouti, B. (2014). Organic photovoltaic cells: Operating principles, recent developments and current challenges–review, *Phys. Chem. News*, 72, pp. 73–84.

[32] Ng, T.-W., Lo, M.-F., Yang, Q.-D. and Lee, C.-S. (2016). *Graphene Science Handbook: Electrical and Optical Properties*, Chapter 22: Direct threat of UV–ozone-treated indium-tin oxide in organic optoelectronics and stability enhancement using graphene oxide as anode buffer layer (CRC Press, Boca Raton) pp. 365–380.

[33] Fung, D. D. S. and Choy, W. C. H. (2013). *Organic Solar Cells: Materials and Device Physics*, W. C. H. Choy (ed.), "Introduction to organic solar cells" (Springer London, London) pp. 1–16.

[34] Dyer-Smith, C., Nelson, J. and Li, Y. (2018). *McEvoy's Handbook of Photovoltaics*, 3rd edn., S. A. Kalogirou (ed.), Chapter I-5-B: Organic solar cells (Academic Press, United Kingdom) pp. 567–597.

[35] Eggenhuisen, T. M., Galagan, Y., Biezemans, A. F. K. V., Slaats, T. M. W. L., Voorthuijzen, W. P., Kommeren, S., Shanmugam, S., Teunissen, J. P., Hadipour, A., Verhees, W. J. H., Veenstra, S. C., Coenen, M. J. J., Gilot, J., Andriessen, R. and Groen, W. A. (2015). High efficiency, fully inkjet printed organic solar cells with freedom of design, *J. Mater. Chem. A*, 3, pp. 7255–7262.

[36] Oliveira, J., Costa, C. M. and Lanceros-Méndez, S. (2018). *Printed Batteries — Materials, Technologies and Applications*, Senentxu Lanceros-Méndez and Carlos Miguel Costa (eds.), Chapter 1: Printed batteries: An overview (John Wiley & Sons).

[37] Beidaghi, M. and Gogotsi, Y. (2014). Capacitive energy storage in micro-scale devices: Recent advances in design and fabrication of micro-supercapacitors, *Energy Environ. Sci.*, 7, pp. 867–884.

[38] Tian, X., Jin, J., Yuan, S., Chua, C. K., Tor, S. B. and Zhou, K. (2017). Emerging 3D-printed electrochemical energy storage devices: A critical review, *Adv. Energy Mater.*, 7, p. 1700127.

[39] Zhang, M., Mei, H., Chang, P. and Cheng, L. (2020). 3D printing of structured electrodes for rechargeable batteries, *J. Mater. Chem. A*, 8, pp. 10670–10694.

[40] Gaikwad, A. M., Arias, A. C. and Steingart, D. A. (2015). Recent progress on printed flexible batteries: Mechanical challenges, printing technologies, and future prospects, *Energy Technol.*, 3, pp. 305–328.

[41] Yue, Y. and Liang, H. (2018). 3D current collectors for lithium-ion batteries: A topical review, *Small Methods*, 2, p. 1800056.

[42] Kong, D., Wang, Y., Huang, S., Zhang, B., Lim, Y. V., Sim, G. J., Valdivia y Alvarado, P., Ge, Q. and Yang, H. Y. (2020). 3D printed compressible quasi-solid-state nickel–iron battery, *ACS Nano*, 14, pp. 9675–9686.

[43] Xu, Y., Wu, X., Guo, X., Kong, B., Zhang, M., Qian, X., Mi, S. and Sun, W. (2017). The boom in 3D-printed sensor technology, *Sensors*, 17, p. 1166.

[44] Ni, Y., Ji, R., Long, K., Bu, T., Chen, K. and Zhuang, S. (2017). A review of 3D-printed sensors, *Appl. Spectrosc. Rev.*, 52, pp. 623–652.

[45] Agarwala, S., Goh, G. L., Yap, Y. L., Goh, G. D., Yu, H., Yeong, W. Y. and Tran, T. (2017). Development of bendable strain sensor with embedded microchannels using 3D printing, *Sens. Actuators, A*, 263, pp. 593–599.

[46] Agarwala, S., Goh, G. L. and Yeong, W. Y. (2018). Aerosol jet printed strain sensor: Simulation studies analyzing the effect of dimension and design on performance (September 2018), *IEEE Access*, 6, pp. 63080–63086.

[47] Goh, G. L., Agarwala, S., Tan, Y. J. and Yeong, W. Y. (2018). A low cost and flexible carbon nanotube pH sensor fabricated using aerosol jet technology for live cell applications, *Sens. Actuators, B*, 260, pp. 227–235.

[48] Macdonald, E., Salas, R., Espalin, D., Perez, M., Aguilera, E., Muse, D. and Wicker, R. B. (2014). 3D printing for the rapid prototyping of structural electronics, *IEEE Access*, 2, pp. 234–242.

[49] Shemelya, C., Cedillos, F., Aguilera, E., Espalin, D., Muse, D., Wicker, R. and MacDonald, E. (2015). Encapsulated copper wire and copper mesh capacitive sensing for 3-D printing applications, *IEEE Sens. J.*, 15, pp. 1280–1286.

[50] Jeranče, N., Bednar, N. and Stojanović, G. (2013). An ink-jet printed eddy current position sensor, *Sensors*, 13, pp. 5205–5219.

[51] Kit-Anan, W., Olarnwanich, A., Sriprachuabwong, C., Karuwan, C., Tuantranont, A., Wisitsoraat, A., Srituravanich, W. and Pimpin, A. (2012). Disposable paper-based electrochemical sensor utilizing inkjet-printed polyaniline modified screen-printed carbon electrode for ascorbic acid detection, *J. Electroanal. Chem.*, 685, pp. 72–78.

[52] Lind, J. U., Busbee, T. A., Valentine, A. D., Pasqualini, F. S., Yuan, H., Yadid, M., Park, S.-J., Kotikian, A., Nesmith, A. P. and Campbell, P. H. (2017). Instrumented cardiac microphysiological devices via multimaterial three-dimensional printing, *Nat. Mater.*, 16, pp. 303–308.

[53] Sauerbrunn, E., Chen, Y., Didion, J., Yu, M., Smela, E. and Bruck, H. A. (2015). Thermal imaging using polymer nanocomposite temperature sensors, *Physica Status Solidi (a)*, 212, pp. 2239–2245.

[54] Gao, J., Sidén, J., Nilsson, H. and Gulliksson, M. (2013). Printed humidity sensor with memory functionality for passive RFID tags, *IEEE Sens. J.*, 13, pp. 1824–1834.

[55] Hoffmann, K. (1989). *An Introduction to Measurements Using Strain Gages* (Hottinger Baldwin Messtechnik GmbH).

[56] NI. (2020). Measuring strain with strain gages. Retrieved from http://www.ni.com/white-paper/3642/en/.

[57] Serway, R. A. and John W. Jewett, J. (2014). *Physics for Scientists and Engineers with Modern Physics*, Chapter 12: Static equilibrium and elasticity, pp. 363–387.

[58] Thompson, B. and Yoon, H. S. (2013). Aerosol-printed strain sensor using PEDOT:PSS, *IEEE Sens. J.*, 13, pp. 4256–4263.

[59] Agarwala, S., Goh, G. L., Dinh Le, T.-S., An, J., Peh, Z. K., Yeong, W. Y. and Kim, Y.-J. (2019). Wearable bandage-based strain sensor for home healthcare: Combining 3D aerosol jet printing and laser sintering, *ACS Sens.*, 4, pp. 218–226.

[60] Borghetti, M., Serpelloni, M. and Sardini, E. (2019). Printed strain gauge on 3D and low-melting point plastic surface by aerosol jet printing and photonic curing, *Sensors*, 19, p. 4220.

[61] Manjakkal, L., Szwagierczak, D. and Dahiya, R. (2020). Metal oxides based electrochemical pH sensors: Current progress and future perspectives, *Progr. Mater. Sci.*, 109, p. 100635.

[62] Qin, Y., Kwon, H.-J., Howlader, M. M. R. and Deen, M. J. (2015). Microfabricated electrochemical pH and free chlorine sensors for water quality monitoring: recent advances and research challenges, *RSC Adv.*, 5, pp. 69086–69109.

[63] Gou, P., Kraut, N. D., Feigel, I. M., Bai, H., Morgan, G. J., Chen, Y., Tang, Y., Bocan, K., Stachel, J., Berger, L., Mickle, M., Sejdić, E. and Star, A. (2014). Carbon nanotube chemiresistor for wireless pH sensing, *Sci. Rep.*, 4, p. 4468.

[64] Fan, C., Pavlidis, S., Papapolymerou, J., Yung Hang, C., Kan, W., Zhang, C. and Ben, W. (2014). Aerosol jet printing for 3-D multilayer passive microwave circuitry, *presented at the 44th European Microwave Conference (EuMC)*, Rome, Italy.

[65] Khan, S., Lorenzelli, L. and Dahiya, R. S. (2015). Technologies for printing sensors and electronics over large flexible substrates: A review, *IEEE Sens. J.*, 15, pp. 3164–3185.

[66] Kunnari, E., Valkama, J., Keskinen, M. and Mansikkamäki, P. (2009). Environmental evaluation of new technology: Printed electronics case study, *J. Cleaner Prod.*, 17, pp. 791–799.

[67] Kamyshny, A. and Magdassi, S. (2019). Conductive nanomaterials for 2D and 3D printed flexible electronics, *Chem. Soc. Rev.*, 48(6):1712–1740.

[68] Huang, Q. and Zhu, Y. (2019). Printing conductive nanomaterials for flexible and stretchable electronics: A review of materials, processes, and applications, *Adv. Mater. Technol.*, 4, p. 1800546.

[69] Khan, Y., Garg, M., Gui, Q., Schadt, M., Gaikwad, A., Han, D., Yamamoto, N. A. D., Hart, P., Welte, R., Wilson, W., Czarnecki, S., Poliks, M., Jin, Z., Ghose, K., Egitto, F., Turner, J. and Arias, A. C. (2016). Flexible hybrid electronics: Direct interfacing of soft and hard electronics for wearable health monitoring, *Adv. Funct. Mater.*, 26, pp. 8764–8775.

[70] Magdassi, S., Grouchko, M., Berezin, O. and Kamyshny, A. (2010). Triggering the sintering of silver nanoparticles at room temperature, *ACS Nano*, 4, pp. 1943–1948.

[71] Wu, W. (2019). Stretchable electronics: Functional materials, fabrication strategies and applications, *Sci. Technol. Adv. Mater.*, 20, pp. 187–224.

[72] Muth, J. T., Vogt, D. M., Truby, R. L., Mengüç, Y., Kolesky, D. B., Wood, R. J. and Lewis, J. A. (2014). Embedded 3D printing of strain sensors within highly stretchable elastomers, *Adv. Mater.*, 26, pp. 6307–6312.

[73] Tan, H., Chua, C., Uttamchand, M. and Tran, T. (2019). Fully 3D printed horizontally polarised omnidirectional antenna, *presented at the Industry 4.0–Shaping The Future of The Digital World*, Manchester, UK.

[74] Espera, A. H., Dizon, J. R. C., Chen, Q. and Advincula, R. C. (2019). 3D-printing and advanced manufacturing for electronics, *Progr. Addit. Manufact.*, 4, pp. 245–267.

[75] Adams, J. J., Duoss, E. B., Malkowski, T. F., Motala, M. J., Ahn, B. Y., Nuzzo, R. G., Bernhard, J. T. and Lewis, J. A. (2011). Conformal printing of electrically small antennas on three-dimensional surfaces, *Adv. Mater.*, 23, pp. 1335–1340.
[76] Joe Lopes, A., MacDonald, E. and Wicker Ryan, B. (2012). Integrating stereolithography and direct print technologies for 3D structural electronics fabrication, *Rapid Prototyping J.*, 18, pp. 129–143.

Problems

1. What is a passive electrical component? What are some of the examples of passive electrical components? Sketch the schematic diagrams of these passive components.
2. What is an active electrical component? Give some examples of active electrical components.
3. Sketch the schematic diagrams of the various OTFTs design architectures.
4. Sketch the schematic diagrams of the sandwich-type printed battery and the interdigitated-type printed battery. Explain briefly the functions of each layer (e.g. electrodes, current collectors and separator with electrolyte) in a 3D sandwich-type printed battery.
5. Describe the main differences between a P-LED and an OPV device.
6. Discuss some of the challenges for 3D printed electronics.
7. What are the application areas that 3D printing of electronics can potentially applied to in the future?
8. Apart from the examples seen in this chapter, which electronic devices you think is suitable for 3D printing and why?

Appendix

List of Companies

BotFactory, Inc.
4334 32nd Place, FL 3, RM 3Rb
Long Island City, NY 11101
USA
Tel: +1 (347) 377 2687
Email: contact@botfactory.co
Website: www.botfactory.co

Enjet, Inc.
45, Saneop-ro 92 Beon-gil
Gwonseon-gu, Suwon-si, Gyeonggi-do
Republic of Korea
Tel: +82 70 4892 8113
Tel: +82 70 4892 8100
E-mail: sales@enjet.co.kr
Website: http://en.enjet.co.kr/

FUJIFILM Dimatix, Inc.
2250 Martin Avenue,
Santa Clara, CA 95050
USA
Tel: +1 (888) DIMATIX or +1 (888) 346-2849 (US toll free)

Tel: +1 (408) 565-9150
E-mail: printinginfo@dimatix.com
Website: www.dimatix.com

Integrated Deposition Solutions, Inc.
5901 Indian School Rd NE. Suite #125
Albuquerque, NM 87110
USA
Tel: +1 (505) 200-9527
E-mail: info@idsnm.com
Website: www.idsnm.com

Nano Dimension Ltd.
13798 NW 4th St., Suite 315
Sunrise, FL 33325-6227
USA
Website: www.nano-di.com

Neotech AMT GmbH
Petzoltstr. 3
90443, Nuremberg
Germany
E-mail: info@neotech-amt.com
Website: www.neotech-amt.com

nScrypt, Inc.
12151 Research Parkway, Suite 150
Orlando, FL 32826
USA
Tel: +1 (407) 275-4720
E-mail: info@nscrypt.com
Website: www.nscrypt.com

Optomec, Inc.
3911 Singer Blvd N.E.
Albuquerque, NM 87109
USA
Tel: +1 (505) 761-8250

E-mail: sales@optomec.com
Website: www.optomec.com

Sonoplot, Inc.
3030 Laura Lane, Suite 120
Middleton, WI 53562
USA
Tel: +1 (608) 824 9311
E-mail: contact@sonoplot.com
Website: www.sonoplot.com

Voltera, Inc.
113 Breithaupt St., Suite 100
Kitchener, ON, N2H 5G9
Canada
E-mail: hello@voltera.io
Website: www.voltera.io

Index